Graduate Texts in Physics

Graduate Texts in Physics publishes core learning/teaching material for graduate- and advanced-level undergraduate courses on topics of current and emerging fields within physics, both pure and applied. These textbooks serve students at the MS- or PhD-level and their instructors as comprehensive sources of principles, definitions, derivations, experiments and applications (as relevant) for their mastery and teaching, respectively. International in scope and relevance, the textbooks correspond to course syllabi sufficiently to serve as required reading. Their didactic style, comprehensiveness and coverage of fundamental material also make them suitable as introductions or references for scientists entering, or requiring timely knowledge of, a research field.

Elena Bannikova · Massimo Capaccioli

Foundations of Celestial Mechanics

 Springer

Elena Bannikova
Department of Astronomy and Space
Informatics
V. N. Karazin Kharkiv National University
Kharkiv, Ukraine

Massimo Capaccioli
Department of Physics "Ettore Pancini"
University of Naples Federico II
Naples, Italy

ISSN 1868-4513   ISSN 1868-4521 (electronic)
Graduate Texts in Physics
ISBN 978-3-031-04578-3   ISBN 978-3-031-04576-9 (eBook)
https://doi.org/10.1007/978-3-031-04576-9

This Springer imprint is published by the registered company Springer Nature Switzerland AG
The registered company address is: Gewerbestrasse 11, 6330 Cham, Switzerland

To our masters
Gérard Henri de Vaucouleurs
Victor Moiseevich Kontorovich

Aristotle

Teaching is the highest form of
understanding.

Mustafa Kemal Atatürk

A good teacher is like a candle that consumes
itself while lighting the way for others.

Introduction

Vladimir Igorevich Arnold

Celestial mechanics is the origin of dynamical systems, linear algebra, topology, variational calculus, and symplectic geometry.

In its most classical sense, celestial mechanics is the discipline that studies the motion of the Solar System bodies such as planets, their satellites, asteroids, and comets, as well as artificial satellites and space probes, in the context of classical mechanics and Newton's theory of gravitation.[1] The nature of the problems addressed and the degree of precision required for the results do not mandate the use of general relativity but in special cases, when Einstein's gravitational theory compensates for a falsification of classical mechanics (for example, in the full explanation of Mercury's perihelion advance; see the last two sections of Chap. 4) or when the measurements are so accurate as to challenge relativity itself (for instance, in experiments designed to verify the Lense–Thirring effect[2]).

The origins or at least the motivations of celestial mechanics are intertwined with the very roots of astronomy, the oldest of all natural sciences (see, for instance, [1] and [2]). Already during Prehistory, the closest and brightest celestial bodies and the stars visible to the naked eye were observed with the aim of describing and then predicting their motions. The reasons for this interest reside in the apparent centrality of mankind

---

[1] The locution celestial mechanics was coined by the French mathematician Pierre Simon de Laplace in the late eighteenth century: from 1798 to 1825, he published the five volumes of his monumental *Traité de mécanique céleste*. The term dynamics, from the Greek word for force, was instead introduced (originally in French) by the German polymath Gottfried Leibniz (1646–1716) right a century earlier.

[2] Joseph Lense (1890–1985) and Hans Thirring (1888–1976) were Austrian physicists. They predicted the relativistic correction to the precession of a gyroscope near a large rotating mass. For example, while classically a satellite orbiting a spherical symmetric body is unaffected by the possible rotation of the primary, relativity requires that the plane of the satellite's orbit suffers a precession in the direction of the primary's rotation axis.

with respect to the firmament and in the manifest perfection of the stars, both fixed (true stars) and wandering (planets, from the ancient Greek term for vagabond). These otherwise unchanging light sources[3] appeared to be subject to periodic movements as well as recurring phenomena (Lunar phases, Solar and Lunar eclipses, Meton cycle[4]). They showed no signs of birth or death,[5] which placed them outside the domain of time, making them eternal. By contrast, anything on Earth appeared corruptible and imperfect. Despite this remarkable diversity, eternity and celestial perfection seemed to be able to tune in well with earthly events (such as seasonal variations) and hence with the rhythms of human life. his reassuring synergy generated the belief of a conspiracy between the two distinct Aristotelian worlds, celestial and terrestrial, one within the other and with a boundary at the sphere of the Moon,[6] and each one with its own composition and physics.[7] In the wake of an anthropocentric reading of the world, the next logical step was to postulate that the whole construction had been

---

[3] Apart from a few novae and supernovae visible to the naked eye, the photometric variability of stars was virtually unknown to ancient astronomers, and in any case ignored. The Moon and the planets did change with time their luminosity and aspect (Moon phases), but these alterations were periodic.

[4] If a phenomenon is produced by a combination of two independent periodic phenomena with periods P_1 and P_2, it is also periodic with period $P = mP_1 = nP_2$, where m and n are the smallest integers satisfying the relation. For instance, about 2700 years ago Chaldean astronomers discovered that every 223 synodic months, equivalent to about 18.03 Julian years (each lasting 365.25 days), the Sun and the Moon resume the same relative positions. Hence, the chronology of the eclipses repeats almost identically every 18.03 years (but Solar eclipses will occur in different places on Earth). This periodicity, christened the Saros cycle in 1686 by the English astronomer Edmund Halley (1656–1742) who derived the name from a Babylonian or Greek term of uncertain meaning, is the consequence of two requirements for an eclipse to occur: (1) the Moon must be in conjunction or opposition with the Sun, which happens with the period of a synodic month, every 29.53 days, and (2) the Moon, whose orbit is inclined by $\sim 5°$, must lie on the ecliptic, otherwise the alignment is incomplete (hence ineffective). This requires that the Moon be transiting through one node of its orbit, which happens twice in a draconic month of 27.21 days. Now, $223 \times 29.53 \simeq 242 \times 27.21 \simeq 6585\,\text{days} = 18.03$ years. The Saros cycle has had great importance since, owing to the extraordinary suggestiveness of eclipses (especially Solar), predicting them gave great power to those who were able to do so.

Another example is provided by the Meton cycle. In the fifth century BC, the Athenian Meton discovered (or learned from the Chaldeans) that the phases of the Moon repeat at approximately the same dates every 19 years. The duration of this cycle comes from the equation $235 \times 29.53 \simeq 19 \times 365.25$ days.

[5] Random appearances of a comet or nova star were treated as meteorological (earthly) phenomena.

[6] According to a model devised by Plato and then accepted by the Greek-Alexandrian culture with a few exceptions and some complications dictated by the need for greater adherence to observational evidence, the world consisted of spheres concentric to the Earth, each assigned to a wandering star according to the sequence; Moon, Mercury, Venus, Sun, Mars, Jupiter, and Saturn.

[7] To the four ultimate elements that make up the terrestrial world (fire, air, water, and earth) according to Empedocles of Akragas, Sicily (5th century BC), Aristotle (384–322 BC) added the ether or quintessence (Latinate name for the Greek ether), filling the celestial spheres. Note that, according to the Stagirite, two are the types of motion, natural and violent. Cause of natural motion is the return

created[8] for the central lodger, the human being, a creature privileged by God(s). Wandering stars were then interpreted as pencils with which the deity represented the messages addressed to men, to guide, admonish, or simply inform them[9]. This claim made astronomy become the instrument of the mysterious astrology, judged as the highest form of knowledge and therefore delivered into the hands of wise priests. A pseudoscience that easily slipped into superstition:

> This is the excellent foppery of the world, that, when we are sick in fortune, often the surfeit of our own behaviour, we make guilty of our disasters the sun, the moon, and the stars; as if we were villains on necessity; fools by heavenly compulsion; knaves, thieves, and treachers by spherical pre-dominance; drunkards, liars, and adulterers by an enforc'd obedience of planetary influence; and all that we are evil in, by a divine thrusting on. An admirable evasion of whore-master man, to lay his goatish disposition to the charge of a star! My father compounded with my mother under the Dragon's Tail, and my nativity was under Ursa Major, so that it follows I am rough and lecherous. Fut! I should have been that I am, had the maidenliest star in the firmament twinkled on my bastardizing.
>
> W. Shakespeare, *King Lear* [Act 1, Sc. 2]

The simple use of celestial phenomena for orientation, time keeping, or practical life, as in the *Works and Days* of the Greek Hesiod (c. 750–650 BC) which regards agriculture, while so important, has long appeared as secondary to astrology.[10] Examples of such attempts to interpret the motion of the stars are some Neolithic constructions dating back to that period, such as Stonehenge on Salisbury Plain in Wiltshire, England, and the menhirs (long stones in Celtic languages). Later, various populations in the Middle East, Egypt, Central America, and China elaborated this relationship with the sky through frequent and well-documented observations of astronomical phenomena and through the construction of huge architectural works such as the Egyptian pyramids and the Mesopotamian ziggurats.

Activities of this kind are widespread in Greece. Pythagoras of Samos (c. 570–495 BC) placed the Sun and the other planets on concentric spheres around a central fire and made this correspond to the harmony of the musical scales. Plato (c. 428–348 BC) was one of the first to formulate a geocentric hypothesis for the Solar System, stating that the celestial spheres carrying the wandering stars rotate around the Earth, seen as a motionless globe. He guessed the sphericity of the Sun, claiming that the Moon

---

of a body to its natural place (up for air and fire, down for water and earth) where it will remain in quiet; the velocity is proportional to the size of the body. The cause of violent motion is an external motor in contact with the body (remote actions are not allowed); the speed is directly proportional to the applied force and inversely to the magnitude of the body. Aristotle also postulated that the natural motions are rectilinear (but not uniform) on Earth, while they are circular and uniform in the sky.

[8] The idea of a genesis is common to most ancient civilizations (but not to the thought of Aristotle), and to modern cosmology too (the Big Bang).

[9] In one of his writings, Johannes Kepler stated: *That the sky does something to man is obvious enough: but what it does specifically remains hidden.*

[10] In the *Harmonies of the World* (1619), where he reports the discovery of his famous Third Law, Kepler writes: *The soul of the newly born baby is marked for life by the pattern of the stars at the moment it comes into the world, unconsciously remembers it, and remains sensitive to the return of configurations of a similar kind.*

does not shine on its own but receives its light from the Sun. Other contributions to the knowledge of the Solar System were made, for example, by the Milesian philosophers Thales (c. 624–548 BC) and Anaximander (c. 610–546 BC), the Pythagorean and Pre-Socratic Philolaus of Croton (470–385 BC), and Heraclides Ponticus (385–322 or 310 BC), philosopher and astronomer, contemporary to Eudoxus.

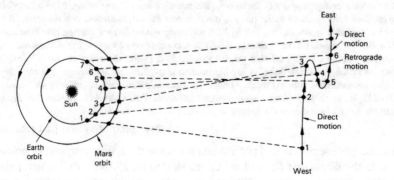

Retrograde motion of Mars. The apparent reversal in the direction of the planet's motion is caused by the different angular velocity of Mars relative to that of the Earth's observer. *Credit*: NASA

The concept of the geocentric system, placing the Earth at the center of the universe, started to spread in the Western World. With various upgrades, it would endure for many centuries. Eudoxus of Cnidus (c. 408–355 BC), student of Plato and the Pythagorean Archytas (428–360 BC), elaborated the model of homocentric spheres: a universe as a clockwork with the Earth at the geometric center. Each sphere carries a planet in a circular and uniform motion, different from that of others. This simple model worked reasonably well; but in order to match the observations (in particular, the so-called retrograde motion, when a planet reverses temporarily its prograde motion with respect to fixed stars), more spheres were needed for some planets, so much so that the overall number grew largely.

The reason why this geocentric conception managed to survive until the time of Copernicus, and even longer in some circles, is mainly threefold. It was able to account for some phenomena such as the lack of an annual parallax (mistakenly assigned to the Earth's immobility and not to the enormous distances of the stars) or the absence of any apparent consequence of a high orbital velocity required for an Earth revolving about the Sun. It was also in keeping with men's innermost need to consider themselves at the center of the universe as proof of a direct connection to God(s). The third reason has been the enormous cultural influence of Aristotle, the first physicist of ancient philosophers, the Master of those who know, as Dante calls him in Canto IV of the *Inferno*. He theorized the geocentric model, discussing all of its consequences. At the end of the Middle Ages, the scholastic philosophers, represented chiefly by the most learned Italian scholar Thomas Aquinas (1225–1274), merged the geocentric cosmology with the Christian theology, which also required an immobile Earth to match the prayer of Joshua to the Lord reported in the Old Testament: *Sun, stand still over Gibeon, and you, moon, over the Valley of*

Aijalon (*Joshua* 10, 13). By now an integral part of Christian doctrine, geocentrism became an absolute dogma.

Note that, to account for improved observations, Aristotle (384–322 BC) had been forced to add more spheres to the original ones of Eudoxus. His celestial clockwork became a complex system of 55 spheres animated by the so-called *primum mobile* (first mover) which allowed the spheres to turn, each one well tuned to the appropriate motion, through friction.

Heliocentric theories, however, were not completely abandoned. For instance, Aristarchus of Samos (c. 310–230 BC) refined the cosmological model of Heraclides Ponticus in which the Sun was at the center of the universe, making the description of planetary motions simpler while not matching better the observations, since the orbits were still circular. Later, Apollonius of Perga (c. 262–192 BC) introduced the system of epicycles and deferents. The motivation for this elegant complication was to save both appearances and the uniform circular motions by combining two such motions in the following way: planets were made to revolve at a constant velocity on circular orbits, called epicycles, whose centers were made to rotate around the Earth on immaterial circles called deferents. This geometrical construction was modified further by Hipparchus of Nicaea (c. 190–120 BC) who, to account for the observed properties of the Moon (change of velocity and angular size during a lunation) and of the planets (change of brightness), placed the Earth off the center of each deferent, in an eccentric position.

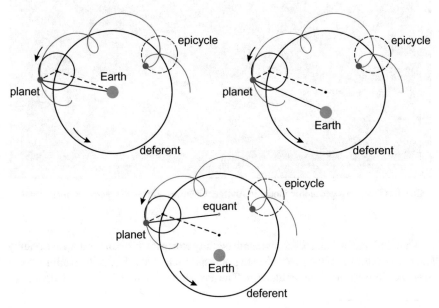

Sketch of the planetary models based on (top-left) epicycle and deferent, (top-right) eccentric, and (bottom) equant

Finally, during Hellenism, Claudius Ptolemy (c. 100–170 AD), who worked in the famous Musaeum[11] of Alexandria, in Egypt, perfected the hypothesis of eccentric motion by shifting, for each planet, the focus of the uniform motion from the geometrical center to a point placed symmetrically with respect to the Earth. With this gimmick, which was never fully accepted by his followers, the Alexandrian philosopher was able to keep up with the Aristotelian principles of circularity of the orbits and uniformity of motions, improving significantly the accuracy of the model in representing the planetary motions (required to compute reliable ephemerides).

Claudius Ptolemy's geocentric model in Andreas Cellarius' *Harmonia macrocosmica* (1660)

Mankind had to wait for the sixteenth century to see the birth of modern astronomy and the withdrawal of the geocentric hypothesis that had required several adjustments over the time to meet the evidences. The Polish astronomer Nicolaus Copernicus

---

[11] The Musaeum (temple of the Muses), which included the Library, was erected in Alexandria by Ptolomy I Soter (c. 367–283 BC), the first successor of Alexander the Great. Conceived as a meeting place for scholars and also for teaching, it represented for centuries the highest cultural institution of the Hellenistic world. It was perhaps destroyed by order of the Emperor Aurelian in 272 AD.

(1473–1543) was the first to build a heliocentric system, then called after his name. He described the motions of planets through circular orbits around the Sun, keeping the epicyclical architecture, including the eccentric. In other words, Copernicus was both a revolutionary and a conservative. In the same direction were the fundamental contributions of the Italian Galileo Galilei 1564–1642) and the German Johannes Kepler (1571–1630).

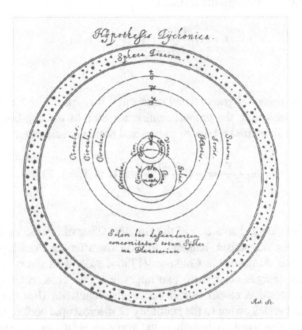

Illustration of the Hypothesis Tychonica from Johannes Hevelius' *Selenographia* (1647)

In particular, by using the precise observations made by the Danish Tycho Brahe (1546–1601), Kepler established that the planets move on elliptical orbits where they accelerate while approaching the Sun and decelerate when moving away. He summarized his speculations into three empirical laws, named after him; the first two published in *Astronomia Nova* (1609) and the last one in *Harmonices Mundi* (1619). Galilei, instead, created the basis to demolish the principle of immutability of the heaven. With his pioneering observations of the sky through a telescope, he discovered the Solar spots, the phases of Venus, and the existence of four large satellites[12] orbiting around Jupiter, all together strongly supporting the heliocentric hypothesis.

However, it was only with Isaac Newton (1642–1726) that celestial mechanics was born. Indeed the English genius (see [3]) managed to elaborate a theory of (universal)

---

[12] The term satellite was coined for the occasion by Kepler drawing on the Latin *satelles*, a word of Etruscan origin which means bodyguard (used in the Middle Ages also in a derogatory sense).

gravitation capable of describing, in the framework of his mechanics, the motions of the Solar System bodies, which later the Anglo-German astronomer William Herschel (1738–1822) would prove to work for the sidereal world too (binary stellar systems). It establishes that two point-like bodies attract each other with a force directed along the line joining them with an intensity f which is proportional to the product of the (gravitational) masses m_i and m_j and inversely to the square of the mutual distance r_{ij}, through the formula:

$$f = G \frac{m_i m_j}{r_{ij}^2} \quad (i \neq j = 1, 2) \tag{1}$$

where G is the universal gravitational constant.[13] It couples with Newton's second law of dynamics stating the proportionality, for each of the two bodies, between acceleration (in an inertial reference frame) and force per unit mass:

$$m_i \frac{d^2 \mathbf{r}_i}{dt^2} = -G \frac{m_i m_j}{r_{ij}^2} \frac{\mathbf{r}_{ij}}{r_{ij}} \quad (i \neq j = 1, 2) \tag{2}$$

This has the identical structure of the law describing of interaction between two charges in electromagnetism, named Coulomb law after the French physicist and engineer Charles Augustin de Coulomb (1736–1806), with the only remarkable difference that charges occur with two opposite signs. Thus, contrary to gravity that is purely attractive, electric forces can be also repulsive. They mutually cancel at astronomical scales owing to the neutrality of macroscopic bodies, leaving a free field to gravity. The equation describing the gravitational action between two celestial bodies is therefore at the foundation of all the phenomena treated by celestial mechanics (motion of Solar System planets, comets, asteroids, artificial satellites) as well as double stars and even the interaction between two galaxies. Newton's gravitational force is an instantaneous action at distance: a concept that would have horrified Aristotle and that Newton himself disliked, accepting it only on a pragmatic basis:

> Hitherto I have not been able to discover the cause of those properties of gravity from phenomena, and I frame no hypotheses [hypotheses non fingo]; for whatever is not deduced from the phenomena is to be called an hypothesis; and hypotheses, whether metaphysical or physical, whether of occult qualities or mechanical, have no place in experimental philosophy. In this philosophy particular propositions are inferred from the phenomena, and

---

[13] G is an empirical physical constant involved by Isaac Newton in the calculation of the dynamical effects of his universal gravitational law and by Albert Einstein in his general theory of relativity, where it quantifies the relation between the geometry of spacetime and the stress–energy tensor. The notation (G) was introduced in 1890 by the British physicist Charles Bays (1855–1944). Its modern value, $G = 6.67430 \times 10^{-11}$ m^{3}kg^{-1}s^{-2}, was first (implicitly) measured with an accuracy of 1% by the English natural philosopher Henry Cavendish (1731–1810) with an experiment aimed at determining the average density of the Earth.

afterwards rendered general by induction. Thus it was that the impenetrability, the mobility,
and the impulsive force of bodies, and the laws of motion and of gravitation, were discovered.
And to us it is enough that gravity does really exist, and act according to the laws which we
have explained, and abundantly serves to account for all the motions of the celestial bodies,
and of our sea.

Letter of Newton to Robert Hooke, 15 February 1676

Newton's model endured for 230 years and was finally reformulated in 1916 by Albert Einstein (1879–1955) with his general relativity, while continuing to represent well the situations of weak gravitational fields and of velocities far away from that of light.

Since the first appearance of Newton's *Philosophiae Naturalis Principia Mathematica* in 1687 (re-published twice, in 1713 and 1726, with corrections and improvements), the mechanical theory of the Lucasian professor, coupled with the gravity law, has collected countless successes in its applications to both terrestrial and celestial phenomena. Probably, the triumph was reached with the discovery of Neptune on the *tip of the pen*, as François Arago[14] (1786–1853), director of the Paris Observatory, described the prediction made by Urbain Le Verrier (1811–1877) on the position on the sky of an unknown planet perturbing the orbit of Uranus (see also pg. 27). This sort of miracle was made possible by the high degree of development reached by celestial mechanics since the eighteenth century owing to the work of Leonhard Euler, Joseph Louis Lagrange, Pierre-Simon Laplace, and many others. After World War II, a new age of the discipline started with *(a)* the coming into operation of electronic computers which allowed to solve numerically those problems where accuracy was conflicting with mathematical difficulties, and *(b)* the development of space flights. For a review of the history of celestial mechanics in the last three centuries, see [4], and for an overview of *Celestial mechanics: past, present, future*, [5].

This book aims at making a mathematical description of the gravitational interaction of celestial bodies. The approach to the problem is purely formal. It allows us to write equations of motion and solve them as much as possible, either exactly or by approximate techniques when there is no other way. The results obtained provide predictions that can be compared with observations. Several appendices have been added to the five main chapters. They review some selected mathematical tools, deepen some questions (so as not to interrupt the logic of the main frame with heavy technicalities or digressions), offer a few examples, and provide an overview of special functions which are useful here as well as in many other fields of physics. We also present, as an example of a modern application, an original investigation of the potential of a torus. General relativity is used only in Sect. 4.14 dealing with the perihelion advance of Mercury; to follow this section, the reader is not required to know deeply about general relativity.

The present version, which corrects, improves, and enlarges the lecture notes written in Italian by one of us some forty years ago, takes advantage of the experience

---

[14] François Arago (1786–1853) was a French mathematician, physicist, astronomer, and politician. He promoted awareness and dissemination of the discovery of a revolutionary photographic technique by Louis-Jacques-Mandé Daguerre (1787–1851).

of many years of lecturing of celestial mechanics by both of us, either at the V. N. Karazin Kharkiv National University (EB) and at the Padua University (MC).

We are grateful to Crescenzo Tortora, then at the Naples University Federico II and now at the INAF-Capodimonte Astronomical Observatory, Naples, who long ago reviewed the Italian manuscript, and to Nina Akerman, now at the Padua University, and the students of V. N. Karazin Kharkiv National University, who helped us edit the current version, checking for both errors and unclear passages. Useful suggestions and criticisms came to us from the colleagues Elbaz Abouelmagd and Niraj Pathak of the National Research Institute of Astronomy and Geophysics of Egypt (NRIAG), Albert Kotvytskiy and Sergey Poslavsky of V. N. Karazin Kharkiv National University, and Ester Piedipalumbo of Naples University Federico II. We are especially grateful to Filippo Zerbi, Scientific Director of the Italian National Institute for Astrophysics (INAF), for the constant support to complete this work, and to the V. N. Karazin Kharkiv National University, the Institute of Radio Astronomy, National Academy of Sciences of Ukraine, and the INAF-Capodimonte Astronomical Observatory for their hospitality.

Kharkiv, Ukraine   Elena Bannikova
Naples, Italy   Massimo Capaccioli
February 2022

References

1. M. Hoskin, *The History of Astronomy: A Very Short Introduction*, Oxford, Oxford University Press, 2003.
2. M. Capaccioli, *The Enchantment of Urania. 25 Centuries of Exploration of the Sky*, Singapore: World Scientific, 2022.
3. R. S. Westfall, *Isaac Newton*, Cambridge, Cambridge University Press, 2007.
4. V. G. Szebehely, H. Mark, *Adventures in Celestial Mechanics*, New York: Wiley, 1998.
5. V. A. Brumberg, *Solar System Research*, Vol. 47, No. 5, pp. 347–358, New York: Pleiades Publishing, Inc., 2013.

Contents

Chapter 1
About the N-Body Problem

John von Neumann (from the podium of the first national
meeting of the Association for Computing Machinery, in 1947)

*If people do not believe that mathematics is simple, it is only
because they do not realize how complicated life is.*

The basic task of classical celestial mechanics, set by the need to model and predict
the motions of Solar System bodies in the context of Newton's mechanics and gravitational law, is to determine the dynamical evolution of systems with a finite[1] number
N of distinct bodies that are subject to mutual gravitational forces. The classic formulation[2] of the problem, given by the system of equations (1.2), is straightforward, but
the exact solution does not exist for $N > 2$ even under oversimplified assumptions,
i.e., when: *i.* the external forces are deliberately disregarded (the system is then said
to be isolated), *ii.* there are no other internal forces but gravitational interactions, and
iii. the bodies in play are all assimilated to massive points.

It was this generalized indetermination that fostered, at the beginning of the 18-th
century,[3] the development of a new science focused on seeking approximate solutions to systems of differential equations such as (1.2). The whole history of celestial
mechanics, enlightened with extraordinary cultural achievements and enhanced by
spectacular successes, developed around the search for approximate methods to overcome the obstacle of having just one exact solution: that of the so-called two-body
problem that we consider in detail in the next chapter. Before getting to the heart of
the matter, a comment on the simplifications listed above is in order.

---

[1] When the number N of bodies becomes large, as it is the case of the stars in a cluster or in a
galaxy, celestial mechanics turns into stellar dynamics. The bodies lose their identities and, in some
applications, systems are idealized by continuous media; cf. [1].

[2] Strong gravitational fields require the theory of general relativity.

[3] The unsolvability of the $N > 2$ problem had been only guessed until the beginning of the 20-th
century, when it was demonstrated by Bruns and Poincaré; see p. 7.

© The Author(s), under exclusive license to Springer Nature Switzerland AG 2022
E. Bannikova and M. Capaccioli, *Foundations of Celestial Mechanics*, Graduate Texts
in Physics, https://doi.org/10.1007/978-3-031-04576-9_1

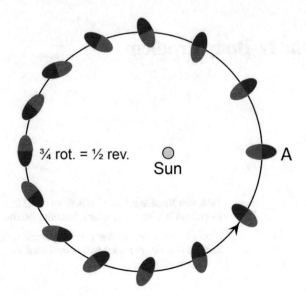

Fig. 1.1 Cartoon for the spin-orbit resonance of Mercury. The planet is sketched as an ellipse to mark the direction of the long axis of its figure. The two colors are used to visualize the effect of diurnal rotation along the elongated (elliptical) orbit. It is apparent that, after one complete revolution (staring from A and moving counterclockwise), at perihelion the planet, whose diurnal angular velocity is 3/2 the orbital velocity, alternately shows to the Sun one of the two opposite faces, thus keeping its long axis always aligned with the center of force when the distance is at a minimum. In this way the torque is also minimal

Bodies considered by celestial mechanics – as small as a Sputnik or as large as a star or a galactic nucleus – are often idealized by massive dimensionless mathematical entities. The introduction of massive points is justified by the known property of the center of mass: any motion, of both a continuous body and a system of *N* particles, can be split into (*1*) the motion of the barycenter, treated as a physical point containing the entire mass, and (*2*) the motion of the finite body (or of the *N* particles) about the barycenter. Dimensionless points allow us to ignore the complications introduced by a motion with respect to the center of mass and no longer require us to specify either the nature or the geometry or the status of the bodies involved. Thus, interactions that are not purely mechanical can be ignored (for instance, exchanges between mechanical and internal energies).[4] Moreover, the potential of a massive point has a much simpler and better tractable analytic expression than a finite body (cf. Chap. 5), at least for distances comparable to the dimensions of the body itself[5]; its expression is identical to the external potential of a finite body with spherically symmetric density.

---

[4] Provided that the sizes of the bodies are small compared with the distances among macroscopic entities.

[5] As the distances increases, the potential of a body tends always to that of a massive point of equal mass, no matter how complicate its shape is.

Fig. 1.2 Cartoon for the
spin-orbit resonance of
Venus. The planet is coded
as in the previous figure. It is
apparent that the period of
the diurnal rotation of Venus
is the same as the time
interval Δt between two
conjunctions, where:
$$\frac{1}{\Delta t} = \frac{1}{P_{\venus}} - \frac{1}{P_{\earth}}, \text{ and}$$
$P_{\venus} = 224.7\,d$ and
$P_{\earth} = 365.25\,d$ are the
periods of revolution of
Venus and Earth
respectively. The reader can
verify that $5 \times \Delta t \simeq 8 \times P_{\earth}$

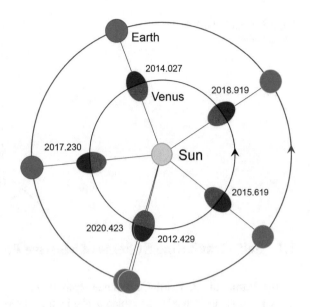

We want to stress that here the term isolation means just the absence of external
forces. In other words, there is no conceptual conflict with the Mach principle.[6] We
also ignore all internal forces other than gravitational. This latter assumption is not
always realistic. Consider, for instance, the consequences of the exchange between
mechanical and thermal energies which is the basis of innumerable evolutionary
phenomena in the Solar System. A classic example is the spin-orbit resonance causing
the Moon to show (on average)[7] the same face to the Earth ($P_{rot} = P_{rev}$). Others such
examples are those of Mercury (Fig. 1.1), which completes three rotations in the time
of two revolutions, and of Venus (Fig. 1.2), which shows always the same face to
the Earth in the lower conjunction, when the planet is aligned with the Earth and the
Sun and it is situated between them.

In the following of this chapter, we analyze the general properties of an isolated,
self-gravitating system of N massive points, leaving the detailed description of some
particular cases to Chaps. 2 and 3.

---

[6] The principle or conjecture of Ernst Mach (1838–1916) originates from the following question:
if every motion is relative, how does one measure inertia in an isolated system? Given an isolated
system of two massive points subject to gravitational forces, an external inertial reference system
is needed to set the dynamical condition for the gravitational equilibrium.

[7] A small lunar libration allows the terrestrial observer to see up to 59% of the surface of the satellite,
in spite of the tidal locking. The causes are the ellipticity of the orbit, the change of view point
because of the finite size of the Earth (parallax), and a physical oscillation.

Fig. 1.3 Sketch of a N-body configuration

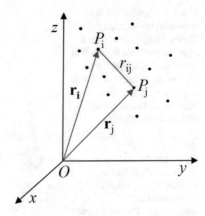

Fig. 1.3 Sketch of a N-body configuration

1.1   Self-Gravitating Systems of Massive Points

In the frame of an inertial reference system with orthogonal Cartesian[8] axes $O[x, y, z]$, let us consider an isolated mechanical system of N massive, gravitationally interacting point-like bodies P_i $(i = 1, \ldots, N)$ of masses m_i (Fig. 1.3). Each point P_i is attracted by the remaining $(N - 1)$ points P_j $(1 \leq j \neq i \leq N)$, each of which exerts on P_i a force $\mathbf{f}_{ij}$ directed as the vector[9] $\overrightarrow{P_i P_j} = \mathbf{r}_{ij} = \mathbf{r}_j - \mathbf{r}_i$ (P_i is pulled by P_j), with an intensity directly proportional to the mass product and inversely to the square of the distance between P_i and P_j (cf. Chap. 5):

$$\mathbf{f}_{ij} = G \, \frac{m_i m_j}{|\overrightarrow{P_i P_j}|^2} \, \frac{\overrightarrow{P_i P_j}}{|\overrightarrow{P_i P_j}|} = G \, \frac{m_i m_j}{r_{ij}^3} \, \mathbf{r}_{ij} \qquad (i \neq j = 1, \ldots, N). \qquad (1.1)$$

According to the third law of dynamics, it must be $\mathbf{f}_{ij} = -\mathbf{f}_{ji}$. The application of the second law assures that the acceleration of P_i has to balance the resultant of the forces (1.1):

$$m_i \frac{d^2 \mathbf{r}_i}{dt^2} = G \sum_{j \neq i = 1}^{N} \frac{m_i m_j}{r_{ij}^3} \, \mathbf{r}_{ij} \qquad (i \neq j = 1, \ldots, N). \qquad (1.2)$$

The system of the N vector equations (1.2) contains all the dynamical information which can be provided to describe the motion of the set of points P_i. It is equivalent to a system of $3N$ second order scalar differential equations:

---

[8] It may be interesting to recall that the first use of geometric coordinates was by the Catholic bishop Nicholas of Oresme (1323–1382), a polymath active in Paris. The French mathematician and philosopher René Descartes, latinized in Renatus Cartesius (1596–1650), took up the studies of Oresme for his work on the fusion of algebra with Euclidean geometry.

[9] Vector comes from the Latin word for 'carrier', since $\overrightarrow{P_i P_j}$ tries to carry the point P_i to P_j.

$$
\begin{cases}
\dfrac{d^2 x_i}{dt^2} = G \displaystyle\sum_{j \neq i = 1}^{N} \dfrac{m_j}{r_{ij}^3} (x_j - x_i) \\[2.5ex]
\dfrac{d^2 y_i}{dt^2} = G \displaystyle\sum_{j \neq i = 1}^{N} \dfrac{m_j}{r_{ij}^3} (y_j - y_i) \qquad\qquad (i \neq j = 1, \dots, N), \\[2.5ex]
\dfrac{d^2 z_i}{dt^2} = G \displaystyle\sum_{j \neq i = 1}^{N} \dfrac{m_j}{r_{ij}^3} (z_j - z_i)
\end{cases}
\tag{1.3}
$$

where the distance $|\overrightarrow{P_i P_j}|$ is:

$$
r_{ij} = r_{ji} = \left[(x_j - x_i)^2 + (y_j - y_i)^2 + (z_j - z_i)^2 \right]^{1/2}. \tag{1.4}
$$

Each equation of the system (1.3) can be transformed in a pair of differential equations of the first order by setting: $X_i = \dot{x}_i$, $Y_i = \dot{y}_i$, and $Z_i = \dot{z}_i$. In this way the system (1.3) is replaced by another set of $6N$ first order scalar differential equations:

$$
\begin{cases}
X_i = \dot{x}_i \\[1.5ex]
\dot{X}_i = G \displaystyle\sum_{j \neq i = 1}^{N} \dfrac{m_j}{r_{ij}^3} (x_j - x_i) \\[2.5ex]
Y_i = \dot{y}_i \\[1.5ex]
\dot{Y}_i = G \displaystyle\sum_{j \neq i = 1}^{N} \dfrac{m_j}{r_{ij}^3} (y_j - y_i) \qquad\qquad (i \neq j = 1, \dots, N). \\[2.5ex]
Z_i = \dot{z}_i \\[1.5ex]
\dot{Z}_i = G \displaystyle\sum_{j \neq i = 1}^{N} \dfrac{m_j}{r_{ij}^3} (z_j - z_i)
\end{cases}
\tag{1.5}
$$

For a generalization of this transformation and of what follows in this section, see Sect. 4.2.

The search for general solutions of the system (1.3) is one of the major steps in the history of celestial mechanics, and is the pivot around which much of the activity of physicists and mathematicians from the 18-th century to the first half of the 20-th century has revolved.

We can perhaps argue that, if the system (1.3) were easy to solve for any value of N (or at least for N large enough), celestial mechanics would have had difficulty acquiring its independence from the other physical, mathematical, and astronomical disciplines. In turn, an important stimulus for building fundamental tools of mathematical analysis and computation would have been lost.

Let us now clarify what we mean with general solution of the system (1.3). The Cauchy[10]-Lipschitz[11] theorem ensures the existence and uniqueness of the solutions for the systems (1.3) or (1.5) only in the neighborhood of a regular point.[12] Therefore, with the help of analytic or numeric methods, it is always possible, although not necessarily easy, to find a solution associated with particular initial conditions, and extend it to an arbitrarily large (but still finite) time interval. We must proceed as nature does. At each epoch t, we search for the infinitesimal displacement relative to each point P_i as a consequence of the particular configuration of positions and velocities assumed at that time by the other particles, iterating the process over a convenient time interval. When the tools of mathematical analysis fail to provide us with a solution, we can resort to numerical integration.

It should be well understood, though, that the direct solution applies only to problems with explicit initial conditions: precisely what happens in nature, where only specific problems are tackled. The direct approach does not provide the general solution we are interested in. It does not inform us, for instance, about the variations that the dynamic configuration of the system undergoes because of minor changes in the initial conditions. Moreover, it is a poor and expensive surrogate to an analytical criterion which, on the ground of the knowledge of the initial conditions, tells us whether a given set of bodies forms a bound system rather than one where the mutual distances grow indefinitely as the motion progresses.

The general solution of the system (1.3) is a set of $3N$ pairs of independent functions,[13] one for each coordinate of position and velocity:

$$\begin{cases} x_i = x_i(t; a_1, a_2, \ldots, a_{6N}) \\ \dot{x}_i = \dot{x}_i(t; a_1, a_2, \ldots, a_{6N}) \\ y_i = y_i(t; a_1, a_2, \ldots, a_{6N}) \\ \dot{y}_i = \dot{y}_i(t; a_1, a_2, \ldots, a_{6N}) \\ z_i = z_i(t; a_1, a_2, \ldots, a_{6N}) \\ \dot{z}_i = \dot{z}_i(t; a_1, a_2, \ldots, a_{6N}) \end{cases} \qquad (i = 1, \ldots, N) , \qquad (1.6)$$

---

[10] Augustin-Louis Cauchy (1789–1857); French mathematician and engineer, one of the fathers of mathematical analysis. Very prolific writer, he was among the first to state and rigorously prove theorems of calculus using the concepts of limit and continuity. He also founded complex analysis.

[11] Rudolph Lipschitz (1832–1903); German mathematician who contributed to mathematical analysis, differential geometry, and classical mechanics.

[12] Cauchy-Lipschitz or the existence and uniqueness theorem ensures that a system of n linear differential equations of the first order, rewritten in compact terms as $\dot{\mathbf{u}} = \mathbf{f}(t, \mathbf{u})$, where $\mathbf{u} = (u_1, \ldots, u_n)$ and $\mathbf{f} = (f_1, \ldots, f_n)$, associated with proper initial conditions at an epoch t_0, $\mathbf{u}(t_0) = \mathbf{u}_0$ (this set of equations is referred to as Cauchy problem), have only one solution in the neighborhood of t_0. The function $\mathbf{f}$ must be defined in a neighborhood of the point $(t_0, \mathbf{u}_0)$, where it must be at least continuous and satisfy Lipschitz condition with respect to $\mathbf{u}$. Systems of the type $\ddot{\mathbf{u}} = \mathbf{f}(t, \mathbf{u})$, with a degree higher than the first, can be transformed into systems of first order equations through the simple position $\mathbf{v} = \dot{\mathbf{u}}$, thus re-entering in the case described above.

[13] Possible dependence inside the solutions would indicate the existence of constraints (cf. Sect. 4.2), which are not considered in the formulation of the problem given in this section.

which satisfy the system identically for any value of the time t. They contain $6N$ integration constants a_i, as many as the first order equations (1.5), that link the solutions to the set of initial conditions.

The independence of the functions (1.6) is also the condition for the inversion of the system, that is for the existence of $6N$ constant functions of coordinates, velocities, and time, independent of each other:

$$a_i = a_i(x_1, y_1, z_1, \ldots, x_N, y_N, z_N, \dot{x}_1, \dot{y}_1, \dot{z}_1, \ldots, \dot{x}_N, \dot{y}_N, \dot{z}_N, t), \qquad (1.7)$$

that are called first integrals or integrals of motion. The existence of $6N$ first integrals for the system of equations (1.3), i.e., of $6N$ constant functions of coordinates and velocity components, is a necessary condition for the existence of the general solution. But, if the system (1.7) is reversed, one obtains again (1.6), that is the general solution of (1.3). So, the existence of the $6N$ first integrals is also a sufficient condition for the existence of the general solution. This means that the search for solutions of the (1.3) can be completely replaced by the search for $6N$ independent first integrals.

1.2 The Fundamental First Integrals

We prove here that, for any $N \geq 2$, the system of equations (1.3) has always at least 10 first (scalar) integrals. Six of them are related to the motion of the center of mass, three to the projections of the total angular momentum on the coordinate axes, and the last one to the total energy. We will see (cf. Chap. 2) that, if $N = 2$, the search for the remaining integral of motion is reduced to simple quadratures. The difficulties arise when $N > 2$.

There are theorems due to Bruns[14] and Poincaré,[15] which prove that, under selected coordinate systems, no further first integrals exist either in the form of algebraic and uniformly transcendent functions [2, 3]. These theorems do not solve completely the problem as they lack exploring all the possible types of integrals and coordinate systems. However, no any method capable of providing the complete solution of the so-called N-body problem is known to date when $N > 2$. Actually, between 1906 and 1909 the Finnish mathematician and astronomer Karl Frithiof Sundman (1873–1949) produced an exact analytical solution of the three-body problem, developing an infinite convergent series of powers of $t^{1/3}$; but the mathematical formulation reaches a level of complexity that makes the solution unusable. The

---

[14] Ernst Heinrich Bruns (1848–1919); German mathematician, astronomer, and geodesist.

[15] Jules Henri Poincaré (1854–1912): French mathematician and physicist, one of the greatest and most influential scientists of his time, active in many fields, from pure to applied mathematics, from mathematical physics to celestial mechanics, where he was the first to discover a deterministic chaotic system, setting the foundations of the modern theory of chaos. His researches on the principle of relativity and the Lorentz transformations have been fundamental for the formulation of the special theory of relativity.

convergence of the series is so slow that nearly 10^7 terms are required to return an astronomically acceptable accuracy.

We shall see in Chap. 4 how to get around the obstacle of the unsolvability of the *N*-body problem though approximate methods based on the knowledge of the exact solution of the two-body problem, which we explore in the next chapter. Here we briefly consider the simplest case of $N = 1$. If the only body in play is isolate (infinitely far from other bodies and not subject to external forces[16]), we say that the reference system in which it remains stationary ($\mathbf{v} = 0$) or moves uniformly in a straight line ($\mathbf{v} = $ constant vector) is inertial. This statement is nothing but the principle of inertia or first principle of dynamics. Formulated by Galileo Galilei[17] following his studies on the motion of bodies on inclined and horizontal planes, this principle opposed the erroneous Aristotelian theory according to which a moving body tends to slow down and stop unless it is supported by an external force. Clearly, the opinion of Aristotle is biased by ignorance of friction. Everyday experience seems to support the philosopher of Stagira, but in an ideal experiment where friction is negligible (similar to that made by Galilei himself), a body which is not subject to forces continues in its motion indefinitely and with constant velocity.

1.2.1 The Conservation of Momentum

Let us now return to a *N*-body system. Adding up all the equations of (1.3) related to the same axis, three identities are obtained:

$$\sum_{i=1}^{N} m_i \ddot{x}_i = 0,$$

$$\sum_{i=1}^{N} m_i \ddot{y}_i = 0, \qquad\qquad\qquad (1.8a)$$

$$\sum_{i=1}^{N} m_i \ddot{z}_i = 0.$$

To be convinced, just apply the third principle of the dynamics to each pair of bodies and sum up all the pairs. By integrating a first time the (1.8a), it is:

---

[16] There are no internal forces since we deal with a massive point.

[17] The inertia principle has been anticipated in a qualitative form in Dante Alighieri's *Divine Comedy* (*Inf.* XVII, vv. 115-117), where the Italian poet describes the descent of the steep cliff from the 7-th to the 8-th circle riding the monster Geryon without feeling the movement (Fig. 1.4: *Onward he goeth, swimming slowly, slowly; / Wheels and descends, but I perceive it only / By wind upon my face and from below.*

Fig. 1.4 Geryon flies
carrying Dante and Virgil on
his back; illustration by
Gustave Doré

$$\sum_{i=1}^{N} m_i \dot{x}_i = p_x,$$

$$\sum_{i=1}^{N} m_i \dot{y}_i = p_y, \qquad (1.8b)$$

$$\sum_{i=1}^{N} m_i \dot{z}_i = p_z,$$

where $\mathbf{p} = (p_x, p_y, p_z)$ has the dimensions of a momentum. By integrating once
more, we obtain 6 first integrals:

$$\sum_{i=1}^{N} m_i x_i = p_x t + q_x,$$

$$\sum_{i=1}^{N} m_i y_i = p_y t + q_y, \qquad (1.8c)$$

$$\sum_{i=1}^{N} m_i z_i = p_z t + q_z.$$

By introducing the coordinates of the barycenter,[18] (x_c, y_c, z_c), it is:

$$
\begin{aligned}
\sum_{i=1}^{N} m_i x_i = x_c \sum_{i=1}^{N} m_i = x_c M &\quad\Rightarrow\quad x_c = \frac{p_x t + q_x}{M}, \\
\sum_{i=1}^{N} m_i y_i = y_c \sum_{i=1}^{N} m_i = y_c M &\quad\Rightarrow\quad y_c = \frac{p_y t + q_y}{M}, \qquad (1.9)\\
\sum_{i=1}^{N} m_i z_i = z_c \sum_{i=1}^{N} m_i = z_c M &\quad\Rightarrow\quad z_c = \frac{p_z t + q_z}{M},
\end{aligned}
$$

where $M = \displaystyle\sum_{i=1}^{N} m_i$ is the total mass of the system. It is apparent that the three sets of (1.8a), (1.8b), and (1.8c) represent the application to an isolated mechanical system of the theorem on the motion of the mass center: in an inertial reference frame this fictitious point moves with rectilinear and uniform motion with a time law of motion given by (1.9).

The existence of the 6 first integrals $(p_x, p_y, p_z, q_x, q_y, q_z)$, corresponding to the motion of the center of gravity, can be inferred more elegantly from the properties of the potential function (see also Chap. 5):

$$\mathcal{U} = G \sum_{i=1}^{N} \sum_{j>i}^{N} \frac{m_i m_j}{r_{ij}}, \qquad (1.10a)$$

or:

$$\mathcal{U} = \frac{1}{2} G \sum_{i=1}^{N} \sum_{j\neq i=1}^{N} \frac{m_i m_j}{r_{ij}}. \qquad (1.10b)$$

---

[18] Given a generic reference system centered on O, the barycenter or center of mass B of a system of N points P_i of mass m_i is the dynamical point satisfying the equation: $\displaystyle\sum_{i=1}^{N} m_i \overrightarrow{OP_i} = \overrightarrow{OB} \sum_{i=1}^{N} m_i$, or $\displaystyle\sum_{i=1}^{N} m_i \overrightarrow{BP_i} = 0$ (for a continuous body replace the summation with an integral). The name barycenter comes from the combination of two words of ancient Greek for heavy and center.

It is easily verified that:

$$m_i \ddot{x}_i = \frac{\partial \mathcal{U}}{\partial x_i},$$
$$m_i \ddot{y}_i = \frac{\partial \mathcal{U}}{\partial y_i}, \qquad (1.11)$$
$$m_i \ddot{z}_i = \frac{\partial \mathcal{U}}{\partial z_i}.$$

For instance, recalling (1.3):

$$\frac{\partial \mathcal{U}}{\partial x_i} = G\, m_i \frac{\partial}{\partial x_i} \sum_{j \neq i=1}^{N} \frac{m_j}{r_{ij}} = G\, m_i \sum_{j \neq i=1}^{N} \frac{m_j}{r_{ij}^3} (x_j - x_i) = m_i \ddot{x}_i. \qquad (1.12)$$

The potential $\mathcal{U}$ tends to zero only when all the distances r_{ij} between the particles tend to infinity, i.e., when the system becomes completely disconnected. Conversely, $\mathcal{U} \to \infty$ when at least one of the mutual distances tends to zero, i.e., at least two particles collide.[19]

The potential function $\mathcal{U}$ depends on the coordinates of the point-like bodies P_i, but only through special combinations of these, corresponding to the mutual distances between pairs of bodies. Therefore $\mathcal{U}$ has to be invariant to translations of the reference system due to the homogeneity of space.[20] This implies that:

$$\sum_{i=1}^{N} \frac{\partial \mathcal{U}}{\partial x_i} = \sum_{i=1}^{N} \frac{\partial \mathcal{U}}{\partial y_i} = \sum_{i=1}^{N} \frac{\partial \mathcal{U}}{\partial z_i} = 0. \qquad (1.13)$$

In fact, let it be for example:

$$x_i' = x_i + x_o \qquad (i = 1, \ldots, N), \qquad (1.14)$$

with $x_o =$ constant. The partial derivative of $\mathcal{U}$ with respect to x_o is:

$$\frac{\partial \mathcal{U}}{\partial x_o} = \sum_{i=1}^{N} \frac{\partial \mathcal{U}}{\partial x_i'} \frac{\partial x_i'}{\partial x_o} = \sum_{i=1}^{N} \frac{\partial \mathcal{U}}{\partial x_i'}, \qquad (1.15)$$

---

[19] This condition is purely formal in that, at small distances, it is no longer possible to ignore the physical nature of the massive points and therefore the existence of a Schwarzschild radius which assigns them a sort of finite dimension (see p. 19). The singularity can also intervene in the numerical calculation and must be properly treated.

[20] Homogeneity of space (from the classical Greek 'same' and 'type') means that no point is special, so the same basic laws of physics govern all of the space.

as $\partial x_i'/\partial x_o = 1$. But it must vanish since $\mathcal{U}$ is independent of x_o. Eliminating the quotes from (1.15), we obtain the first of the (1.13). Finally, using (1.11), we find again the (1.8a).

1.2.2 The Conservation of Angular Momentum

We now prove the existence of three first integrals connected to components of the total angular momentum by resorting to another characteristic of the potential (1.10a). This function is invariant to rotations of the reference system since classical space is not only homogeneous but also isotropic.[21] Consider, as an example, a rotation of the reference system by an angle ϕ around the z axis, set by the transformation equations:

$$
\begin{aligned}
x_i' &= x_i \cos \phi - y_i \sin \phi \\
y_i' &= x_i \sin \phi + y_i \cos \phi \qquad (i = 1, \dots, N). \\
z_i' &= z_i
\end{aligned}
\tag{1.16}
$$

If the angle ϕ is small, we have $\sin \phi \sim \phi$, $\cos \phi \sim 1$, and, in the differential form:

$$
\begin{aligned}
dx_i &= x_i' - x_i = -y_i d\phi, \\
dy_i &= y_i' - y_i = x_i d\phi, \\
dz_i &= 0.
\end{aligned}
\tag{1.17}
$$

Being invariant to rotations, $\mathcal{U}$ does not depend on the coordinates but on the mutual distances only; thus, using (1.17), the total differential is:

$$
\begin{aligned}
d\mathcal{U} = \sum_{i=1}^{N} \left(\frac{\partial \mathcal{U}}{\partial x_i} dx_i + \frac{\partial \mathcal{U}}{\partial y_i} dy_i + \frac{\partial \mathcal{U}}{\partial z_i} dz_i \right) = \\
= \sum_{i=1}^{N} \left(-\frac{\partial \mathcal{U}}{\partial x_i} y_i d\phi + \frac{\partial \mathcal{U}}{\partial y_i} x_i d\phi \right) = \\
= d\phi \sum_{i=1}^{N} \left(-\frac{\partial \mathcal{U}}{\partial x_i} y_i + \frac{\partial \mathcal{U}}{\partial y_i} x_i \right) = 0.
\end{aligned}
\tag{1.18}
$$

By repeating the exercise for rotations about x and y and combining the results with the (1.11), we obtain the system:

---

[21] Isotropy of space (from the classical Greek 'equal' and 'turn') means that no direction is special.

$$\sum_{i=1}^{N} m_i \left(y_i \frac{d^2 z_i}{dt^2} - z_i \frac{d^2 y_i}{dt^2} \right) = 0,$$

$$\sum_{i=1}^{N} m_i \left(z_i \frac{d^2 x_i}{dt^2} - x_i \frac{d^2 z_i}{dt^2} \right) = 0, \qquad (1.19)$$

$$\sum_{i=1}^{N} m_i \left(x_i \frac{d^2 y_i}{dt^2} - y_i \frac{d^2 x_i}{dt^2} \right) = 0,$$

which integrates in:

$$\sum_{i=1}^{N} m_i \left(y_i \frac{dz_i}{dt} - z_i \frac{dy_i}{dt} \right) = h_x,$$

$$\sum_{i=1}^{N} m_i \left(z_i \frac{dx_i}{dt} - x_i \frac{dz_i}{dt} \right) = h_y, \qquad (1.20)$$

$$\sum_{i=1}^{N} m_i \left(x_i \frac{dy_i}{dt} - y_i \frac{dx_i}{dt} \right) = h_z.$$

The three relations (1.20) are first integrals. Their left terms express the projections of the total angular momentum $\mathbf{h} = (h_x, h_y, h_z)$ on each of the coordinated axes. It is convenient to rewrite (1.20) in the vector form: $\sum_{i=1}^{N} \mathbf{r}_i \times m_i \dot{\mathbf{r}}_i = \mathbf{h}$. The constancy of the individual components requires the total angular momentum vector to be also constant and hence the existence of a single privileged plane, called invariable plane by Laplace.[22] Since it contains the center of gravity and is orthogonal to the total moment $\mathbf{h}$, its position is independent of time.

1.2.3   The Conservation of Energy

The last of the 10 first integrals listed above is a consequence of the fact that the gravitational potential does not depend explicitly on time. Multiplying the (1.11) by

---

[22] French mathematician, physicist, and astronomer, Pierre Simon de Laplace (1749–1827) gave a great contribution to the establishment of determinism and, through his famous *Treatise on Celestial Mechanics* (five volumes published between 1799 and 1825), to the transformation of Newtonian mechanics from a geometric science to an analytical discipline. He was the one who coined the term 'celestial mechanics' for that part of classical mechanics that applies to the motions of celestial objects. He was also one of the founders of the probability theory. Made count by the Napoleon, of whom he was also a minister of the interior (quickly dismissed as "*he brought the spirit of the infinitely small into the government*"), after the Restoration he received the title of marquis by the Bourbons; a genius in science and in politics a man for all seasons.

$\dot{x}_i$, $\dot{y}_i$, and $\dot{z}_i$ respectively, then summing up for the index i, and finally adding all the results:

$$\sum_{i=1}^{N} m_i \left(\frac{d^2 x_i}{dt^2} \frac{dx_i}{dt} + \frac{d^2 y_i}{dt^2} \frac{dy_i}{dt} + \frac{d^2 z_i}{dt^2} \frac{dz_i}{dt} \right) = \frac{d\mathcal{U}}{dt}, \qquad (1.21)$$

or:

$$\frac{d}{dt} \left(\sum_{i=1}^{N} \frac{1}{2} m_i \left[\left(\frac{dx_i}{dt} \right)^2 + \left(\frac{dy_i}{dt} \right)^2 + \left(\frac{dz_i}{dt} \right)^2 \right] \right) = \frac{d\mathcal{U}}{dt}. \qquad (1.22)$$

By integrating over time, we obtain the 10-th first integral:

$$T = \mathcal{U} + \mathcal{E}_{\circ}, \qquad (1.23)$$

where:

$$T = \frac{1}{2} \sum_{i=1}^{N} m_i \left[\left(\frac{dx_i}{dt} \right)^2 + \left(\frac{dy_i}{dt} \right)^2 + \left(\frac{dz_i}{dt} \right)^2 \right], \qquad (1.24)$$

is the total kinetic energy,[23] evaluated with respect to the adopted inertial reference system, and the integration constant $\mathcal{E}_{\circ} = T - \mathcal{U}$ is the total energy of the system; in fact, $-\mathcal{U}$ is the total potential energy. This statement is verified by calculating the work done by the mutual attractive force between two bodies P_i and P_j for a displacement which changes the mutual distance from r_{ij}° into $r_{ij} < r_{ij}^{\circ}$:

$$W_{ij} = -G\, m_i m_j \int_{r_{ij}^{\circ}}^{r_{ij}} \frac{dr'_{ij}}{r_{ij}'^2} = G\, m_i m_j \left(\frac{1}{r_{ij}} - \frac{1}{r_{ij}^{\circ}} \right). \qquad (1.25)$$

By assuming that the two bodies are originally at an infinite distance ($r_{ij}^{\circ} \to +\infty$), the (1.25) becomes[24]:

$$W_{ij} = G \frac{m_i m_j}{r_{ij}}, \qquad (1.26)$$

---

[23] Some philosophers of antiquity, such as Thales of Miletus, had already sensed the concept of energy conservation. The first to give it a mathematical formulation was Leibniz. He noted that, in a dynamical system, the sum of the products of the mass of each particle times the squared velocity, $\sum_i m_i v_i^2$, is conserved, and he named this quantity *vis viva*, from the Latin 'living force', as opposed to *vis mortua*, 'dead force', which is our inertia. This principle of conservation of the kinetic energy, also called theorem of the 'living forces', had a coeval rival in another fundamental physical principle due to the French natural philosopher Descartes and to Newton: the conservation of the momentum, i.e., of the vector quantity $\sum_i m_i \mathbf{v}_i$. Later, the two principles became complementary. With the advent of thermodynamics, in the 18-th century, it was understood that the kinetic energy is not conserved as such, since it is partly converted into heat.

[24] Notice the sign in (1.25): the work done by gravity on the system in going from infinity to a finite distance is positive because the field is attractive.

which provides the work needed for the pair of bodies to go from infinity to the mutual distance r_{ij}. By the definition of potential (1.10b), for a N-body system it is:

$$\mathcal{U} = \frac{1}{2} \sum_{i=1}^{N} \sum_{j=1}^{N} W_{ij}, \qquad (i \neq j). \qquad (1.27)$$

For instance, the potential energy of the Sun tells us how much gravitational energy the star was able to extract from its own self-gravity since the time it started its collapse as a protostar. This value is far lower than needed to fuel the Sun during its lifetime of 4.6 billion years (an overall budget that we deduce by assuming that the Solar luminosity has been constant over time),[25] and induces to look for other, more efficient sources (nuclear energy). Note however that gravitational energy is still an important resource for stars, which they draw on in the protostellar phase (before the ignition of the nuclear source) and in all those phases in which nuclear sources are insufficient or even extinct.

The expression (1.23) shows that the total energy, $\mathcal{E}_o$, of an isolated self-gravitating system of N massive points is an undefined constant. In fact, both the kinetic and the potential energy have arbitrary zero points.

The total kinetic energy consists of the sum of two terms. The first is given by the motion of the N bodies relative to center of gravity (for this reason we call it specific kinetic energy); the second by the motion of the center of gravity treated as a point holding the entire mass M of the system. Indeed, said B the center of gravity, it is:

$\overrightarrow{OP_i} = \overrightarrow{OB} + \overrightarrow{BP_i}$, and thus, indicating the total mass with $M = \sum_{i=1}^{N} m_i$:

---

[25] The gravitational potential energy of a spherically symmetric body of density $\rho(r)$ and radius R is obtained by integrating the contributions $d\text{W} = -\dfrac{GM_r}{r} 4\pi r^2 \rho \, dr$ by the thin spherical layers of radius $0 \leq r \leq R$, where $M_r = 4\pi \int_0^r \rho r^2 dr$. It is $\text{W} = -4\pi G \int_0^R M_r \, \rho r \, dr$, which integrates in $\text{W} = -\dfrac{3GM^2}{5R}$ if $\rho = \rho_o = $ const, and $M = \dfrac{4}{3}\pi\rho_o R^3$. From the virial theorem (see Sect. 1.5) we learn that a gravitational system passing from one equilibrium state to another must turn half of its available energy into internal kinetic energy. The other half is radiated away. Therefore, from its birth as a protostar, the Sun has radiated $\Delta\mathcal{E}_r = \dfrac{3GM_\odot^2}{10R_\odot} = 1.14 \times 10^{41}$ J at the expenses of its potential energy. Since our star shines at a rate of $L_\odot = 3.832 \times 10^{26}$ J s^{-1}, whether this flux rate has been preserved over time, the gravitational energy has lasted for not more than 2.9×10^{14} s $\simeq 10^7$ yr.

$$T = \frac{1}{2} \sum_{i=1}^{N} m_i v_i^2 = \frac{1}{2} \sum_{i=1}^{N} m_i \left(\frac{d\overrightarrow{OP_i}}{dt} \right)^2 = \frac{1}{2} \sum_{i=1}^{N} m_i \left(\frac{d\overrightarrow{OB}}{dt} + \frac{d\overrightarrow{BP_i}}{dt} \right)^2 =$$

$$= \frac{1}{2} M \left(\frac{d\overrightarrow{OB}}{dt} \right)^2 + \frac{1}{2} \sum_{i=1}^{N} m_i \left(\frac{d\overrightarrow{BP_i}}{dt} \right)^2 + \frac{d\overrightarrow{OB}}{dt} \sum_{i=1}^{N} m_i \frac{d\overrightarrow{BP_i}}{dt} =$$

$$= \frac{1}{2} M \left(\frac{d\overrightarrow{OB}}{dt} \right)^2 + \frac{1}{2} \sum_{i=1}^{N} m_i \left(\frac{d\overrightarrow{BP_i}}{dt} \right)^2, \qquad (1.28)$$

since, according to the definition of barycenter, it is $\sum_{i=1}^{N} m_i \frac{d\overrightarrow{BP_i}}{dt} = 0$.

The indetermination about T is therefore a consequence of the Galilean equivalence principle of reference systems in relative translatory and uniform motion, and it is eliminated considering a system of coordinates connected to the center of gravity of the mechanical system, which is allowed in the absence of external forces.

The indetermination about the potential energy derives formally from the definition of the scalar potential as a primitive function of a vector force field; integration involves the introduction of an arbitrary constant that can be fixed only by imposing some initial conditions. This is exactly what we did by assuming, in (1.26), that the gravitational potential between two bodies at infinite distance is zero.

In conclusion, lacking proper conventions, it makes little sense to talk of total energy, although it stays constant. In the following we stipulate that, unless otherwise stated, the total energy is always the specific one, with motions relative to the gravity center and $\mathcal{U} = 0$ at infinity.

1.3   The N-Body Problem in Barycentric and in Relative Systems

Consider the motion of the particle P_i in the barycentric reference system. Being $\mathbf{r}_i$ the radius vector of P_i, the equation of motion writes as:

$$\frac{d^2 \mathbf{r}_i}{dt^2} = G \sum_{j=1}^{N} m_j \frac{\mathbf{r}_j - \mathbf{r}_i}{r_{ij}^3}, \qquad (i \neq j = 1, \ldots, N). \qquad (1.29)$$

From the definition of the barycenter:

$$\sum_{i=1}^{N} m_i \, \mathbf{r}_i = 0, \qquad (1.30)$$

it follows that:

$$\mathbf{r}_1 = -\frac{1}{m_1} \sum_{i=2}^{N} m_i \mathbf{r}_i, \tag{1.31}$$

where we have isolated the coordinates of one of the masses, arbitrarily choosing that with the index $i = 1$ with no loss of generality. By putting in evidence the terms related to m_1 in (1.29), we have:

$$\frac{1}{G} \ddot{\mathbf{r}}_i = \frac{m_1 \mathbf{r}_1}{r_{i1}^3} - \frac{m_1 \mathbf{r}_i}{r_{i1}^3} + \sum_{i=2}^{N} m_j \frac{\mathbf{r}_j - \mathbf{r}_i}{r_{ij}^3} =$$

$$= -\frac{1}{r_{i1}^3} \sum_{j \neq i=2}^{N} m_i \mathbf{r}_i - \frac{m_1 \mathbf{r}_i}{r_{i1}^3} + \sum_{j \neq i=2}^{N} m_j \frac{\mathbf{r}_j - \mathbf{r}_i}{r_{ij}^3}. \tag{1.32}$$

If we divide the first sum of (1.32) in two terms with i and $j \neq i$, with simple manipulations we obtain:

$$\ddot{\mathbf{r}}_i + G(m_1 + m_i) \frac{\mathbf{r}_i}{r_{i1}^3} = G \sum_{j \neq i=2}^{N} m_j \left[\frac{\mathbf{r}_j - \mathbf{r}_i}{r_{ij}^3} - \frac{\mathbf{r}_j}{r_{i1}^3} \right]. \tag{1.33}$$

The first term on the right side of (1.33) is the sum of the forces exerted on the i-th mass by all the others of the system. The second term is the sum of the inertial forces due to the acceleration of the mass m_1 under the influence of others. The barycentric reference system is inertial since, in the absence of external forces, the center of gravity is stationary or moves by a uniform rectilinear motion.

In the event that one of the masses largely exceeds all the others, i.e., when $m_1 \gg m_i$, where the choice of the index for the dominant mass does not invalidate the generality of the discussion, it may be convenient to place the origin of the system on it. We will call this system a relative reference system.

Let then assume that P_1 coincides with the origin, so that $\mathbf{r}_i' = \overrightarrow{P_1 P_i}$ is the radius vector of point P_i. The relations of the new coordinates to the old ones are:

$$\mathbf{r}_i' = \mathbf{r}_i - \mathbf{r}_1. \tag{1.34}$$

Isolating in (1.30) the terms relative to the masses m_1 and m_i, we write:

$$m_1 \mathbf{r}_1 + m_i \mathbf{r}_i + \sum_{j \neq i=2}^{N} m_j \mathbf{r}_j = 0, \tag{1.35}$$

which, substituted in (1.34) to eliminate $\mathbf{r}_1$, gives:

$$m_1 \mathbf{r}'_i = (m_1 + m_i)\mathbf{r}_i + \sum_{j \neq i = 2}^{N} m_j \mathbf{r}_j. \tag{1.36}$$

Let us now derive (1.34) twice:

$$\ddot{\mathbf{r}}'_i = \ddot{\mathbf{r}}_i - \ddot{\mathbf{r}}_1; \tag{1.37}$$

using (1.33) for $\ddot{\mathbf{r}}_i$ and (1.29) with $i = 1$ for $\ddot{\mathbf{r}}_1$, after some simplifications we obtain the equation of motion for the relative reference system:

$$\ddot{\mathbf{r}}'_i + G(m_1 + m_i)\frac{\mathbf{r}'_i}{r'^3_i} = G \sum_{j \neq i = 2}^{N} m_j \left[\frac{\mathbf{r}'_j - \mathbf{r}'_i}{r'^3_{ij}} - \frac{\mathbf{r}'_j}{r'^3_j} \right], \tag{1.38}$$

where $r_{1i} = r'_i = \sqrt{(x'_i)^2 + (y'_i)^2 + (z'_i)^2}$. Again, the second term on the left side of (1.38) is the result of the gravitational influence of m_1 on the i-th mass. The first term on the right side of this equation is the sum of the forces exerted on the i-th mass by all the others of the system. The second term is the sum of the inertial forces due to the acceleration of the mass m_1 under the influence of others.

Remember that in this case the coordinate system is no longer inertial. This implies that the first motion integral for the center of mass does not exist. We can rewrite the (1.38) in the form:

$$\ddot{\mathbf{r}}_i + G(m_1 + m_i)\frac{\mathbf{r}_i}{r^3_i} = \mathbf{F}_i, \tag{1.39}$$

omitting the prime. When $m_1 \gg m_i$ for all $i = 2$ to N, it happens that $\mathbf{F}_i \ll G(m_1 + m_i)\frac{\mathbf{r}_i}{r^3_i}$. In this case the N-body problem can be replaced by $(N - 1)$ two-body problems, as many as the direct interactions of each particle P_i with P_1. This is what we pretend happens in the Solar System, when we consider the motion of each planet around the Sun using the two-body problem approach ($\mathbf{F}_i = 0$). The gravitational influence of the other planets obviously remains, but it is only a perturbation. It can be tackled by the methods illustrated in the following chapters.

1.4 On the Solution of the N-Body Problem

Let us now try to understand why the N-body problem does not admit a general solution for $N \geq 3$. To this end, we must first realize that, due to the properties of the center of gravity, the actual number of independent bodies is $(N - 1)$. Take any one of the bodies, for example that with index N, and isolate it in (1.9):

$$x_N = \frac{1}{m_N} \left(x_c M - \sum_{i=1}^{N-1} m_i x_i \right),$$

$$y_N = \frac{1}{m_N} \left(y_c M - \sum_{i=1}^{N-1} m_i y_i \right), \qquad (1.40)$$

$$z_N = \frac{1}{m_N} \left(z_c M - \sum_{i=1}^{N-1} m_i z_i \right).$$

The coordinates of P_N, and its velocities as they result by differentiating the (1.40), are functions of the coordinates of the other $(N-1)$ points and of the center of gravity, which we can assume identically null since the system is isolated. This means that, in a two-body problem, only one of the two points is independent. If $N > 2$, the number of independent bodies increases to $(N-1) \geq 2$ and there are two or more of them in play. They can exchange energy one with the other by means of collisions (close approaches), which can be very effective as the gravitational potential diverges when the distance between two points, r_{ij}, tends to zero. What does it happens in this case? Just a little difference in the approach trajectory is enough to cause enormous differences in the subsequent development of the motion. In other words, the motions in a system with $N \geq 3$ bodies for two sets of initial conditions differing by infinitesimals can, at a given moment, diversify greatly due to a close encounter between (at least) two of the bodies in play. Therefore, the continuity between initial conditions and the characteristics of the motion of a system with $N \geq 3$ bodies is not guaranteed. This is the reason why it is normally impossible to find the general solution to the N-body problem (i.e., a complete set of first integrals), while it is always possible to search for particular solutions that are sufficiently accurate within an ample time interval or even analytic solutions to special problems.[26] We will resume this question in Chap. 4.

Before going any further, let us look attentively at the divergence of the potential of two massive points as their distance $d = r_{12}$ reduces. While formally d may tend to zero, thus allowing any loss of potential energy and an equally unlimited production of kinetic energy, from a physical point of view d can at most reduce to the sum of the Schwarzschild[27] radii r_g of the two massive points, where the gravitational (Schwarzschild) radius for a mass m is [5]:

---

[26] These considerations form the basis of the chaos theory, i.e., the study of dynamical systems that exhibit exponential sensitivity to initial conditions [4].

[27] Karl Schwarzschild (1873–1916) was a German mathematician, physicist, and astronomer in the years when general relativity, quantum mechanics, and astrophysics were born. He died young due to a disease aggravated by the hard living conditions at the Russian front during the First World War. His son Martin, American citizen, became a famous theoretical astrophysicists. The gravitational radius is named after him since it is the spherically symmetric solution of Einstein's equations that he found in 1916. We will apply his solution – Schwarzschild's metric (4.370) – to find the relativistic advance of Mercury perihelion in Sect. 4.14.

$$r_g = \frac{2Gm}{c^2} = 1.485 \times 10^{-27} \frac{m}{1\,\mathrm{kg}} \mathrm{m} = 2.95 \times \left(\frac{m}{M_\odot}\right) \mathrm{km}, \qquad (1.41)$$

with c the speed of light and $M_\odot = 1.989 \times 10^{33}$ g the mass of the Sun. For one Solar mass, $r_g \simeq 3$ km. It is easy to verify that under these conditions the maximum potential energy removable from a pair of points with equal mass m is of the order of mc^2. From (1.26):

$$W_{1,2} = \Delta \mathcal{U} = G\frac{m_1 m_2}{r_{12}} = G\frac{m^2}{r_g} = \frac{mc^2}{2}. \qquad (1.42)$$

In the so-called N-body simulations, applied to study the evolution of sets of many points (up to $N \simeq 10^9$ and beyond[28]) by solving numerically the system of differential equations of motion, the divergence of the potential as r_{ij} tends to zero is controlled by an ad hoc modification of the expression of the potential of each massive point. A minimum spatial scale, ϵ, called softening length, is introduced to force the regularization [6]:

$$\mathcal{U} = G \sum_{i=1}^{N} \sum_{j \neq i=1}^{N} \frac{m_i m_j}{\sqrt{r_{ij}^2 + \epsilon^2}}; \qquad (1.43)$$

so that the force acting on each particle P_i:

$$\mathbf{f}_i = \nabla \left(\sum_{j \neq i=1}^{N} \frac{m_i m_j}{\sqrt{r_{ij}^2 + \epsilon^2}} \right) = m_i \sum_{j \neq i=1}^{N} \frac{m_j}{\left(r_{ij}^2 + \epsilon^2\right)^{3/2}} \mathbf{r}_{ij}, \qquad (1.44)$$

does not diverge for small values of r_{ij}. Most of the spectacular simulations of galaxies encounters downloadable from the web have been made with this 'gimmick'.

1.5 The Virial Theorem: Classical Formulation

This paragraph is devoted to a statistical relation between the potential and the kinetic energies of a dynamical system. Born within the thermodynamics, it has found useful applications in several branches of physics, from astrophysics to classical mechanics.

Given a system of N massive points and an inertial reference frame, consider the explicit function of time:

$$G(t) = \sum_{i=1}^{N} \mathbf{p}_i \cdot \mathbf{r}_i, \qquad (1.45)$$

---

[28] See, for instance, https://www.illustris-project.org, the site of the cosmological simulation project named *Illustris*.

where $\mathbf{p}_i = m_i \dot{\mathbf{r}}_i$ is the momentum of the i-th point. The function $G(t)$ has the dimensions of an action. Apart from an arbitrary constant of integration, its primitive is the moment of inertia of the system relative to the origin of the reference system:

$$\int^t G(t)\,dt = \frac{1}{2}\sum_{i=1}^N m_i\,\mathbf{r}_i \cdot \mathbf{r}_i. \tag{1.46}$$

Take now the time derivative of G:

$$\frac{dG(t)}{dt} = \sum_{i=1}^N \mathbf{p}_i \cdot \dot{\mathbf{r}}_i + \sum_{i=1}^N \dot{\mathbf{p}}_i \cdot \mathbf{r}_i. \tag{1.47}$$

The first term at the right side equals twice the total kinetic energy, $T = \frac{1}{2}\sum_{i=1}^N m_i \dot{\mathbf{r}}_i^2$.

By denoting with $\mathbf{f}_i$ the total force acting on the i-th point, the second term can be written as:

$$\sum_{i=1}^N \dot{\mathbf{p}}_i \cdot \mathbf{r}_i = \sum_{i=1}^N \mathbf{f}_i \cdot \mathbf{r}_i. \tag{1.48}$$

If the force field depends on the potential $\mathcal{U}$, then:

$$\sum_{i=1}^N \mathbf{f}_i \cdot \mathbf{r}_i = \sum_{i=1}^N \nabla_i\,\mathcal{U} \cdot \mathbf{r}_i, \tag{1.49}$$

where the differential operator (cf. Appendix C) is calculated with reference to the point P_i. Finally, if the potential $\mathcal{U}$ is a homogeneous function of degree k, the Euler[29] theorem (see p. 228) ensures that:

$$\sum_{i=1}^N \nabla_i\,\mathcal{U} \cdot \mathbf{r}_i = k\,\mathcal{U} = -k\,\mathcal{W}, \tag{1.50}$$

where $\mathcal{W} = -\mathcal{U}$ is the total potential energy. In conclusion, the (1.47) reduces to:

$$\frac{dG(t)}{dt} = 2\,T - k\,\mathcal{W}. \tag{1.51}$$

---

[29] A student of Johann Bernoulli, the Swiss Leonhard Euler (1707–1783) is considered the most important mathematician of the Enlightenment. Extraordinarily prolific, he made fundamental contributions to infinitesimal analysis, special functions, rational and celestial mechanics, and to the theory of numbers. His name is linked to an impressive number of theorems and mathematical objects. He spent 14 year in Saint Petersburg and the following 25 years in Berlin. In 1766 he returned to Russia where he stayed up to his death.

Suppose now that the system is subject to a periodical motion with period P. In this case the value of dG/dt averaged over one period:

$$\frac{1}{P}\int_t^{t+P}\frac{dG(t)}{dt}dt = \frac{1}{P}\Big[G(t+P) - G(t)\Big],\qquad(1.52)$$

is null. Thus, the mean value at the right-hand side of (1.51) is also null:

$$\langle 2T - kW\rangle = 0.\qquad(1.53)$$

This important relation, called classical Virial theorem or Virial of Clausius,[30] works even under hypotheses larger than the simple periodicity of the system. Actually, the average value of (1.52) is null as long as the system is limited in the phase space ($|\mathbf{r}_i| \le r_i^\circ$, and $|\mathbf{p}_i| \le p_i^\circ$, at any time) and P, which now has just the meaning of a time interval, is large enough (formally, when it tends to infinity).

It can also be shown that the condition (1.53) is satisfied even in the presence of dissipative forces, provided that they do not cancel the motion in the time interval in which the average value is calculated.

For elastic forces ($k = 2$), the Virial (1.53) returns the well-known property $\langle T\rangle = \langle W\rangle$, telling us, for example, that the kinetic energy of a simple pendulum is equal, on average, to the potential energy. For the gravitational potential, $k = -1$, and then:

$$\langle 2T + W\rangle = 0,\qquad(1.54)$$

with the usual convention $W(\infty) = 0$.

The Virial theorem can be generalized to include the contribution of electric and magnetic fields and becomes useful for analyzing the effect of magnetic fields in plasmas. In this case, its form becomes more complex than (1.53). For the tensor version of this theorem, fundamental in stellar dynamics, see the book of Binney and Tremaine [1].

References

1. J. Binney, S. Tremaine, *Galactic Dynamics* (Princeton University Press, Princeton, 1987)
2. E.T. Whittaker, *A Treatise on the Analytical Dynamics of Particles and Rigid Bodies: With an Introduction to the Problem of Three Bodies* (Dover, New York, 1944)
3. A. Chenciner, *Poincaré and the Three-Body Problem*, in Poincaré, 1912-2012, Séminaire Poincaré XVI (2012), pp. 45–133. http://www.bourbaphy.fr/chenciner.pdf

[30] Rudolf Clausius (1822–1888); German physicist and mathematician, one of the founders of thermodynamics. He coined the term 'virial' inspired by the Latin *vis viva*; see note 23 in this chapter.

4. R. Badii, A. Politi, *Complexity: Hierarchical Structures and Scaling Physics* (Cambridge University Press, Cambridge, 1997)
5. S. Weinberg, *Gravitation and Cosmology* (Wiley, New York, 1972)
6. H. Zhan, Optimal softening for N-body halo simulations. Astrophys. J. **639**, 617 (2006)

Chapter 2
On the Two-Body Problem

Albert Einstein, *address to*
Prussian Academy of Sciences (1921)

As far as laws of mathematics refer to reality, they are not
certain; and as far as they are certain, they do not refer to reality.

The modern formulation of the two-body problem appeared for the first time in the third and last book of the *Philosophiae Naturalis Principia Mathematica*, the treaty that Isaac Newton published owing to a decisive stimulus and material help by Edmund Halley: unanimously considered one of the most important contributions to the scientific thought. A physical theory and an interaction model were needed to interpret the phenomena occurring on Earth and on the sky. Inspired by Galilei, Newton postulated that the laws of mechanics are universal (they hold below and above the sphere of the Moon, to use the Aristotelian scheme of the world) and that the same force causing the mythical apple fall on the Earth keeps the Moon tied to its primary.[1]

In this sense, Newton's law of gravitational force, the first of the four fundamental interactions of physics to be discovered and modeled mathematically,[2] gained the character of a universal law: an attribute that marks another important victory over

---

[1] From a letter that in mature age Newton wrote, perhaps to the Huguenot scholar Pierre Des Maizeaux, to tell of his discoveries made in the years of the plague 1665–1666: *"I deduced that the forces which keep the planets in their orbits must be reciprocally as the squares of their distances from the centers about which they revolve; and thereby compared the force requisite to keep the Moon in her orbit with the force of gravity at the surface of the Earth; and found them answer pretty nearly"*. As for the myth of the apple, it was invented by Voltaire, *nom de plum* of the French philosopher François-Marie Arouet (1694–1778).

[2] In decreasing order of strength there are the strong nuclear forces, the weak interactions, the electromagnetic forces, and gravity.

© The Author(s), under exclusive license to Springer Nature Switzerland AG 2022
E. Bannikova and M. Capaccioli, *Foundations of Celestial Mechanics*, Graduate Texts in Physics, https://doi.org/10.1007/978-3-031-04576-9_2

the Aristotelian dichotomy between terrestrial and celestial worlds, each with the its own $\varphi \upsilon \sigma \iota \kappa \acute{\eta}$ (physics). It is worth remembering how this law of force postulates the existence of the gravitational mass as a characteristic property of a body, enduring as long as the body keeps its integrity: a concept that is dramatically updated in Einstein's theory of relativity.

The postulated existence of a force directly proportional to the mass is also the theoretical key to close once for ever the millennial cosmological issue about who is at the center of the world (intended as just the Solar System): either the Earth or the Sun. This role obviously belongs to the Sun because its mass is far larger than the Earth's.

To tackle the problem of the motion of two bodies under the mutual attractive force generated by their gravitational masses, a complete mechanical theory was also needed. Newton exposed it in the first two books of the *Principia*. Here, in addition to introducing the inertial mass as a degree of resistance of a body to dynamical changes, quantified by the ratio between the applied force and the acceleration, Newton postulated the existence of a Euclidean absolute space and an absolute time, each one independent of the other (at variance with Descartes' and Leibniz's thoughts), and measurable. These hypotheses[3] survived for over two centuries, during which Newton's mechanics collected extraordinary and countless successes (besides the scientific applications, it was the drive to the industrial revolutions and to the birth of modern technological society), up to the critical revision made by Einstein with the special (in 1905) and the general (in 1915) theory of relativity, with the unification of space and time and the equivalence principle between gravity and inertia.

Newton's gravitational theory gained readily the attention of the scientific community as it was able to reproduce all the three planetary laws found empirically by Johannes Kepler from the year 1609 to 1619. It also provided a convincing explanation of the tides: phenomena that even Galileo Galilei had unsuccessfully tried to interpret using inertia only. Nonetheless, the idea of instant actions at distance implied by the gravitational force was hard to digest. Newton himself found uncom-

---

[3] See this translation from Latin to English of the *Scholium*, the reasoned summary Newton made of his overall theory: "*Hitherto I have laid down the definitions of such words as are less known, and explained the sense in which I would have them to be understood in the following discourse. I do not define time, space, place, and motion, as being well known to all. Only I must observe, that the common people conceive those quantities under no other notions but from the relation they bear to sensible objects. And thence arise certain prejudices, for the removing of which it will be convenient to distinguish them into absolute and relative, true and apparent, mathematical and common*".

fortable to accept it,[4] though he had shown that the instantaneousness of the action was implied by the constancy of the angular momentum required by Kepler's laws.

In astronomy, the Newtonian theory reached the top of the success in the middle of the 19th century when Le Verrier predicted the existence of a new planet, then named Neptune, in order to account for the perturbations exhibited by Uranus, remaining strictly within the Newtonian orthodoxy. Discovered in 1782 by the German-born British musician and astronomer William Herschel (1738–1822), Uranus showed a projected velocity at variance with the predictions made by applying celestial mechanics methods to the system of the main bodies of the Solar System. Le Verrier and, independently, the English student John Couch Adams (1819–1892), asked themselves the following question: could these inconsistencies between theory and observations be due to an unknown planet perturbing the orbit of Uranus? If so, where should this ghost presence be and with which mass in order to produce the wanted effect? A very difficult enterprise indeed for the time, where all calculations had to be done by hand. The iterated solution worked out by Le Verrier provided a position of the assumed perturber with a small error box. Pointing there the refractor of the Berlin Observatory, Johann Gottfried Galle (1821–1910) and Heinrich Louis d'Arrest (1822–1875) readily discovered Neptune. It was the year 1846.

Note that the explanation of Uranus' perturbations could have been also sought in a failure, at large scales, of the Newtonian formulation of gravity and/or law of inertia.[5]

A similar problem arose in the second half of the 20th century with the discovery of the so-called flat rotation curves of spiral galaxies which implied the presence of a force additional to that generated by the distribution of luminous matter (in turn traced by the light density distribution under the assumption of a constant mass-to-light ratio). The canonical solution to this conundrum has lead to postulate the existence of a cosmic Dark Matter, a new and dominant material ingredient, whose particle has not yet been detected though. It has also stimulated attempts of falsifying the

---

[4] From a letter that Newton wrote to Rev. Richard Bentley (1662-1742) in 1692/93 in response to pressing requests for clarification on the deeper meanings of the theory set forth in the last book of *Principia*: "*It is inconceivable that inanimate brute matter should, without the mediation of something else which is not material, operate upon and affect other matter without mutual contact, as it must be, if gravitation in the sense of Epicurus, be essential and inherent in it. And this is one reason why I desired you would not ascribe innate gravity to me. That gravity should be innate, inherent, and essential to matter, so that one body may act upon another at a distance through a vacuum, without the mediation of anything else, by and through which their action and force may be conveyed from one to another, is to me so great an absurdity that I believe no man who has in philosophical matters a competent faculty of thinking can ever fall into it. Gravity must be caused by an agent acting constantly according to certain laws; but whether this agent be material or immaterial, I have left open to the consideration of my readers*".

[5] A priori, there were other ways to justify the discrepancy. For instance, one might have called for a revision of either the Newton's second law or the behavior of the gravitational force in a 'weaker field', as it is the gravitational field of the Sun in the (then unexplored) outskirts of the Solar System. Actually, the falsification of the theory, rather than the existence of a new planet as suspected by Le Verrier, was Einstein's solution to the difficulties met by classical celestial mechanics to account for the total value of Mercury's perihelion precession; *cf.* Sect. 4.14.

Fig. 2.1 Two-body
problem: vectors of the two
massive points P_1 and P_2 in
an inertial reference system
centered either in O or in B,
where B is the barycenter

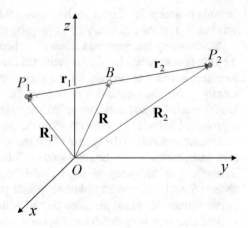

Fig. 2.1 Two-body problem: vectors of the two massive points P_1 and P_2 in an inertial reference system centered either in O or in B, where B is the barycenter

Newtonian dynamics in the presence of weak gravitational fields (MOND = Modified Newtonian Dynamics; see the review in [1]).

In this chapter we present the classical treatment of the problem of the motion of two massive points using for now the Cartesian formalism. The same problem will be proposed again in Chap. 4 with the Hamiltonian formalism and in Sect. 4.14 for the relativistic case. A classical application to the orbital motion of visual binaries will be given in Appendix J.

2.1 Motion Relative to the Center of Mass

We reconsider the isolated self-gravitating system of point-like masses defined at the beginning of Chap. 1, with $N = 2$. If $\mathbf{r}_i = \overrightarrow{OP_i}$ is the radius vector of the point P_i $(i = 1, 2)$ from the origin O of the inertial reference system, and $\mathbf{r}_{ij} = -\mathbf{r}_{ji}$ is the vector $\overrightarrow{P_iP_j}$ of the point P_j relative to P_i, the motion of the two massive points P_1 and P_2 is described by the solutions, $\mathbf{r}_i = \mathbf{r}_i(t)$, of the system of vector equations (see (1.2)):

$$\begin{cases} m_1\,\ddot{\mathbf{r}}_1 = -G\,\dfrac{m_1 m_2}{r_{12}^2}\,\dfrac{\mathbf{r}_{21}}{r_{21}}, \\[2mm] m_2\,\ddot{\mathbf{r}}_2 = -G\,\dfrac{m_2 m_1}{r_{21}^2}\,\dfrac{\mathbf{r}_{12}}{r_{12}}. \end{cases} \tag{2.1}$$

As we know, the pair of vector equations (2.1) is equivalent to a system of 6 scalar differential equations of the second order or to a system of 12 first order equations. The general solution of the problem requires 12 integration constants, i.e., 12 first integrals of motion.

The problem can be simplified by referring the motion to the center of mass B. As already observed (*cf.* Sect. 1.4), this allows to reduce the order from N to $(N - 1)$.

To this end we replace the two Eqs. (2.1) with their sum and their difference:

$$\begin{cases} m_1\ddot{\mathbf{r}}_1 + m_2\ddot{\mathbf{r}}_2 = 0, \\ \ddot{\mathbf{r}}_2 - \ddot{\mathbf{r}}_1 = -G\left(m_1 + m_2\right)\dfrac{\mathbf{r}_{12}}{r_{12}^3}. \end{cases} \tag{2.2}$$

The solution of the first of these new equations:

$$m_1\mathbf{r}_1 + m_2\mathbf{r}_2 = \left(m_1 + m_2\right)\left(\dot{\mathbf{r}}_B^\circ t + \mathbf{r}_B^\circ\right), \tag{2.3}$$

where $\dot{\mathbf{r}}_B^\circ$ and $\mathbf{r}_B^\circ$ are constant vectors, can be written in the following form:

$$\mathbf{r}_B = \frac{m_1\mathbf{r}_1 + m_2\mathbf{r}_2}{m_1 + m_2} = \dot{\mathbf{r}}_B^\circ t + \mathbf{r}_B^\circ. \tag{2.4}$$

The vectors $\mathbf{r}_B$ and $\dot{\mathbf{r}}_B$ give position and velocity of the center of gravity. The (2.4) expresses the usual law of motion of the center of gravity of an isolated mechanical system. Thus, without losing the generality, the barycenter B of the system can be chosen as the origin O of an inertial reference system. In this case, $\mathbf{r}_i = \overrightarrow{BP_i}$.

Now we express the radius $\mathbf{r}_2$ from the first equation in (2.2) and replace it in the second (2.2), considering that: $\mathbf{r}_{12} = \mathbf{r}_2 - \mathbf{r}_1 = -(m_1 + m_2)\mathbf{r}_1/m_2$. By repeating the operation for $\mathbf{r}_1$, we obtain the equations of motion for each of the two masses in the barycentric reference system:

$$\begin{aligned} \ddot{\mathbf{r}}_1 + G\frac{m_2^3}{(m_1 + m_2)^2}\frac{\mathbf{r}_1}{r_1^3} &= 0, \\ \ddot{\mathbf{r}}_2 + G\frac{m_1^3}{(m_1 + m_2)^2}\frac{\mathbf{r}_2}{r_2^3} &= 0. \end{aligned} \tag{2.5}$$

Equations (2.5) are independent of each other. They have the same form for both masses and differ only by a constant in the second term; it means that the shape of the two orbits is the same but the scales can be different (Fig. 2.2).

The right panel in Fig. 2.2 represents, for instance, the situation of a massive planet in a circular orbit around a star. It is easy to prove that the first of (2.5) is satisfied if the circular velocity of P_1 is $V_1^c = \sqrt{G\mu_1/r_1}$, where $\mu_1 = m_2^3/(m_1 + m_2)^2$ is a fictitious mass placed in the barycenter, and the velocity vector is perpendicular to $\mathbf{r}_1$. The small oscillation of the star either (a) in its astrometric position or (b) in its radial velocity offers the way to reveal the presence of the planetary companion, even when it is not possible to observe it directly because of the dominant glow of the star. It was by measures of this type that (a) in 1844 the German Friedrich Bessel disclosed the presence of a companion star orbiting around Sirius and (b) in 1995 the Swiss astronomers Michel Mayor and Didier Queloz found the first exoplanet around the star 51 Pegasi [2]: a discovery prized with the Nobel in 2019.

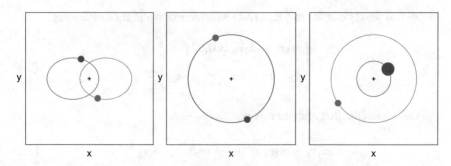

Fig. 2.2 The two-body problem in the barycentric system. Trajectories for three cases: elliptical and circular orbits for $m_1 = m_2$, and circular orbits for $m_2 = 2.5\,m_1$

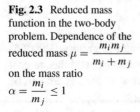

Fig. 2.3 Reduced mass function in the two-body problem. Dependence of the reduced mass $\mu = \dfrac{m_i m_j}{m_i + m_j}$ on the mass ratio $\alpha = \dfrac{m_i}{m_j} \leq 1$

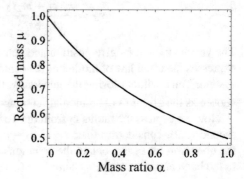

If $m_2 \gg m_1$, it is convenient to center the reference system on the larger mass. This is just what we are going to do now.

Placing $\mathbf{r} = \mathbf{r}_2 - \mathbf{r}_1 = \overrightarrow{P_1 P_2}$ (note the arbitrary choice of one of the two possible directions of the vector $\mathbf{r}$), and:

$$\mu = \frac{m_1 m_2}{m_1 + m_2}, \qquad (2.6)$$

the second equation of the system (2.2) may be rewritten in the following way:

$$\mu \ddot{\mathbf{r}} + G\, m_1 m_2 \frac{\mathbf{r}}{r^3} = 0. \qquad (2.7)$$

The parameter μ, which has the dimension of a mass, is called reduced mass[6] as its value tends to that of the smaller of the two masses. If $m_i \leq m_j$, or $\alpha = m_i/m_j \leq 1$, then $\mu = m_i(1 + \alpha)^{-1}$, implying $\mu \to m_i$ for $\alpha \to 0$ and $\mu = m_i/2$ for $\alpha = 1$ (Fig. 2.3).

---

[6] The reduced mass is an harmonic mean of the masses: $\dfrac{1}{\mu} = \dfrac{1}{m_1} + \dfrac{1}{m_2}$.

Finally, multiplying both members of (2.7) by $\dot{\mathbf{r}}$ and integrating, we obtain the expression of the total specific energy[7]:

$$\mathcal{E}_s = T_s + \mathcal{W} = \frac{1}{2}\mu\dot{r}^2 - G\frac{m_1 m_2}{r} = \text{const.} \tag{2.8}$$

Notice how the initial problem has been reduced to one requiring the knowledge of 6 first integrals only, as many as there are additive constants in the general solution of (2.7). The latter is identical to the equation of motion of a point P of (reduced) mass $\mu = m$, subject to an acceleration $\mathbf{a} = -G\left(m_1 + m_2\right)\dfrac{\mathbf{r}}{r^3}$ due to a force coming from the center C in which the whole mass of the system is concentrated, identified by the radius vector $\mathbf{r} = \overrightarrow{CP}$. The identity between $\mathbf{r} = \overrightarrow{P_1 P_2}$ and $\mathbf{r} = \overrightarrow{CP}$ ensures that, at any time, position and velocity of the dummy body P coincide with those of P_2 (or of P_1, for a different choice of the direction of the vector $\mathbf{r}$), if the center of force C is thought to coincide with P_1 (or with P_2). In other words, the transformation of the problem following the new interpretation of Eq. (2.7) allows us to treat P_1 (or P_2) as a fixed point,[8] which removes the typical complications of relative motions (that is, of non-inertial reference frames). It also allows us to exploit the properties of the central motions, herewith briefly reviewed.

A force field $\mathbf{f}$ is said to be central if there is a fixed point C, called center of force, so that $\mathbf{f}(P) \times \overrightarrow{CP} = 0$. The motion of a point-like body P in a central field is plane and has a constant areal velocity[9] (central-force theorem). A theorem due to Bertrand[10] states that the orbit of P is closed if and only if the central force depends either linearly on the distance of P from C (Hooke's elastic force) or on the inverse square (Newtonian force); see [3] and [4].

Given a reference system centered in C and indicated with α, β, and γ, the direction cosines of $\mathbf{r} = \overrightarrow{CP}$, the three scalar equations of motion of the point-like body P of mass m are:

$$m\ddot{x} = |\mathbf{f}|\,\alpha, \qquad m\ddot{y} = |\mathbf{f}|\,\beta, \qquad m\ddot{z} = |\mathbf{f}|\,\gamma, \tag{2.9a}$$

that is:

$$m\ddot{x} = |\mathbf{f}|\,\frac{x}{r}, \qquad m\ddot{y} = |\mathbf{f}|\,\frac{y}{r}, \qquad m\ddot{z} = |\mathbf{f}|\,\frac{z}{r}. \tag{2.9b}$$

---

[7] Remember that the specific energy is the sum of the kinetic energy in the motion about the center of gravity when we place $T_B = \dfrac{1}{2}(m_1 + m_2)\,\dot{r}_B^2 = 0$, and of the potential energy with a zero point chosen so that $\mathcal{W} \to 0$ for $r \to +\infty$.

[8] It is essential to emphasize that P_1 is to be considered fixed only from the point of view of writing the equations of motion of P_2 of reduced mass μ, which are those characteristic of an inertial reference system. Actually, P_1 orbits around the barycenter of the system.

[9] Rate at which the area is swept out by the vector $\mathbf{r}$ of a point moving along a curve: $|\mathbf{v_A}| = \dfrac{1}{2}\lim\limits_{\Delta t \to 0}\dfrac{|\mathbf{r}(t) \times \mathbf{r}(t + \Delta t)|}{\Delta t}$. Note that $\mathbf{v_A}$ is normal to the instantaneous plane of motion.

[10] Joseph Betrand (1822–1900): French mathematician who studied number theory, differential geometry, probability theory, thermodynamics, and history of science. He translated some of Gauss's works into French.

It is easy to prove that central motions are planar and imply a constant areal velocity. Indeed, multiplying the first of (2.9b) by $-y$ and the second by $+x$, and summing the results, it is $m(\ddot{x}y - \ddot{y}x) = 0$, whose integral is $(\dot{x}y - \dot{y}x) = h_z$, where h_z is a constant. Therefore, the projection of the orbital angular momentum $\mathbf{J}$ of P on the z axis is constant. It is similarly proven that the other two projections of $\mathbf{J}$ are also constant, and therefore $\mathbf{J} = \mathbf{r} \times m\dot{\mathbf{r}} = \mathbf{r}_o \times m\dot{\mathbf{r}}_o = \mathbf{c}$, where $\mathbf{c}$ is a constant vector, and $\mathbf{r}_o$ and $\dot{\mathbf{r}}_o$ are the initial values of the radius vector and of the velocity. The vectors $\mathbf{r}_o$ and $\dot{\mathbf{r}}_o$ lie on a plane orthogonal to $\mathbf{J}$, which has a constant position and contains the fixed point C, and is therefore invariable. This proves that the motion develops on a plane.

On this fixed plane we introduce a system of orthogonal Cartesian axes x and y centered in C. Indicated with θ the angle that the radius r forms with the positive semi-axis x, so that $\theta = \arctan \dfrac{y}{x}$, the module of the angular momentum is $|\mathbf{J}| = mr^2\dot{\theta} =$ const. This relation proves that the areal velocity $A_\theta = \dfrac{1}{2}r^2\dot{\theta}$ is also constant.

2.2   Reduction to the Plane

Equation (2.8) can be formally interpreted as the energy integral in the motion of a point P of mass μ (see Eq. (2.7)) around P_1, which is now fixed and at the center of a gravitational field with potential:

$$\mathcal{U} = G \frac{m_1\, m_2}{\mu} \frac{1}{r} = G\,(m_1 + m_2)\frac{1}{r}. \tag{2.10}$$

The fact that P_2 coincides with P, and that P_1 appears as a fixed point around which P_2 moves, is a consequence of the arbitrary orientation of the vector $\mathbf{r} = \overrightarrow{P_1 P_2}$. Since the motion of P is central (and therefore plane), the motion of P_2 around P_1 is also plane. The plane of motion must contain the center, i.e., P_1, and therefore also the center of gravity B of the system.

In conclusion, the two-body problem can be always reduced to a plane, where we now introduce a polar coordinate system with pole in P_1 and polar axis coincident with the x axis of a generic Cartesian reference system $O[x, y]$, also contained on the plane. It is (Fig. 2.4):

$$\mathbf{r} = r\cos\theta\,\mathbf{i} + r\sin\theta\,\mathbf{j}, \tag{2.11a}$$

where $\mathbf{i}$ and $\mathbf{j}$ represent the unit vectors. The polar angle θ is called true anomaly. The time derivative of (2.11a) is:

$$\dot{\mathbf{r}} = \dot{r}\Big[\cos\theta\,\mathbf{i} + \sin\theta\,\mathbf{j}\Big] + r\,\dot{\theta}\Big[-\sin\theta\,\mathbf{i} + \cos\theta\,\mathbf{j}\Big]. \tag{2.11b}$$

Fig. 2.4 Reference system
for the orbit of P_2 relative to
P_1 on the plane of motion

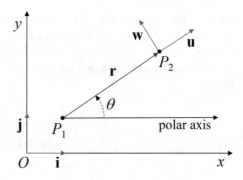

By placing:

$$\cos\theta\,\mathbf{i} + \sin\theta\,\mathbf{j} = \mathbf{u},$$
$$-\sin\theta\,\mathbf{i} + \cos\theta\,\mathbf{j} = \mathbf{w}, \tag{2.12}$$

where the unit vectors $\mathbf{u}$ and $\mathbf{w}$ are one parallel and the other orthogonal to $\mathbf{r}$, Eqs. (2.11a) and (2.11b) assume the compact form:

$$\mathbf{r} = r\,\mathbf{u},$$
$$\dot{\mathbf{r}} = \dot{r}\,\mathbf{u} + r\,\dot{\theta}\,\mathbf{w}, \tag{2.13}$$

and Eq. (2.8) rewrites as:

$$\mathcal{E} = \frac{1}{2}\mu\left(\dot{r}^2 + r^2\dot{\theta}^2\right) - G\frac{m_1\,m_2}{r} = \text{const.} \tag{2.14}$$

Observe now that, since the central-force theorem ensures the constancy of the angular momentum of P_2 in motion around P_1, then:

$$\mu\mathbf{r} \times \dot{\mathbf{r}} = \mathbf{c}, \tag{2.15}$$

with $\mathbf{c}$ constant vector.[11] From (2.12) and (2.13):

$$\mu\mathbf{r} \times \dot{\mathbf{r}} = \mu r^2\,\dot{\theta}\,\mathbf{u} \times \mathbf{w} = \mathbf{c}, \tag{2.16}$$

that is:

$$r^2\,\dot{\theta} = h, \tag{2.17}$$

where the constant h has the dimensions of an areal velocity. Since the only non-null component of the angular momentum per unit mass h is that perpendicular to the

---

[11] One can obtain (2.15) by vector multiplying by $\mathbf{r}$ the Eq. (2.7), adding the null term $\dot{\mathbf{r}} \times \dot{\mathbf{r}}$, and then simplifying.

orbital plane, we can reduce the problem by two further degrees. We will use the first two integrals (2.14) and (2.17) to obtain the equation of the trajectory and the time equation of motion of P_2 around P_1. Writing (2.17) as $d\theta = (h/r^2)dt$, the expression (2.14), which is actually a first integral (cf. Sect. 4.6), can be placed in two different forms:

$$dt = \sqrt{\frac{\mu}{2}} \frac{dr}{\sqrt{\mathcal{E} + G\dfrac{m_1 m_2}{r} - \dfrac{\mu h^2}{2r^2}}}, \tag{2.18a}$$

$$d\theta = \sqrt{\frac{\mu h^2}{2}} \frac{dr/r^2}{\sqrt{\mathcal{E} + G\dfrac{m_1 m_2}{r} - \dfrac{\mu h^2}{2r^2}}}. \tag{2.18b}$$

These equations establish the functional dependence of the modulus of $\mathbf{r}$ on the time t and on the true anomaly θ respectively. We will firstly integrate equation (2.18b) in order to derive the relation between θ and r, i.e., the equation of the trajectory. Before that, though, we want to analyze directly the type of motion considering the so-called effective potential energy.

2.3   The Effective Potential Energy

We obtain $\dot{\theta}$ from the (2.17) and replace it in the expression of the total energy (2.14) for the extreme case $\mu \to m_2$, when $m_1 \gg m_2$:

$$\mathcal{E} = \frac{1}{2}m_2\,\dot{r}^2 + \frac{m_2 h^2}{2r^2} - G\frac{m_1 m_2}{r}. \tag{2.19}$$

We call effective potential energy the quantity:

$$\mathcal{E}_{eff} = \mathcal{E} - \frac{1}{2}m_2\,\dot{r}^2 = \frac{m_2 h^2}{2r^2} - G\frac{m_1 m_2}{r}. \tag{2.20}$$

It equals the total energy when the radial velocity is zero, i.e., when r has an extreme. Therefore the trajectory (we will see in the next section that it has to be a conic in the two-body problem) must be confined between the values of the intercepts of the $\mathcal{E}_{eff}$ curve (given for an assigned value of the angular momentum per unit mass h) with the constant value of the total energy $\mathcal{E}$ (Fig. 2.5). The minimum $d\mathcal{E}_{eff}/dr = 0$ corresponds to the circular orbit since this is the case in which the centrifugal force equals the gravitational pull, as it can be verified by direct derivation. What do we learn from that?

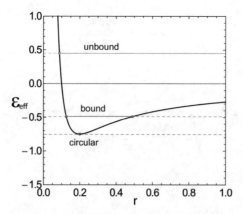

Fig. 2.5 Example of effective potential energy, $\mathcal{E}_{eff}$, for a given value of the angular momentum per unit mass h. The intercepts with a line of constant total energy ($\mathcal{E}$) with $\mathcal{E}_{eff} \leq 0$ give the range of radial distances allowed to the motion (*solid blue line*). Notice that the circular orbit (*red dot*) occurs there where $\mathcal{E}_{eff}$ has a minimum. For $\mathcal{E}_{eff} > 0$ there is only a lower limit to r (*solid green line*)

1. First of all, not all negative values of $\mathcal{E}$ are allowed for a given value of h; only those when $\mathcal{E} \geq \min(\mathcal{E}_{eff})$.
2. The minimum of the allowed total energy corresponds to a circular orbit with radius $r_c = Gm_1/h^2$ and circular velocity $V_c^2 = Gm_1/r_c$. This implies that the circular velocity is that with the highest stability (you must always spend energy to change it).
3. For all values of $\min(\mathcal{E}_{eff}) < \mathcal{E} < 0$, the trajectory is confined between a pericentric (r_{min}) and an apocentric (r_{max}) value, so it is limited in space.
4. For $\mathcal{E} \geq 0$ (there is no upper limit to the values of the total energy), the trajectories have only a lower boundary (r_{min}), i.e., they are open.

2.4 The Trajectory

We now solve Eq. (2.18b). The structure of the right-hand side suggests to choose the reciprocal of r as the integration variable. We put:

$$x = \frac{1}{r}\sqrt{\frac{\mu h^2}{2}} + x_\circ, \qquad (2.21a)$$

where $x_\circ$ is a constant to be determined in such a way that the expression under the root at the right side of (2.18b) turns into a difference of squares. It is:

$$x_\circ = -G \frac{m_1 \, m_2}{\sqrt{2 \, \mu \, h^2}}, \tag{2.21b}$$

from where it follows:

$$d\theta = \frac{-\,dx}{\sqrt{\mathcal{E} + \dfrac{(G m_1 \, m_2)^2}{2 \, \mu \, h^2} - x^2}} = \frac{-\,dz}{\sqrt{1 - z^2}}, \tag{2.22}$$

with:

$$z = \frac{x}{\sqrt{\mathcal{E} + \dfrac{(G m_1 \, m_2)^2}{2 \, \mu \, h^2}}}. \tag{2.23a}$$

The latter integrates into:

$$\theta - \theta_\circ = \arccos z = \arccos \frac{x}{\sqrt{\mathcal{E} + \dfrac{(G m_1 \, m_2)^2}{2 \, \mu \, h^2}}}, \tag{2.23b}$$

where $\theta_\circ$ is an integration constant. Re-inserting the source variable r by the (2.21a) and (2.21b), after some manipulation we obtain the equation of the trajectory of P_2 in its motion relative to P_1:

$$r = \frac{\dfrac{\mu \, h^2}{G \, m_1 \, m_2}}{1 + \sqrt{1 + \mathcal{E} \dfrac{2 \, \mu \, h^2}{(G \, m_1 \, m_2)^2}} \; \cos(\theta - \theta_\circ)} = \frac{p}{1 + e \, \cos(\theta - \theta_\circ)}, \tag{2.24}$$

with the parameters:

$$p = \frac{\mu \, h^2}{G \, m_1 \, m_2} \geq 0, \tag{2.25a}$$

$$e = \sqrt{1 + \mathcal{E} \frac{2 \mu \, h^2}{(G \, m_1 \, m_2)^2}} \geq 0. \tag{2.25b}$$

Whatever the (non-negative) values of the parameters are, Eq. (2.24) represents a real non-degenerate conic in a coordinate system whose pole, coincident with P_1, also contains the focus of the conic itself. According to one of the possible definitions for this family, a conic is in fact any plane curve being the geometric locus of the points whose distances r and k from a fixed point (focus) and from a fixed line (directrix) respectively have a constant ratio e (eccentricity). Then: $e = \dfrac{r}{k - r \cos\theta}$, identical to (2.24) if we place $p = ke$ (see [5]).

In summary, so far we have been able to prove that, under the action of the mutual gravitational forces only, the massive point P_2 (or P_1) moves with constant areal velocity around P_1 in a plane orbit that is a conic with one focus in P_1. Note the great generality of this simple result, and remember the fact that neither P_1 nor P_2 define an inertial system.

2.5 Laplace-Runge-Lenz Vector

We now find the vector that is constant during the motion. We will see that this additional first integral is not independent, being related to two fundamental ones (momentum and total energy). Let us multiply vectorially each term of the equation of motion (2.7) by the momentum vector $\mathbf{J} = \mathbf{r} \times \mu\,\dot{\mathbf{r}}$:

$$\mu\ddot{\mathbf{r}} \times \mathbf{J} + Gm_1m_2\,\frac{\mathbf{r} \times \mathbf{J}}{r^3} = 0. \tag{2.26}$$

Since $\mathbf{J}$ is constant, the first term turns to be: $\mu\dfrac{d(\dot{\mathbf{r}} \times \mathbf{J})}{dt}$. For the second term we have[12]:

$$\frac{\mathbf{r} \times \mathbf{J}}{r^3} = \mu\frac{\mathbf{r} \cdot (\mathbf{r} \cdot \dot{\mathbf{r}}) - \dot{\mathbf{r}} \cdot (\mathbf{r} \cdot \mathbf{r})}{r^3} = \mu\frac{\mathbf{r} \cdot \dot{r} - r \cdot \dot{\mathbf{r}}}{r^2} = -\mu\frac{d}{dt}\left(\frac{\mathbf{r}}{r}\right). \tag{2.27}$$

After integration we obtain:

$$\mathbf{A} = \dot{\mathbf{r}} \times \mathbf{J} - Gm_1m_2\frac{\mathbf{r}}{r}, \tag{2.28}$$

where the constant vector $\mathbf{A}$ is named after Laplace-Runge-Lenz.[13] It is simple to prove that $\mathbf{A}$ is orthogonal to the momentum vector $\mathbf{J}$. In fact:

$$\mathbf{J} \cdot \mathbf{A} = \mathbf{J} \cdot (\dot{\mathbf{r}} \times \mathbf{J}) - G\,m_1\,m_2\frac{\mathbf{r} \cdot (\mathbf{r} \times \mu\,\dot{\mathbf{r}})}{r} = 0. \tag{2.29}$$

---

[12] The vector triple product develops into dot products according to the following formula:

$$\mathbf{a} \times (\mathbf{b} \times \mathbf{c}) = \mathbf{b}\,(\mathbf{a} \cdot \mathbf{c}) - \mathbf{c}\,(\mathbf{a} \cdot \mathbf{b}).$$

[13] Carl Runge (1856-1927) and Wilhelm Lenz (1888-1957): German mathematicians and physicists.

This means that, for a central Keplerian motion, $\mathbf{A}$ lies always on the plane of motion. Let us now square the (2.28):

$$A^2 = (\dot{\mathbf{r}} \times \mathbf{J})^2 - 2\,G\,m_1\,m_2 \frac{\mathbf{r} \cdot (\dot{\mathbf{r}} \times \mathbf{J})}{r} + (G\,m_1\,m_2)^2. \qquad (2.30)$$

Using the permutation rule[14] in the second term at the right-hand side and the orientation of the vectors:

$$\mathbf{r} \cdot (\dot{\mathbf{r}} \times \mathbf{J}) = \mathbf{J} \cdot (\mathbf{r} \times \dot{\mathbf{r}}) = \frac{\mathbf{J} \cdot \mathbf{J}}{\mu} = \frac{J^2}{\mu}, \qquad (2.31)$$

with which the (2.30) becomes:

$$A^2 = \dot{r}^2\,J^2 - \frac{2G\,m_1\,m_2}{\mu} \frac{J^2}{r} + (G\,m_1\,m_2)^2. \qquad (2.32)$$

Taking into account the energy conservation law given by Eq. (2.8), we derive the second relation between the first integrals (the first relation is (2.29)):

$$A^2 = (G\,m_1\,m_2)^2 + \frac{2\,\mathcal{E}\,J^2}{\mu}. \qquad (2.33)$$

We remind here that the two-body system of differential equations is of 6-th order. Thus, at least 5 first integrals of motion are needed for the solution. Four of them concern the angular momentum (3 scalars) and the total energy. Due to an existence of the additional first integral (the Laplace-Runge-Lenz vector), the two-body problem is solved. This first integral, as we could see, establishes a relations between the others two. Note that we finally have exactly 5 first integrals taking into account the two provided by the relations (2.29) and (2.33). This allows us to obtain the equation of orbit as we can see immediately. By developing the scalar product:

$$\mathbf{A} \cdot \mathbf{r} = A\,r \cos\theta = \mathbf{r} \cdot (\dot{\mathbf{r}} \times \mathbf{J}) - G\,m_1\,m_2\,r, \qquad (2.34)$$

we obtain:

$$r = \frac{\dfrac{J^2}{G\,m_1\,m_2\,\mu}}{1 + \dfrac{A}{G\,m_1\,m_2} \cos\theta}, \qquad (2.35)$$

---

[14] The cyclic permutation rule of the cross product states that:

$$\mathbf{a} \cdot (\mathbf{b} \times \mathbf{c}) = \mathbf{c} \cdot (\mathbf{a} \times \mathbf{b}) = \mathbf{b} \cdot (\mathbf{c} \times \mathbf{a}).$$

Fig. 2.6 Orbital system of coordinates. The center coincides with one focus of the conic orbit (sketched as an ellipse here). The axes x' and y' lie on the orbital plane and the x' is chosen to pass through the pericenter. We have indicated the directions of **J**, **A**, and **J** $\times$ **A**, without tracing the corresponding vectors

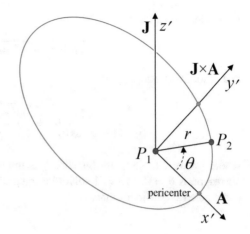

which is the equation of a conic with eccentricity and focal parameter given by:

$$e = \frac{A}{G\,m_1\,m_2},$$
$$p = \frac{J^2}{G\,m_1\,m_2\,\mu}. \tag{2.36}$$

We obtain in this way the result already acquired (see (2.24) with (2.25a), (2.25b)) that a particle in a central Keplerian force field describes a conic orbit focused on the center of motion.

Since the two-body problem is plane, it is convenient to use a coordinate system with one of its principal planes containing that of motion (the so-called orbital system). Indeed, we can construct this coordinate system $[x'y'z']$ with origin at the massive point P_1 and axes parallel to three orthogonal vectors. One of them is the total momentum **J** which is always orthogonal to the orbital plane. So, we can choose the z' axis along the direction of **J**. The second axis, x', is set coincident with the line through the point P_1 and the pericenter, with positive direction being the same of the Laplace-Runge-Lenz vector **A**. Finally, the third axis y' is the vector along **J** $\times$ **A** (see Fig. 2.6).

In this system the coordinates of the point P_2 are: $x' = r\cos\theta$, $y' = r\sin\theta$, where r is given by (2.35). The orbital system of coordinates is appropriate for dealing with the motion of the massive point on its orbit without paying attention to the orientation of the orbit itself. We will consider in Sect. 2.8 the system accounting for the orientation of the orbit in space.

We can now find the law of the motion $\theta(t)$. From the conservation of momentum: $\mathbf{J} = \mathbf{r} \times \mu\dot{\mathbf{r}} = \overrightarrow{\text{const}} = J_{z'}\mathbf{k}$, where **k** is the unit vector of the z' axis, we have:

$$J_{z'} = \mu h = \mu(x'\dot{y}' - y'\dot{x}') = \mu r^2\dot{\theta}, \tag{2.37}$$

from which it follows:

$$\frac{p^2}{(1 + e \cos \theta)^2} \frac{d\theta}{dt} = h,$$

(2.38)

and, after integration:

$$\int_0^\theta \frac{d\theta}{(1 + e \cos \theta)^2} = \sqrt{\frac{G m_1 m_2}{\mu}} \frac{1}{p^{3/2}} (t - t_o),$$

(2.39)

where we used (2.25a). In this case t_o is the time of passage at the pericenter and corresponds $\theta = 0$. We will solve the integral in (2.39) for each conic orbit after some general considerations.

2.6   Geometry of Conic Orbits

Multiplying the expression for Laplace-Runge-Lenz vector (2.28) by $\mathbf{r}$ we obtain:

$$\mathbf{r} \cdot \mathbf{A} = \frac{J^2}{\mu} - G m_1 m_2 r,$$

(2.40)

or:

$$x A_x + y A_y + z A_z - \frac{J^2}{\mu} + G m_1 m_2 \sqrt{x^2 + y^2 + z^2} = 0,$$

(2.41)

which is the equation of a conic surface. The equation of the plane of the moving mass:

$$x J_x + y J_y + z J_z = 0,$$

(2.42)

is given by $\mathbf{r} \cdot \mathbf{J} = 0$. The orbit of the mass is the result of the intersection of plane (2.42) with the surface (2.41) and is determined by the initial conditions. Since the conic surface is of the second order, the resulting intersection curve is of the second order too. We remind that in a Cartesian reference system, the family of the conics is analytically represented by a generic equation of second degree in two variables:

$$c_1 x^2 + c_2 y^2 + c_3 x y + c_4 x + c_5 y + c_6 = 0.$$

(2.43)

In this sense, conics are second in the hierarchy of complexity of plane curves, immediately after the straight lines, represented by linear equations [5]. A general property of the quadratic forms (2.43) is that they usually allow a coordinate transformation able to simplify their expression in:

$$c_1' x^2 + c_2' y^2 + c_3' = 0.$$

(2.44)

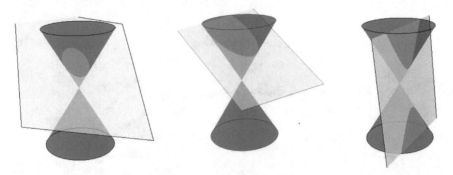

Fig. 2.7 Conic sections: ellipse, parabola, hyperbola

When this happens (in the case of the ellipse and of the hyperbola), the origin of the new coordinate system is also the center of the conic, its axes contain the main axes (which are also symmetry axes), and Eq. (2.44) is said to be canonical.

The conics[15] derive their name from the fact that they can be obtained as sections of a (double) cone by a plane (see Fig. 2.7). If α is the semi-aperture of the cone (angle between the axis and one generatrix) and β is the angle that the normal to the secant plane forms with the axis of the cone, it is:

an elliptical section if $0 < \beta < \alpha$;

a parabolic section if $\beta = \alpha$;

a hyperbolic section if $\alpha < \beta \leq \pi/2$.

2.6.1 Ellipse

Figure 2.8 shows an ellipse in rectangular coordinates with central origin and coordinate axes coinciding with the main axes. The corresponding (canonical) equation:

$$\frac{x^2}{a^2} + \frac{y^2}{b^2} = 1, \tag{2.45}$$

---

[15] Conic sections were known in ancient Greece. They were discovered by Menaechmus (380–320 BC), a Greek mathematician and philosopher, student of Eudoxus. He found these sections by solving the problem of doubling the cube and was the first to prove that ellipse, parabola, and hyperbola can be obtained by cutting a cone with an inclined plane. There is a famous answer of Menaechmus to Alexander the Great who asked him the best way to learn geometry: *O king, to travel through the country there are private roads and royal roads, but in geometry there is one road for all.* Conic sections became important in mechanics only after the Newtonian law of gravity.

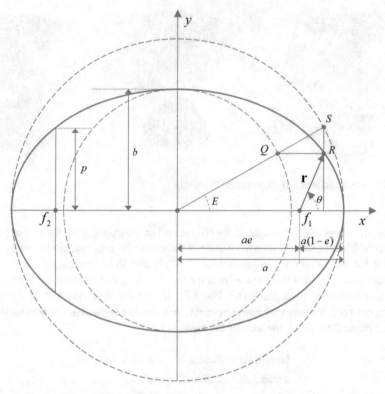

Fig. 2.8 Geometrical properties of the ellipse. The outer circle is named pedal: it is the locus of the orthogonal projections of one focus on the tangent lines to the ellipse

contains the lengths a and b of the major and minor semi-axes respectively. The following quantities are defined:

$$\text{axial ratio} \quad 0 < \frac{b}{a} \leq 1, \tag{2.46a}$$

$$\text{ellipticity} \quad \epsilon = 1 - \frac{b}{a}, \tag{2.46b}$$

$$\text{eccentricity} \quad e = \sqrt{1 - \frac{b^2}{a^2}}. \tag{2.46c}$$

The product ae measures the distance of each focus from the center. Note that the ellipticity is not widely used by mathematicians, but it is common among astronomers.[16]

---

[16] For example, the apparent flattening of elliptical galaxies in the Hubble morphological classification is given as $10 \times (1 - b/a)$, where a and b are the lengths of the average semi-axes of the elliptical images of these objects.

Let us now introduce, on the plane of the ellipse, a system of polar coordinates with center at one focus and polar axis coincident with the focal axis (usually oriented towards the pericenter). This coordinate system coincides with the orbital system (see Sect. 2.5). Through the transformation equations:

$$x = ae + r \cos \theta,$$
$$y = r \sin \theta, \qquad (2.47)$$

the (2.45) turns into:

$$r = \frac{p}{1 + e \cos \theta}, \qquad (2.48)$$

where the constant:

$$p = a (1 - e^2) > 0 \qquad (2.49)$$

is named focal parameter. The angle θ_o in (2.24) has the geometrical meaning of true anomaly of the focal axis.

The parametric equations:

$$x = a \cos E,$$
$$y = b \sin E, \qquad (2.50)$$

provide another representation of the ellipse in a centered system of rectangular coordinates with one of the axes coincident with the focal line, particularly convenient for numerical sampling of the ellipse (for example, when you want to plot an ellipse via a broken line). It is easy to prove that these equations satisfy the (2.45).

The geometric meaning of the parameter E is readily understood looking at Fig. 2.8. It is the angle at the center that is common to the points Q of the minor (inscribed) circle and S of the major (circumscribed) circle,[17] given by horizontal or vertical projections of the point R on the ellipse. You can be soon convinced of this geometrical construction by noting that the ellipse is the orthogonal projection (see Appendix J) of the pedal circle on a plane with inclination $i = \arcsin e$, and that the major axis is parallel to the line of nodes (intersection of the two planes).

In order to find the relation between the true anomaly θ and the angle E, called eccentric anomaly, it is enough to compare the coordinates of the point R in Fig. 2.8.

$$a \cos E = ae + r \cos \theta,$$
$$b \sin E = r \sin \theta. \qquad (2.51a)$$

---

[17] Called pedal, Latinate word for 'foot', because there the tangents to the ellipse encounter their perpendicular lines passing for one of the foci; see also Sect. 2.6.1.

Eliminating r through (2.48), it is:

$$\sin\theta = \frac{\sqrt{1-e^2}\,\sin E}{1-e\cos E},$$

$$\cos\theta = \frac{\cos E - e}{1 - e\cos E},$$

(2.51b)

or, by inverting the equations,

$$\sin E = \frac{\sqrt{1-e^2}\,\sin\theta}{1+e\cos\theta},$$

$$\cos E = \frac{\cos\theta + e}{1 + e\cos\theta}.$$

(2.51c)

Further manipulations give new useful relations. In particular, introducing into (2.48) the expression for $\cos\theta$ given by the second of the (2.51b), we obtain:

$$r = a\,(1 - e\cos E). \tag{2.52}$$

This expression for the focal radius r of (2.48) makes it easy to calculate area integrals. In fact, the area swept by the focal radius when the true anomaly grows from 0 to θ, is:

$$A(\theta) = \frac{1}{2}\int_0^\theta r^2\,d\theta = \frac{a^2}{2}\int_0^\theta \left(1 - e\cos E\right)^2 d\theta =$$

$$= \frac{ab}{2}\int_0^E \left(1 - e\cos E\right) dE, \tag{2.53a}$$

where $\dfrac{d\theta}{dE} = \dfrac{\sqrt{1-e^2}}{1-e\cos E}$ is found by differentiating the first of the (2.51b) and using the second. Expressed as a function of the eccentric anomaly, the area integral is simply found:

$$A(\theta) = A(E) = \frac{ab}{2}\left(E - e\sin E\right). \tag{2.53b}$$

A natural consequence of this equation is the introduction of a third angular parameter, the mean anomaly:

$$M = 2\pi\frac{A}{\pi ab} = \frac{2A}{ab} = E - e\sin E. \tag{2.54}$$

The angle M measures the opening of a circular sector of area A in a circle having the same total area πab as the ellipse. In other words, the transcendental equation (2.54), named after Kepler, establishes a correspondence between points of an ellipse and those of a circumference (thus among M, E, and θ) by equating the areas of the corresponding sectors. Its importance in astronomy is remarkable because it helps finding the position of celestial bodies through time (*cf.* Sect. 2.7.1) and justifies the efforts made to solve it. Some analytic expansions of the Kepler equation and a method to solve it numerically are presented in Appendix F.

2.6.2 Parabola

Same as the circle (obtained by placing $e = 0$), the parabola is a special case of the conic section when $e = 1$. Thus it is also a uniparametric curve. Here the parameter, which for the circle is the radius, is the distance p of the focus from the directrix (see Fig. 2.9). Formally, from (2.48):

$$r = \frac{p}{1 + \cos \theta}. \tag{2.55}$$

Note that, in order to preserve the expression $p = a(1 - e^2)$, the parameter a, not needed for the parabola, must be thought equal to $+\infty$. The Cartesian form of Eq. (2.55) is $y^2 = 2px$. The integration of the area follows with no difficulty:

Fig. 2.9 Geometric properties of the parabola

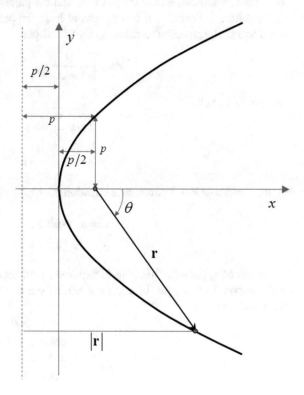

$$A(\theta) = \frac{1}{2} \int_0^\theta r^2 \, d\theta = \frac{p^2}{2} \int_0^\theta \frac{d\theta}{(1 + \cos\theta)^2} = \frac{p^2}{8} \int_0^\theta \frac{d\theta}{\cos^4 \dfrac{\theta}{2}} =$$

$$\quad (2.56a)$$

$$= \frac{p^2}{4} \int_0^\theta \left(1 + \tan^2 \frac{\theta}{2}\right) d\left(\tan \frac{\theta}{2}\right),$$

from where:

$$A(\theta) = \frac{p^2}{4} \left(\tan \frac{\theta}{2} + \frac{1}{3} \tan^3 \frac{\theta}{2}\right). \qquad (2.56b)$$

2.6.3 Hyperbola

The Cartesian equation of the hyperbola, equivalent to (2.45) for the ellipse, is:

$$\frac{x^2}{a_h^2} - \frac{y^2}{b_h^2} = 1, \qquad (2.57)$$

where we added the suffix "h" to remark the difference between the geometric semi-axes of the hyperbola, which are positive, and the parameter $a = -a_h < 0$ that we will introduce in Sect. 2.7. If θ is measured from the pericenter (see Fig. 2.10), the polar form of the equation remains, as for the ellipse:

$$r = \frac{p}{1 + e \, \cos\theta}, \qquad (2.58)$$

where in this case:

$$e = \sqrt{1 + \frac{b_h^2}{a_h^2}} > 1, \qquad (2.59)$$

$$p = a_h \, (e^2 - 1).$$

For the hyperbola too, a parametric equation of the type (2.50) can be defined:

$$x = a_h \, \cosh F, \qquad (2.60)$$
$$y = b_h \, \sinh F.$$

The use of hyperbolic functions is required by the imaginary intercepts of (2.57) on the y-axis. In this case the eccentric anomaly is designated with F. In analogy with the ellipse, it is:

$$\sinh F = \frac{\sqrt{e^2 - 1} \, \sin\theta}{e \, \cos\theta + 1}, \qquad (2.61)$$

$$\cosh F = \frac{e + \cos\theta}{e \, \cos\theta + 1},$$

Fig. 2.10 Property of the
hyperbole. The suffix "h"
added to the symbol for the
major axis helps to
distinguish it from the
parameter $a = -a_h < 0$

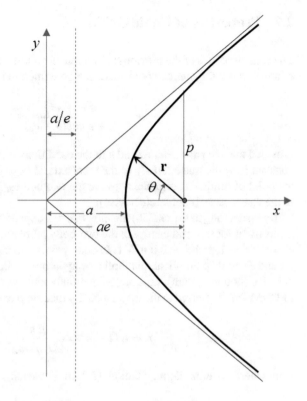

and the equation of the focal radius is:

$$r = a_h \left(e \cosh F - 1 \right). \tag{2.62}$$

Finally, the area integral:

$$A(\theta) = \frac{1}{2} \int_0^\theta r^2 \, d\theta = \frac{a_h^2}{2} \int_0^\theta \left(e \cosh F - 1 \right)^2 d\theta =$$

$$= \frac{a_h \, b_h}{2} \int_0^F \left(e \cosh F - 1 \right) dF, \tag{2.63a}$$

gives:

$$A(\theta) = \frac{a_h \, b_h}{2} \left(e \sinh F - F \right), \tag{2.63b}$$

which is the equivalent of (2.53b) and suggests to introduce a mean anomaly $M = e \sinh F - F$.

2.7 Properties of Conic Orbits

The considerations of the previous paragraph allow us to state that the focal equation of any real not degenerate conic curve can be written in the following form:

$$r = \frac{a(1 - e^2)}{1 + e\,\cos(\theta - \theta_\circ)}, \tag{2.64}$$

provided that the parameters a and e fit the conditions summarized in Table 2.1. The constant $\theta_\circ$ is the true anomaly of the focal axis. If $\theta_0 = 0$, angles are counted from the point of minimum distance between the two bodies, i.e., the pericenter, usually in the direct sense (counterclockwise).

By associating to Eq. (2.25b) the intervals of e appearing in Table (2.1), we obtain a constraint for the total energy in the two-body problem; it must be $\mathcal{E} < 0$ if $e < 1$, $\mathcal{E} = 0$ if $e = 1$, and $\mathcal{E} > 0$ if $e > 1$. In other words, the trajectory of P_2 in its motion around P_1 will be elliptical, parabolic, or hyperbolic whether the specific energy is negative (bound system), null or positive (unbound system). This property is even more evident if referred to the major axis. By making p explicit in (2.25a), we obtain:

$$p = a(1 - e^2) = \frac{\mu h^2}{G\,m_1\,m_2}, \tag{2.65}$$

from where, by eliminating e though (2.25b), it results:

$$a = -G\,\frac{m_1\,m_2}{2\,\mathcal{E}}. \tag{2.66}$$

This important relation shows that the characteristic length a of the conic is completely determined by the specific energy, and vice versa, confirming the classification in Table 2.1. From (2.65) we also see that, other things being equal, eccentricity is fixed by the value of the angular momentum per unit mass h. Based on (2.25b), the latter is upper limited by the specific energy only in the case of the elliptical orbit ($\mathcal{E} < 0$).

We want to visualize these results considering an elliptical orbit. In order to change its maximum size, it will be necessary and sufficient to vary the specific energy, obviously keeping it negative. To change instead the flattening while preserving the maximum size (semi-major axis), we must act on the value of h.

Table 2.1 Limits for the parameters a and e

Conic type	Eccentricity	Semi-major axis	Total energy
Ellipse	$0 \le e < 1$	$a > 0$	$\mathcal{E} < 0$
Parabola	$e = 1$	$a = \pm\infty$	$\mathcal{E} = 0$
Hyperbola	$e > 1$	$a < 0$	$\mathcal{E} > 0$

Replacing in (2.8) the value of $\mathcal{E}$ given by (2.66), we have:

$$\mathcal{E} = \frac{1}{2}\mu\dot{\mathbf{r}}^2 - G\frac{m_1 m_2}{r} = -G\frac{m_1 m_2}{2a}, \tag{2.67}$$

from which it comes out immediately the expression of the modulus of the velocity of P_2 as a function of the focal radius and of the parameter a (Table 2.1):

$$\dot{\mathbf{r}}^2 = 2\,G\,(m_1 + m_2)\left(\frac{1}{r} - \frac{1}{2a}\right). \tag{2.68}$$

The expression (2.68) secures the velocity at any position along the trajectory. For an elliptical orbit, which is periodical, the maximum of velocity over a period is reached at $r_{\min} = a(1 - e)$, i.e., at the minimum value of the separation between P_1 and P_2:

$$V_{\max}^2 = \frac{G\,(m_1 + m_2)}{a}\frac{1 + e}{1 - e}; \tag{2.69}$$

conversely, for $r_{\max} = a(1 + e)$:

$$V_{\min}^2 = \frac{G\,(m_1 + m_2)}{a}\frac{1 - e}{1 + e}. \tag{2.70}$$

So, the ratio of the velocities in the two apsidal points is:

$$\frac{V_{\max}}{V_{\min}} = \frac{1 + e}{1 - e} = \frac{r_{\max}}{r_{\min}}. \tag{2.71}$$

Let us now consider the point B at one end of the minor axis (Fig. 2.11). From (2.46c) it is easily shown that its distance from the focus F is equal to the major axis length: that is, there $r = a$. Consequently, from the (2.68) the velocity in B is:

$$V_B = \sqrt{\frac{G\,(m_1 + m_2)}{a}}. \tag{2.72}$$

Fig. 2.11 Radius vector and velocity at the minor axis extremes

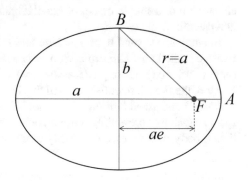

So B satisfies the condition $2T_B + W_B = 0$ given by (1.54), i.e., the velocity V_B and the semi-major axis a are average values in the sense of the virial theorem (Sect. 1.5). One can arrive at the same conclusion by observing that V_B is the modulus of the circular velocity for an orbital radius $r = a$; but, unless $e = 0$, its vector is not perpendicular to the radius vector.

For parabolic motion ($\mathcal{E} = 0$), the law of energy conservation gives immediately:

$$V_e^2(r) = 2\,G\,\frac{m_1 + m_2}{r}, \tag{2.73}$$

which is consistent with (2.68) when $a \to +\infty$. V_e is called escape velocity: it represents the minimum velocity required to P_2 (starting at the distance r from P_1) to reach an asymptotically infinite distance with a zero velocity. In fact, for $r \to +\infty$, it is $V_e \to 0$, i.e., the velocity of P_2 vanishes. Note that the escape velocity at a given distance r is $\sqrt{2}$ times of the circular velocity at the same distance.

The velocity for the hyperbolic motion (positive total energy) is given by (2.68) with the change $a_h = -a > 0$, since now the parameter a is negative (Table 2.1):

$$\dot{r}^2 = 2\,G\,(m_1 + m_2)\left(\frac{1}{r} + \frac{1}{2\,a_h}\right). \tag{2.74}$$

At any distance r, the velocity of the point P_2 is greater than the corresponding escape velocity V_e, and for $r \to +\infty$ it is:

$$V_{\min}^2 = G\,\frac{m_1 + m_2}{a_h}. \tag{2.75}$$

This non-null velocity is called hyperbolic excess.

It is worth noticing that, in the two-body problem, there are three types of the straight motions (degenerate cases) corresponding to negative, positive and zero total energy. They correspond to the three types of universe models in Newtonian cosmology: closed $\mathcal{E} < 0$, open $\mathcal{E} > 0$, and flat $\mathcal{E} = 0$ (see [6] and Footnote 38).

These results are quite general. Further developments require to treat separately the three types of orbits corresponding to the negative, null, and positive values of the specific energy $\mathcal{E}$. Some results obtained in the previous sections are important for the elaboration of a simple model of gravitational interaction, which will be described in Appendix L.

In passing, we want to review here an intriguing use of the formula (2.73) that Laplace made in his *Exposition du Système du Monde* (1796) to introduce the concept of dark body (naive predecessor of what we now call 'black hole'). Following Newton, the French scientist believed that light consisted of tiny particles of negligible mass[18] and asked himself which should be the minimum size of a homogeneous sphere capable of preventing those particle from leaving the surface and flying away (i.e., a body with an escape velocity at the surface equal to the speed of light). If

----

[18] We call them photons after the American physicist Gilbert N. Lewis (1875–1946).

$m_1 = \dfrac{4}{3}\pi\rho_o R^3$ is the mass of the body of density ρ_o and the escape velocity is set equal to the speed of the light,[19] $V_e = c$, the radius of the 'monster' writes[20]:

$$R_{DB} = \frac{2Gm_1}{c^2} = \sqrt{\frac{3c^2}{8\pi G\rho_o}}, \tag{2.76}$$

which, for $\rho_o = 5.5$ gr/cm^3 (density of the Earth) and $c = 3 \times 10^{10}$ cm/s (modern value, very close to that known to Laplace owing to estimates based on the light aberration discovered in 1729 by the English astronomer and priest James Bradley (1693–1762)), turns out to be $R_{DB} = 1.7 \times 10^{13}$ cm $= 246\ R_\odot$. It is self-evident that such a body cannot exist in nature as it could not be hydrostatically supported.

2.7.1 Elliptical Orbit ($\mathcal{E} < 0$)

Let us consider the motion of an isolated and bound system of two massive points ($\mathcal{E} < 0$). The orbit of P_2 around P_1 must be an ellipse with focus in P_1. The time equation of motion (2.18a) can be written as:

$$dt = \sqrt{\frac{\mu}{-2\mathcal{E}}}\,\frac{r\,dr}{\sqrt{\dfrac{\mu h^2}{2\mathcal{E}} + 2\,ar - r^2}}, \tag{2.77}$$

having grouped $-\mathcal{E}$ at the denominator of the second member and made use of (2.66). Adding and subtracting a^2 in the root at the denominator of the second term and noticing that, from the combination of (2.65) and (2.66), it is:

$$\frac{1}{2}\frac{\mu h^2}{\mathcal{E}} = a^2(e^2 - 1), \tag{2.78}$$

we obtain:

$$dt = \sqrt{\frac{a}{G\,(m_1 + m_2)}}\,\frac{r\,dr}{\sqrt{a^2 e^2 - (r - a)^2}}. \tag{2.79}$$

By using Eq. (2.52), we introduce the eccentric anomaly:

---

[19] The value of the speed of light was known since the Danish Ole Rœmer (1644–1710) had measured it at the end of 17-th century (although with a large error: 211,000 instead of 300,000 km/s), using the eclipses of Jupiter's satellites discovered by Galilei in 1609.

[20] The expression for R_{DB} coincides with the gravitational radius (e.g. the radius of a black hole) which appears in the spherically symmetric solution of Einstein's equation in empty space. We will use this Schwarzschild metric to find the relativistic effect in Sect. 4.14.

$$r - a = -ae \cos E,$$
$$dr = ae \sin E \, dE. \tag{2.80}$$

It is:

$$dt = \sqrt{\frac{a}{G(m_1 + m_2)}} \left(a - ae \cos E\right) dE, \tag{2.81a}$$

that is easily integrated in:

$$t - t_o = \sqrt{\frac{a^3}{G(m_1 + m_2)}} \left(E - e \sin E\right), \tag{2.81b}$$

where t_o is a constant. In order to clarify the meaning of this equation, we compare it to (2.53b). It must be:

$$\frac{2A}{ab} = \sqrt{\frac{G(m_1 + m_2)}{a^3}} \left(t - t_o\right), \tag{2.82}$$

where A is the area swept by the focal radius r when the true anomaly varies from zero at the time t_o to the value θ at the time t. For $\theta = 2\pi$, i.e., for a complete revolution, the area of the ellipse is $A = \pi ab$, and the time interval $t - t_o$ is equal to the period P. So, placing:

$$n = \frac{2\pi}{P} = \sqrt{\frac{G(m_1 + m_2)}{a^3}}, \tag{2.83}$$

where n is called mean motion, Eq. (2.81b) becomes:

$$\frac{2\pi}{P}\left(t - t_o\right) = n(t - t_o) = E - e \sin E. \tag{2.84}$$

This expression clarifies the meaning of the mean anomaly:

$$M = \frac{2\pi}{P}\left(t - t_o\right) = n(t - t_o) = \frac{\sqrt{G(m_1 + m_2)}}{a^{3/2}}(t - t_o), \tag{2.85}$$

introduced in Sect. 2.6.1; it represents the true anomaly of P_2 if it moves in a circular orbit with the same period P. Finally, Kepler's equation takes the form:

$$E - e \sin E = M. \tag{2.86}$$

Equations (2.52), (2.84), and (2.85), describe the dependence on time of the azimuth of the focal radius through the auxiliary variables M and E. A convenient expression for the dependence on time of the true anomaly θ is finally found through the relations (2.51b):

$$\tan \frac{\theta}{2} = \sqrt{\frac{1+e}{1-e}} \tan \frac{E}{2}.$$ (2.87)

In fact:

$$\tan \frac{\theta}{2} = \frac{\sin \theta}{1 + \cos \theta} = \sqrt{1 - e^2} \frac{\sin E}{1 - e \cos E} \left(1 + \frac{\cos E - e}{1 - e \cos E}\right)^{-1} =$$

$$= \frac{\sqrt{1 - e^2} \sin E}{1 - e \cos E + \cos E - e} = \frac{\sqrt{1 - e^2} \sin E}{(1 - e)(1 + \cos E)} =$$

$$= \sqrt{\frac{1+e}{1-e}} \tan \frac{E}{2}.$$ (2.88)

By using this change of variable in the integral at the left side of (2.39), we obtain the Kepler's equation (2.86). The above results can be compared with the empirical laws obtained by observing the motions of the planets around the Sun, known as Kepler's laws.[21]

First law: all planets move around the Sun in elliptical orbits, with the Sun at one of the foci.

Second law: the radius vector joining a planet to the Sun sweeps out equal areas in equal lengths of time.

Third law: the square of the sidereal period, P, of any planet in its motion around the Sun is proportional to the cube of the major axis, a, of the orbit.

In the framework of the two-body problem, which represents a first approximation of the real motion of a planet around the Sun, the first two laws are the natural consequence of the fact that the planets are gravitationally bound to the Sun ($\mathcal{E} < 0$). The third law is an approximation of the expression (2.83). The latter, in fact, written in the form:

$$\frac{P^2}{a^3} = \frac{4\pi^2}{G(m_1 + m_2)},$$ (2.89)

shows that P^2 and a^3 are indeed proportional, but for the same total mass ($m_1 + m_2$). In the case of the Solar System, which is dominated by the Sun, the sum ($m_1 + m_2$)

---

[21] German mathematician and astronomer, Johannes Kepler discovered three laws for planetary motions while investigating the accurate astrometric measurements collected by the Danish astronomer Tycho Brahe; the best data of pre-telescope era. In 1599, Tycho, while in Prague serving the Emperor Rudolph II, hired the young Kepler, already famous for his mathematical skills, and charged him to interpret in a Ptolemaic (geocentric) key his measurements of the planetary positions (especially Mars), mostly collected at the observatory of Uraniborg, located in an island of the Northern Sea. But soon after Tycho died, leaving clear field to the young man of Copernican faith. Kepler published his first two laws in the *Astronomia Nova* in 1609 and only 10 years later the third one in the *Harmonices mundi*, the most difficult to discover and the least accurate, but also that of greater physical content. Remarkable is also the work of Kepler in the field of geometric optics, where among other things has coined much of the nomenclature still in use today.

equals to the mass of the Sun $M_\odot$ at better than 0.1% in the least favorable case of the most massive planet Jupiter, and is therefore almost constant. The same would not be true if the bodies in play were of comparable masses, as it happens, for instance, in binary stars.[22]

For a planet of negligible mass ($m_P/M_\odot \to 0$), the equality (2.89) provides:

$$a = \left(\frac{G\, M_\odot}{4\,\pi^2}\, P^2 \right)^{\frac{1}{3}}. \tag{2.90}$$

With $P = 365\overset{d}{.}256365 = 31558149\overset{s}{.}9$, terrestrial sidereal year, and $M_\odot = 1.989 \times 10^{33}$ g, the semi-major axis takes the value $a = 1.4961 \times 10^{13}$ cm. This quantity is called Astronomical Unit (AU) and corresponds to the semi-major axis of the orbit of a massless point in motion around the Sun with sidereal period equal to that of the Earth.

At the end of this paragraph we make some considerations on the velocity of P_2 in the orbital motion. The modulus of $\dot{\mathbf{r}}$ is given by (2.68). The radial and transverse[23] components (parallel and orthogonal to the vector radius) are:

$$\begin{aligned} V_r &= \dot{\mathbf{r}} \cdot \mathbf{u} = \dot{r}, \\ V_t &= \dot{\mathbf{r}} \cdot \mathbf{w} = r\,\dot{\theta}. \end{aligned} \tag{2.91}$$

Let us examine the radial component $\dot{r}$. Differentiating (2.48) with respect to time, we have:

$$\dot{r} = \frac{p\, e\, \dot{\theta}\, \sin\theta}{(1 + e\, \cos\theta)^2}, \tag{2.92a}$$

or, with (2.17) and (2.48):

$$\dot{r} = \frac{e\, h}{p}\, \sin\theta. \tag{2.92b}$$

The transverse component of the velocity is:

$$V_t = r\,\dot{\theta} = r\frac{h}{r^2} = \frac{h}{p}(1 + e\cos\theta). \tag{2.93}$$

So, we see that during motion the module of the velocity varies in the following way:

$$V = \sqrt{V_r^2 + V_t^2} = \frac{h}{p}\sqrt{1 + 2e\cos\theta + e^2}. \tag{2.94}$$

---

[22] Calling P_o the period of a massless particle orbiting around a point of mass m_1, that of a point of mass m_2 is: $P = P_o \times (1 + \alpha)^{-1/2}$, with $\alpha = m_2/m_1$.

[23] Do not mistake the transverse with the tangential component, which is just the velocity itself.

Fig. 2.12 Components of
the orbital velocity in the
elliptical motion of P_2
relative to P_1

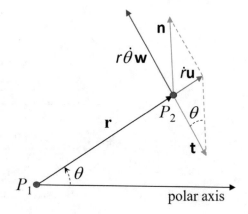

It is again apparent that the extremal values of the velocity occur at the apsidal
points ($\theta = 0, \pi$). We now decompose the radial component V_r into two vectors, one
transverse with modulus equal to:

$$|\mathbf{t}| = \frac{e\,r^2\dot{\theta}}{p}\cos\theta, \qquad (2.95)$$

and the other perpendicular to the main axis (Fig. 2.12). The latter is a constant vector
$\mathbf{n}$ as its modulus is:

$$|\mathbf{n}| = \frac{|\dot{r}\,\mathbf{u}|}{\sin\theta} = \frac{e\,h}{p}, \qquad (2.96)$$

and the direction is also invariable. Summing up the transverse component of the
velocity $\dot{\mathbf{r}}$ of P_2, we obtain a vector whose modulus:

$$|r\,\dot{\theta}\,\mathbf{w} + \mathbf{t}| = |r\,\dot{\theta}\,\mathbf{w}| - |\mathbf{t}| = r\,\dot{\theta} - \frac{e\,r^2\dot{\theta}}{p}\cos\theta = \frac{h}{p}, \qquad (2.97)$$

is also constant. In conclusion, we have proven (Fig. 2.13) that, at any epoch during
the elliptical motion, the velocity of P_2 consists of two components, normal and
transverse, both constant in modulus, the first being also constant in direction. So,
the motion of P_2 consists of a circular motion (but not necessarily uniform) with
velocity h/p, and of a uniform translational with velocity eh/p normal to the focal
axis of the ellipse.

The result just obtained can be used with advantage in many circumstances; for
example in the study of the light aberration, where it allows us to neglect the component of the orbital velocity of the Earth that is constant in direction and orientation,
since it produces an apparent displacement of the position of the stars that is independent of time.

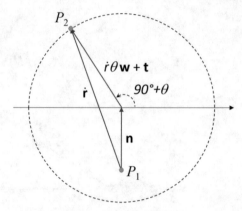

Fig. 2.13 Decomposition of the orbital velocity vector $\dot{\mathbf{r}}$ in the Keplerian motion of P_2 about P_1. The curve described by the velocity vector $\dot{\mathbf{r}}$ applied to a fixed point (P_1) takes the name of hodograph of motion (from the Greek '*hodós*', meaning 'path'). Note that the tangential velocity rotates with the angular velocity $\dot{\theta}$, which is constant only for circular motions. This is the key to understand why the constant velocity vector $\mathbf{n}$ does not generate a secular drift of the orbit

2.7.2 Average Velocity on an Elliptical Orbit

The development in series (N.56) allows us to calculate the average velocity along an elliptical orbit. From the third Kepler law (2.89) we obtain the period:

$$P = a^{3/2} \sqrt{\frac{4\pi^2}{G\,(m_1 + m_2)}}, \tag{2.98}$$

so the average velocity writes as:

$$V_m = \frac{L}{P} = \frac{\sqrt{G\,(m_1 + m_2)}}{2\pi} \frac{L}{a^{3/2}}. \tag{2.99}$$

From (N.49) to (N.56) it is:

$$L = 2\pi\,a \left[1 - \frac{1}{4}e^2 - \frac{3}{64}e^4 - \frac{5}{256}e^6 - \cdots \right], \tag{2.100}$$

and thus:

$$
\begin{aligned}
V_m &= \sqrt{\frac{G\,(m_1 + m_2)}{a}} \left[1 - \frac{1}{4}e^2 - \frac{3}{64}e^4 - \frac{5}{256}e^6 - \cdots \right] = \\
&= V_m^c \left[1 - \frac{1}{4}e^2 - \frac{3}{64}e^4 - \frac{5}{256}e^6 - \cdots \right],
\end{aligned} \tag{2.101}
$$

where $V_m^c = \sqrt{\dfrac{G\,(m_1 + m_2)}{a}}$ is the velocity on the circular orbit with radius a. From this equation we see that the average velocity on an elliptical orbit with the same major axis decreases with ellipticity. This property tell us quantitatively an obvious property since, while the period depends only on the size of mayor axis, the length of the orbit decreases with ellipticity, at least as fast as $1 - e^2/4$.

We can now ask which are the values of the radius in the elliptical orbit where the instantaneous velocity (see (2.68)):

$$V = \sqrt{G\,(m_1 + m_2)}\sqrt{\frac{2}{r} - \frac{1}{a}} = V_m^c \sqrt{\frac{2a}{r} - 1}, \tag{2.102}$$

equals the average velocity (2.101), assuming that we may truncate the expansion to the second order in e. From (2.102) to (2.101) it is:

$$\frac{V}{V_m} = \frac{\sqrt{\dfrac{2a}{r} - 1}}{\left(1 - \dfrac{e^2}{4}\right)} = 1, \tag{2.103}$$

from where:

$$r \approx \frac{4a}{4 - e^2}. \tag{2.104}$$

Remembering (2.64) with $\theta_o = 0$ and using (2.102), we may find the value of the true anomaly θ_m from which $V(\theta_m) = V_m$:

$$\theta_m = \arccos\left(\frac{a(1 - e^2)}{e\,r} - \frac{1}{e}\right) \approx \arccos\left(-\frac{5}{4}e\right), \tag{2.105}$$

provided that e remains small. This tell us that the transition point where the instantaneous velocity equals the mean velocity is beyond the *latus rectum* intersection with the ellipse in the direction of the apocenter and its anomaly grows as the ellipticity increases.

2.7.3 Parabolic Orbit ($\mathcal{E} = 0$)

If the orbit of P_2 is parabolic, it must be $\mathcal{E} = 0$. Thus Eq. (2.18a) becomes:

$$dt = \sqrt{\frac{\mu}{G\,m_1 m_2}}\,\frac{r\,dr}{\sqrt{2r - p}}, \tag{2.106}$$

and, replacing r with the expression (2.55):

$$dt = \sqrt{\frac{p^3}{G\,(m_1 + m_2)}}\,\frac{\sin\theta\,d\theta}{(1 + \cos\theta)^2\sqrt{1 - \cos^2\theta}} =$$

$$= \sqrt{\frac{2q^3}{G\,(m_1 + m_2)}}\,\frac{d(\theta/2)}{\cos^4(\theta/2)}, \tag{2.107}$$

having set, as usual, $p = 2q$. The integral of (2.107), with the introduction of the mean anomaly M, provides the time equation of motion:

$$M = \sqrt{\frac{G\,(m_1 + m_2)}{2q^3}}\,(t - t_\circ) = \tan\frac{\theta}{2} + \frac{1}{3}\tan^3\frac{\theta}{2}. \tag{2.108}$$

We want to prove that this cubic equation in $\theta/2$ has a single real root for each value of t. For the fundamental theorem of algebra, the (2.108) must admit either three real roots or just one real and two complex conjugate roots. It can be proven that, being the coefficients of the same sign and missing the second degree term, two solutions must be complex conjugate. In fact, let $a_1 = \tan(\theta_1/2)$ and $a_2 = \tan(\theta_2/2)$ be two solutions. Replaced in (2.108), by subtraction we obtain:

$$a_1^3 - a_2^3 = -3(a_1 - a_2), \tag{2.109a}$$

from which, developing the difference of the cubes and simplifying:

$$a_1^2 + a_1 a_2 + a_2^2 = -3 < 0, \tag{2.109b}$$

that is:

$$a_1^2 + a_2^2 < -a_1 a_2. \tag{2.109c}$$

But this is absurd if a_1 and a_2 are both real and at least one is different from zero. In fact, if $a_1 \le a_2$, it is $a_1^2 + a_2^2 \ge a_2^2 \ge a_1 a_2$. Therefore, all the three solutions are null, or one is complex and thus there must be its conjugate. Let a_1 be the only real solution. It is $\theta_1 = 2\arctan a_1 + 2n\pi$, which shows that the solution has a unique value in the interval $[0, 2\pi]$.

Regarding the velocity on the parabolic orbit, the same considerations made for the elliptical orbit at the end of Sect. 2.8 apply here too. In this case though, being $e = 1$, the normal and transverse components are equal in modulus (Fig. 2.13; see also Eqs. (2.96) and (2.97)).

2.7.4 Hyperbolic Orbit ($\mathcal{E} > 0$)

The hyperbolic orbit, which occurs when the two points are unbound ($\mathcal{E} > 0$), has a reduced importance in the classical planetary celestial mechanics. It acquires instead full relevance when dealing with meteors, comets, space flight, and the treatment of gravitational collisions. Equation (2.18a) becomes:

$$dt = \sqrt{\frac{\mu}{2\mathcal{E}}} \frac{r \, dr}{\sqrt{-\frac{\mu h^2}{2\mathcal{E}} - 2ar + r^2}}, \tag{2.110}$$

having used (2.66). Proceeding as in Sect. 2.7.1 for the elliptical orbit, through (2.62) and some manipulation we obtain:

$$dt = \sqrt{\frac{a_h{}^3}{G(m_1 + m_2)}} \left(e \cosh F - 1 \right) dF, \tag{2.111}$$

which, by direct integration, gives:

$$t - t_\circ = \sqrt{\frac{a_h{}^3}{G(m_1 + m_2)}} \left(e \sinh F - F \right). \tag{2.112}$$

Equation (2.112) is equivalent to Kepler's equation (2.81b), and the considerations made for this latter hold too.

2.8 Keplerian Elements of the Orbit

Starting from Sect. 2.2, we have analyzed the two-body problem in the plane of motion. Now we need to find the set of parameters to fix this plane with respect to an arbitrary Cartesian reference frame and the position of the orbit on it. Remember that, in order to solve the two-body problem, we must find six constants of motion which are called Keplerian orbital elements.

Three such elements have already been identified: a, e, and $t_\circ$. The first two give size and shape of the orbit (and are equivalent to the total specific energy and to the angular momentum). The third is simply an origin in the time axis. We call it time of passage through the pericenter because, owing to the conventions made, at the time $t = t_\circ$ it is $\theta - \theta_\circ = 0$, i.e., the radius reaches the minimum value; see the generic expression of the radius vector (2.64).

We are now left with the orientation of the plane of motion, i.e., of the reference system already introduced on it, relatively to an external concentric inertial Cartesian reference frame, which requires three additional parameters (Euler angles). Consider for this purpose an orthonormal reference frame $P_1[x, y, z]$ (Fig. 2.14). In the following, the plane xy will be conventionally named reference plane. For heliocentric orbits, this plane is made usually coincide with the ecliptic plane, and the direction of x with that of the point Υ (also called first point of Aries or vernal point), which is precisely one of the two equinoctial points in which the celestial equator intersects

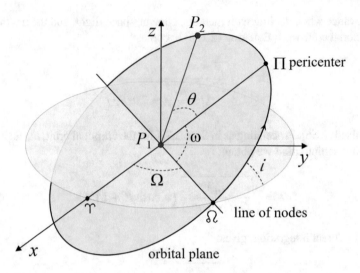

Fig. 2.14 Geometrical Keplerian elements. The figure does not show the orbit but only its plane and the reference plane xy

the ecliptic.[24] As usual, we call line of nodes the intersection of the reference plane with the one containing the orbit, identified in turn by the plane $x'y'$ of the reference system P_1 $[x', y', z']$ where the z' axis is chosen parallel to the angular momentum vector (see also Sect. 2.5 and Fig. 2.6). Three angles are enough to orient the system P_1 $[x', y', z']$ with respect to the reference system P_1 $[x, y, z]$. Usually one chooses (with conventions to be established in order to avoid ambiguity):

$i = \widehat{y\Pi}$ inclination,
$\Omega = \widehat{\Upsilon\Omega}$ longitude to the ascending node,
$\omega = \widehat{\Omega\Pi}$ argument of pericenter.

Note that the argument of pericenter is bound to the angle $u = \widehat{\Omega P_2}$, called argument of latitude, through the relation:

$$u = \omega + \theta, \qquad (2.113)$$

and this justifies the introduction of ω instead of u. Furthermore, the relation (2.113) shows that ω, $\theta_\circ$, and u, are not independent. Indeed, the true anomaly θ, measured relatively to x', is $\theta = \theta_\circ$ when $t = t_\circ$. Therefore, once u is known, ω is also determined. Sometimes, the longitude of the pericenter:

$$\pi = \omega + \Omega. \qquad (2.114)$$

---

[24] In its apparent annual motion, the Sun crosses this point at the spring equinox, passing from the southern to the northern celestial hemisphere, starting the astronomical spring. It is called the point of Aries because, in the ancient times when this terminology was created, during the spring equinox the Sun was in the constellation of Aries. Due to the precession of equinoxes, today the event occurs when the Sun is in the constellation of Pisces.

is used in place of the argument of pericenter. This is convenient when the inclination i tends to zero, that is, when the line of nodes becomes undefined: a situation that cannot be avoided by a proper choice of the reference system when, for example, the inclination of the orbit varies over time (but in this case the motion cannot be a simple Keplerian one).

In conclusion, six Keplerian elements completely identify the orbit in space. The elements a, e, t_o are called dimensional, while i, Ω, ω (or u) are orientation elements. They appear in the direction cosines a_{ij} of the relations, easily invertible, between the two coordinate systems, as shown in the next section.

2.9 Calculation of the Ephemerides

The knowledge of the Keplerian elements makes it possible to calculate a table of positions, or ephemerides,[25] of a body in the context of the hypotheses of this chapter. As an example, let us consider the case of the elliptical orbit. We first calculate the average motion from the semi-major axis:

$$n = \frac{2\pi}{P} = \sqrt{\frac{G\,(m_1 + m_2)}{a^3}}, \qquad (2.115)$$

from where, knowing the time of the passage at pericenter, t_o, the mean anomaly at time t is obtained:

$$M = n\,(t - t_o). \qquad (2.116)$$

By knowing the eccentricity e, we solve the Kepler equation (*cf.* Appendix F):

$$M = E - e \sin E, \qquad (2.117)$$

and, through (2.87), we obtain the true anomaly θ. As we will see below, to obtain the coordinates of the body it is sufficient to known the eccentric anomaly E without passing through θ. By means of (2.52) we can now calculate the radius r from the value of the eccentric anomaly E at the time t.

In order to compute the coordinates in a generic reference system (hereafter called external), we must find the direction cosines (α, β, γ) of the radius vector $\mathbf{r}$ which are related to the orbital elements i, Ω, and ω. Let us indicate with (x', y', z') the coordinates of the point P_2 in the orbital system and with (x, y, z) those in the external reference system. From Fig. 2.15 it is apparent that:

---

[25] Word coming from the Greek $\dot{\varepsilon}\varphi\eta\mu\varepsilon\rho\dot{\iota}\varsigma$, meaning 'daily'. Mesopotamian and pre-Columbian populations used the ephemerides to record the acts of their sovereigns. Currently the use is purely scientific and literary. In particular, they are tables that contain values, calculated over a particular time interval, of various variable astronomical quantities, such as positions.

$$\widehat{z\Omega} = \frac{\pi}{2},\tag{2.118a}$$

$$\widehat{x\Omega} = \Omega,\tag{2.118b}$$

$$\widehat{\Omega y} = \frac{\pi}{2} - \Omega,\tag{2.118c}$$

$$\widehat{\Omega P_2} = u,\tag{2.118d}$$

$$\widehat{zz'} = i.\tag{2.118e}$$

With that, by applying the cosine formula to suitable spherical triangles of Fig. 2.15, we have:

triangle $x\Omega P_2$: $\alpha = \cos(\widehat{xP_2}) = \cos\Omega\,\cos u - \sin\Omega\sin u\,\cos i,\tag{2.119a}$

triangle $y\Omega P_2$: $\beta = \cos(\widehat{yP_2}) = \sin\Omega\,\cos u + \cos\Omega\,\sin u\,\cos i,\tag{2.119b}$

triangle $z\Omega P_2$: $\gamma = \cos(\widehat{zP_2}) = \sin u\,\sin i.\tag{2.119c}$

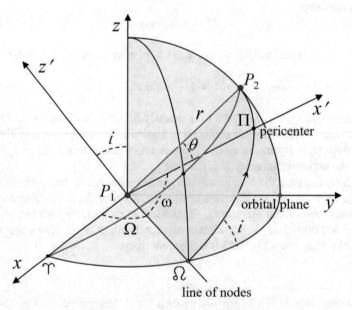

Fig. 2.15 Relations between orbital elements and direction cosines

Thus, the coordinates of the point P_2 are:

$$
\begin{aligned}
x &= r\,\alpha = r\,(\cos\Omega\,\cos u - \sin\Omega\,\sin u\,\cos i), \\
y &= r\,\beta = r\,(\sin\Omega\,\cos u + \cos\Omega\,\sin u\,\cos i), \\
z &= r\,\gamma = r\,\sin u\,\sin i.
\end{aligned}
\tag{2.120}
$$

The differentiation of (2.120) with respect to time t give the components of the velocity:

$$
\begin{aligned}
\dot{x} &= \frac{h}{p}\big(\alpha e \sin\theta + \alpha'(1 + e\cos\theta)\big), \\
\dot{y} &= \frac{h}{p}\big(\beta e \sin\theta + \beta'(1 + e\cos\theta)\big), \\
\dot{z} &= \frac{h}{p}\big(\gamma e \sin\theta + \gamma'(1 + e\cos\theta)\big),
\end{aligned}
\tag{2.121}
$$

where:

$$
\begin{aligned}
\alpha' &= \frac{d\alpha}{du}, \\
\beta' &= \frac{d\beta}{du}, \\
\gamma' &= \frac{d\gamma}{du}.
\end{aligned}
\tag{2.122}
$$

It is simple to find the direction cosines of orbital system of coordinates using the expressions (2.119a), (2.119b), (2.119c). The axis x' corresponds to the condition $\theta = 0$ ($u = \omega$); the axis y' to $\theta = \pi/2$ ($u = \omega + \pi/2$), and the axis z' to $u = \pi/2$ and to the change $i \rightarrow i + \pi/2$. Finally, the direction cosines of orbital system ($[x', y', z']$) relative to the external system ($[x, y, z]$) are:

$$
\begin{aligned}
a_{11} &= \cos(\widehat{xx'}) = \cos\Omega\,\cos\omega - \sin\Omega\,\sin\omega\,\cos i, & (2.123a) \\
a_{12} &= \cos(\widehat{yx'}) = \sin\Omega\,\cos\omega + \cos\Omega\,\sin\omega\,\cos i, & (2.123b) \\
a_{13} &= \cos(\widehat{zx'}) = \sin\omega\,\sin i, & (2.123c) \\
a_{21} &= \cos(\widehat{xy'}) = -\cos\Omega\,\sin\omega - \sin\Omega\,\cos\omega\,\cos i, & (2.123d) \\
a_{22} &= \cos(\widehat{yy'}) = -\sin\Omega\,\sin\omega + \cos\Omega\,\cos\omega\,\cos i, & (2.123e) \\
a_{23} &= \cos(\widehat{zy'}) = \cos\omega\,\sin i, & (2.123f) \\
a_{31} &= \cos(\widehat{xz'}) = \sin\Omega\,\sin i, & (2.123g) \\
a_{32} &= \cos(\widehat{yz'}) = -\cos\Omega\,\sin i, & (2.123h) \\
a_{33} &= \cos(\widehat{zz'}) = \cos i. & (2.123i)
\end{aligned}
$$

Since the motion is plane, it is enough to use only the two orbital coordinates x', y', that is:

$$\begin{pmatrix} x \\ y \\ z \end{pmatrix} = \begin{pmatrix} a_{11} \\ a_{12} \\ a_{13} \end{pmatrix} x' + \begin{pmatrix} a_{21} \\ a_{22} \\ a_{23} \end{pmatrix} y'. \tag{2.124}$$

The expressions for $x'(E)$ and $y'(E)$ are given by (2.51b) and (2.52):

$$x' = a(\cos E - e),$$
$$y' = a\sqrt{1 - e^2} \sin E. \tag{2.125}$$

The velocity components:

$$\begin{pmatrix} \dot{x} \\ \dot{y} \\ \dot{z} \end{pmatrix} = \frac{na^2}{r} \left[-\begin{pmatrix} a_{11} \\ a_{12} \\ a_{13} \end{pmatrix} \sin E + \begin{pmatrix} a_{21} \\ a_{22} \\ a_{23} \end{pmatrix} \sqrt{1 - e^2} \cos E \right], \tag{2.126}$$

where we used Kepler equation (2.117) with (2.116) to express $\dot{E} = dE/dt$ through the mean motion n, are:

$$\dot{x}' = -a\dot{E} \sin E = -\frac{a^2 n}{r} \sin E,$$
$$\dot{y}' = a\sqrt{1 - e^2}\, \dot{E} \cos E = \frac{a^2 n}{r}\sqrt{1 - e^2} \cos E. \tag{2.127}$$

The formulas above allow us to obtain positions and velocities of the celestial body. It is important for the observations to obtain ephemeris in terms of right ascension α and declination δ. In this case we need to make the transition from the ecliptic $[x, y, z]$ to the equatorial $[x_e, y_e, z_e]$ system by a rotation of an angle ϵ. The direction cosines are:

$$b_{11} = a_{11}, \tag{2.128a}$$
$$b_{12} = a_{12} \cos \epsilon - a_{13} \sin \epsilon, \tag{2.128b}$$
$$b_{13} = a_{12} \sin \epsilon + a_{13} \cos \epsilon, \tag{2.128c}$$
$$b_{21} = a_{21}, \tag{2.128d}$$
$$b_{22} = a_{22} \cos \epsilon - a_{23} \sin \epsilon, \tag{2.128e}$$
$$b_{23} = a_{22} \sin \epsilon + a_{23} \cos \epsilon. \tag{2.128f}$$

We also need to make the transformation from heliocentric to geocentric coordinates:

$$x_e = x_\odot + b_{11}x' + b_{21}y',$$
$$y_e = y_\odot + b_{12}x' + b_{22}y', \qquad (2.129)$$
$$z_e = z_\odot + b_{13}x' + b_{23}y',$$

where $x_\odot$, $y_\odot$, $z_\odot$ are the coordinates of the Sun at the time t. Finally we obtain right ascension and declination for the given time:

$$\alpha = \arctan \frac{y_e}{x_e}, \qquad (2.130a)$$

$$\delta = \arctan \frac{z_e}{\sqrt{x_e^2 + y_e^2}}. \qquad (2.130b)$$

Note that the function arctan returns an angle[26] in the interval 0 to π, while α varies in the range $[0, 2\pi]$; this requires us to take into account the signs of both the numerator and the denominator in (2.130a). These formulas can be used to obtain the position of asteroids or planets on the celestial sphere at any time t with a good accuracy by knowing the orbital elements $a, e, \Omega, \omega, i, t_o$ (or M_o), and the coordinates of the Sun from a catalog.

In the following paragraph we deal with the inverse problem of determining the orbital elements of a Solar System body through the minimum number of astrometric observations (projected positions of the body in the sky). This problem has enthralled illustrious mathematicians and astronomers of the past such as Newton, Laplace, and Olbers.[27] We will consider here only the solution given by Laplace, unfriendly in practical use but highly educational. For the other famous solution proposed by Gauss,[28] which allowed to pick up Ceres[29] a year after its discovery, the reader is

---

[26] Right ascension is given in hours, where $1\,\text{h} = 15°$; but all calculations must be done in degrees or better radiants.

[27] German astronomer and medical doctor, Heinrich Wilhelm Olbers (1758–1840) was a discoverer of comets and small planets, and a convinced supporter of the pluralistic idea. His name is associated to the famous paradox: in a Euclidean, transparent, homogeneous, infinite, and eternal universe, the sky should shine like the surface of the Sun. With modern cosmology, the paradox is solved by the finite age of the universe.

[28] The German Carl Fiedrich Gauss (1777–1855) has been one of the greatest mathematicians of the modern age. Born in a humble family, he was an early genius. In a few years he revolutionized various fields of mathematics. In spite of a meticulous caution in communicating only results of which he was absolutely certain (*Pauca sed matura*), he made innovative and decisive contributions in the areas of mathematical analysis, number theory, numerical calculation, differential geometry, geodesy, magnetism, and optics. In 1801 he devised a method to calculate the orbit of the dwarf planet Ceres based on the astrometric observations of Piazzi, by conceiving the method of the 'least squares': a result which in 1807 earned him the directorship of the Göttingen Astronomical Observatory. He had prestigious students such as Richard Dedekind (1831–1916) and Bernhard Riemann (1822–1866). He also fostered the career of the young Friedrich Bessel.

[29] The small planet Ceres, the first "small planet" to be discovered, was noticed by the director of the Palermo Observatory, the Theatine father Giuseppe Piazzi (1746–1826), on the first night of the new year 1801. The astronomer was forced to interrupt the observations early because of an illness and bad weather conditions, well before he could realize which was the trajectory of the

referred to [7] and [8]. The method of the German mathematician is not very intuitive, but it is the most convenient to be translated into a computer program.

2.10   The Laplace Method for Recovering the Orbital Elements from Observations

Let us consider a body P of mass $m \ll M_\odot$ orbiting about the Sun and assume the total absence of perturbations so that we can safely use the results of the two-body problem. Suppose we are able to measure the geocentric positions of P with respect to a given Cartesian absolute reference frame with origin at the center of the Earth (to obviate the diurnal parallax). Such measures have become possible owing to radar techniques,[30] that adds the geocentric distance to the traditional astronomical angular coordinates. It can be proven that radar plus astrometric observations made at two different epochs completely solve (in general) the problem of determining the elements of the heliocentric orbit of P, although only formally as the data are insufficient to account for errors.

More complex is the case in which, as it normally happens, only angular coordinates and no distances are available. Following Laplace, we prove that it is generally possible to determine the orbital elements of P when its geocentric astrometric positions are known at three distinct epochs (for details, see Chap. VI in [7] and Chaps. V–VII in [9]). In this case too, the solution is merely formal because measurement errors are not accounted.

Let be ℓ_i, b_i ($i = 1, 2, 3$) the ecliptical geocentric coordinates[31] of P at the epochs t_i. If these measurements were made in a different astronomical coordinate system, for example in the equatorial system, they can be always transformed to the ecliptical system, which is the most convenient for our purposes (as the orbit of the Earth also enters the game and it lies precisely on the ecliptic plane).

In the geocentric reference system of Fig. 2.16, with plane xy containing the ecliptic, the direction cosines of the line TP joining the Earth T to the point P are:

---

new body among the stars. When the mysterious object disappeared in the daylight (Piazzi did not know yet what it was), it seemed to be irremediably lost. But the few observations accumulated were sufficient to the young Gauss for calculating the orbit. On the basis of his indications, the baron von Franz Xaver Zach (1754–1832) and Olbers were able to trace Ceres when it reappeared in the dark night. The detection was announced on December 31, 1801, exactly one year after the first discovery.

[30] Radar transforms distance determinations d in flight-time measurements $2\Delta t$ of wave packets shut against a target capable of reflecting them in the direction of the observer. The precision of the method increases with the compactness of the wave packet (δt turns into $\delta s = c\delta t$) and the accuracy of *1)* the model for the medium which the light beam will pass through (for example, the Earth atmosphere) for its effects on the speed of the light, and *2)* the reconstruction of the target geometry. Note that the distances cannot be too large since the reflected signal is a factor d^4 weaker when reaching back the observer.

[31] Reduced to the Earth center, after correction for diurnal parallax and light aberration.

Fig. 2.16 Ecliptical
geocentric coordinates

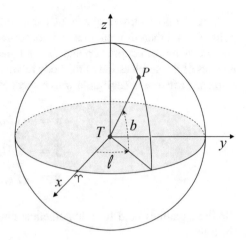

$$\alpha = \cos \ell \cos b,$$
$$\beta = \sin \ell \cos b, \qquad (2.131)$$
$$\gamma = \sin b,$$

where ℓ and b are the geocentric ecliptic coordinates at the same epoch t. Being functions of time, the direction cosines can be developed in a power series in a neighborhood of the epoch t_2. That is why we adopt it as the origin of time, placing $t_2 = 0$; this convention just simplifies the formalism. We have:

$$\alpha(t) = \alpha(t_2) + \sum_{n=1}^{\infty} \left(\frac{d^n \alpha}{dt^n} \right)_{t_2} \frac{(t - t_2)^n}{n!} = \alpha(0) + \sum_{n=1}^{\infty} \left(\frac{d^n \alpha}{dt^n} \right)_{o} \frac{t^n}{n!}, \qquad (2.132)$$

and similar relations for β and γ. If the time interval $\Delta = (t_3 - t_1)$ is small enough (compared to period of P), the developments exemplified in (2.132) can be arrested with some confidence at the terms of the second order. In other words, we assume that, in the interval $t_1 \leq t \leq t_3$, it is:

$$\alpha \approx \alpha_o + \dot{\alpha}_o t + \ddot{\alpha}_o \frac{t^2}{2}, \qquad (2.133)$$

where the suffix indicates that the direction cosines and their derivatives are calculated at epoch $t_2 = 0$. The values of these cosines are immediately derived from the set of the positional measures at the three epochs, solving the system of three equations of the type (2.133). In doing so, we deliberately disregard the presence of measurement errors in ℓ and b. Many more than the three observations, distributed in the interval $-\Delta \leq t \leq \Delta$, are required to account for the uncertainties on the unknowns in (2.133) as well as the consequences of the truncation of the power series at the second order.

Consider now a triplet of ecliptic axes, with origin at the Sun, x-axis towards the point Υ (and therefore parallel and concordant to the geocentric reference frame), and y-axis on the ecliptical plane. Being r the heliocentric distance of P, the components of the acceleration of the body relative to the axes of this ecliptic system are (remember that m is negligible with respect to $M_\odot$):

$$
\begin{aligned}
\ddot{x} &= -G\,M_\odot \frac{x}{r^3}, \\
\ddot{y} &= -G\,M_\odot \frac{y}{r^3}, \\
\ddot{z} &= -G\,M_\odot \frac{z}{r^3}.
\end{aligned}
\tag{2.134}
$$

Similar equations hold for a hypothetical unperturbed motion of the Earth around the Sun:

$$
\begin{aligned}
\ddot{x}_\oplus &= -GM_\odot \frac{x_\oplus}{R_\oplus^3}, \\
\ddot{y}_\oplus &= -GM_\odot \frac{y_\oplus}{R_\oplus^3}, \\
\ddot{z}_\oplus &= -GM_\odot \frac{z_\oplus}{R_\oplus^3} = 0,
\end{aligned}
\tag{2.135}
$$

where $R_\oplus$ is the distance of the Earth to the Sun and $(x_\oplus, y_\oplus, z_\oplus)$ its components. Subtracting (2.135) from (2.134) it is:

$$
\begin{aligned}
\ddot{x} - \ddot{x}_\oplus &= GM_\odot \left(\frac{x_\oplus}{R_\oplus^3} - \frac{x}{r^3} \right), \\
\ddot{y} - \ddot{y}_\oplus &= GM_\odot \left(\frac{y_\oplus}{R_\oplus^3} - \frac{y}{r^3} \right), \\
\ddot{z} &= GM_\odot \left(\frac{z_\oplus}{R_\oplus^3} - \frac{z}{r^3} \right) = -GM_\odot \frac{z}{r^3}.
\end{aligned}
\tag{2.136}
$$

Being ρ the distance of P from the Earth, we have:

$$
\begin{aligned}
x - x_\oplus &= \rho\,\alpha, \\
y - y_\oplus &= \rho\,\beta, \\
z &= \rho\,\gamma,
\end{aligned}
\tag{2.137}
$$

which, differentiated twice with respect to time and substituted in (2.136), gives:

$$\ddot{\rho}\alpha + 2\dot{\rho}\dot{\alpha} + \rho\ddot{\alpha} = GM_\odot \left(\frac{x_\oplus}{R_\oplus^3} - \frac{x_\oplus + \rho\alpha}{r^3} \right),$$

$$\ddot{\rho}\beta + 2\dot{\rho}\dot{\beta} + \rho\ddot{\beta} = GM_\odot \left(\frac{y_\oplus}{R_\oplus^3} - \frac{y_\oplus + \rho\beta}{r^3} \right), \qquad (2.138)$$

$$\ddot{\rho}\gamma + 2\dot{\rho}\dot{\gamma} + \rho\ddot{\gamma} = -GM_\odot \frac{\rho\gamma}{r^3}.$$

We solve the system of the three linear equations (2.138) with the rule of Cramer[32] for the unknowns ρ, $\dot{\rho}$, and $\ddot{\rho}$ at the time $t_2 = 0$, remembering that the direction cosines and their first and second derivatives are known. Meanwhile:

$$\rho = GM_\odot \left(\frac{1}{R_\oplus^3} - \frac{1}{r^3} \right) \frac{D_3}{D}, \qquad (2.139)$$

for[33] $D \neq 0$. The determinants D and D_3 are[34]:

$$D = \begin{vmatrix} \alpha_\circ & 2\dot{\alpha}_\circ & \ddot{\alpha}_\circ + GM_\odot \dfrac{\alpha_\circ}{r^3} \\ \beta_\circ & 2\dot{\beta}_\circ & \ddot{\beta}_\circ + GM_\odot \dfrac{\beta_\circ}{r^3} \\ \gamma_\circ & 2\dot{\gamma}_\circ & \ddot{\gamma}_\circ + GM_\odot \dfrac{\gamma_\circ}{r^3} \end{vmatrix} = 2 \begin{vmatrix} \alpha_\circ & \dot{\alpha}_\circ & \ddot{\alpha}_\circ \\ \beta_\circ & \dot{\beta}_\circ & \ddot{\beta}_\circ \\ \gamma_\circ & \dot{\gamma}_\circ & \ddot{\gamma}_\circ \end{vmatrix}, \qquad (2.140)$$

$$D_3 = 2 \begin{vmatrix} \alpha_\circ & \dot{\alpha}_\circ & x_\oplus \\ \beta_\circ & \dot{\beta}_\circ & y_\oplus \\ \gamma_\circ & \dot{\gamma}_\circ & 0 \end{vmatrix}. \qquad (2.141)$$

With reference to the epoch of the second observation, they contain all known quantities. In fact, the heliocentric coordinates $x_\oplus$, $y_\oplus$, and $z_\oplus (\equiv 0)$, of the Earth at the time t_2 of the second observation, as well as the distance $R_\oplus$, can be derived from ephemeris tables.[35] Equation (2.139) still contains two unknowns, r and ρ.

Another equation that contains the same unknowns is derived from the triangle formed by the Earth, the Sun, and the point P, for which the following relation holds (Fig. 2.17):

---

[32] Swiss mathematician, Gabriel Cramer (1704–1752) was a pupil of Johann Bernoulli. He dealt with algebraic curves and determinants. He discovered the rule for the solutions of systems of n linear equations in n unknowns, which generalizes that of the Scotsman Colin Maclaurin.

[33] $D = 0$ represents an 'unfortunate' set of data: the three observations are contained in the same great circle. Then a new astrometric set is needed, choosing different epochs.

[34] These two properties of determinants are useful to understand the passage.

1. The multiplication by t of all elements of any row (or column) of a matrix changes the determinant by the same factor t. For instance: $\begin{vmatrix} ta & tb \\ c & d \end{vmatrix} = t \begin{vmatrix} a & b \\ c & d \end{vmatrix}$.

2. The determinant acts as a linear function on the rows of the matrix:
$\begin{vmatrix} a+a' & b+b' \\ c & d \end{vmatrix} = \begin{vmatrix} a & b \\ c & d \end{vmatrix} + \begin{vmatrix} a' & b' \\ c & d \end{vmatrix}.$

Fig. 2.17 Graphical representation of the elongation E

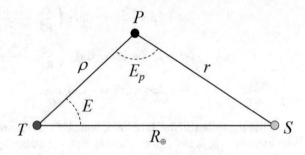

Fig. 2.18 Elongation angle: calculation

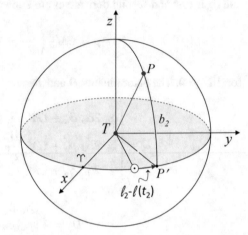

$$r^2 = R_\oplus^2 + \rho^2 - 2R_\oplus\, \rho\, \cos E, \qquad (2.142)$$

where the angle E, named elongation[36] of P, measures the angular distance of P from the Sun with respect to a terrestrial observer (obviously corrected for diurnal parallax). It can be determined by means of geometric considerations based on the measure of ℓ_2 and b_2, or can be measured directly at t_2. Geometrically it results (Fig. 2.18):

$$\cos E = \cos b_2\, \cos\left(\ell_2 - \ell_\odot(t_2)\right), \qquad (2.143)$$

where $\ell_\odot(t_2)$ is the ecliptic longitude of the Sun at t_2, tabulated in the almanacs. From (2.139) to (2.142) we therefore obtain the geocentric and heliocentric distances, ρ and r, of P at the epoch of the second observation.

---

[35] See, for example, *The Astronomical Almanac*, published by the US Naval Observatory and by the Nautical Almanac Office of Her British Majesty, containing the ephemerides of the Solar System and a catalog of selected star and extragalactic objects.

[36] Not to be confused with the eccentric anomaly having the same symbol.

The solution of the system of two equations mentioned above is not trivial. It leads to an algebraic equation of 8-th degree in $r/R_\oplus$, known as the Lagrange equation. Rather than trying to solve it, we take a different approach. Indicated with E_p the relative elongation for an observer in P, from the triangle of Fig. 2.17 we get:

$$\frac{R_\oplus}{\sin E_p} = \frac{r}{\sin E} = \frac{\rho}{\sin(E + E_p)}, \qquad (2.144)$$

from where, through (2.139), it is:

$$R_\oplus \sin E \cos E_p + \left(R_\oplus \cos E - GM_\odot \frac{A}{R_\oplus^3} \right) \sin E_p = $$
$$= -GM_\odot \frac{A}{R_\oplus^3} \frac{\sin^4 E_p}{\sin^3 E}, \qquad (2.145)$$

with $A = D_3/D$. Let then be:

$$N \sin m = R_\oplus \sin E,$$
$$N \cos m = R_\oplus \cos E - GM_\odot \frac{A}{R_\oplus^3}, \qquad (2.146)$$
$$M = -N \frac{R_\oplus^3}{A} \frac{1}{GM_\odot} \sin^3 E.$$

The first two equations, with the second terms consisting of known quantities, show that N and m are uniquely determined once the sign of N is fixed. We choose the latter so that $M > 0$. By replacing (2.146) in (2.145) and simplifying, we have:

$$\sin^4 E_p = M \sin\left(E_p + m \right). \qquad (2.147)$$

This equation, where M and m are known, can be solved for E_p by successive approximations. For example, if E_p° is an approximate value, it is easy to prove that, at the first order, the correction $\Delta E_p^\circ = E_p - E_p^\circ$ is given by:

$$\Delta E_p^\circ = -\frac{\sin^4 E_p^\circ - M \sin\left(E_p^\circ + m \right)}{4\sin^3 E_p^\circ \cos E_p^\circ - M \cos\left(E_p^\circ + m \right)}. \qquad (2.148)$$

Since the position of the terrestrial observer must satisfy the set of equations, the relative solution $E_p = \pi - E$ must be discarded. The real solutions of (2.147) are the intersections of the curves $y_1 = \sin^4 E_p$, and $y_2 = M \sin\left(E_p + m \right)$. Generally they are three (see Fig. 2.19); when there is only one, the problem related to P is not solvable as this solution corresponds to the terrestrial observer.

Consider the case in which $A = D_3/D > 0$. Then from (2.142) $r > R_\oplus$ and, since $\sin E \geq 0$, then $N < 0$ and, from the first of the (2.146), $\pi < m < 2\pi$. In practice, however, if we want 3 solutions, it must be $(3/2)\pi < m < 2\pi$. If instead $A < 0$, it must be $0 < m < \pi/2$.

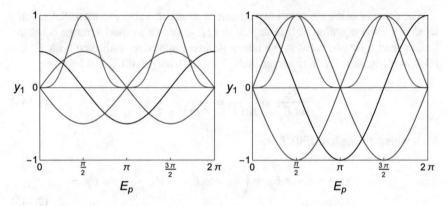

Fig. 2.19 Graphical solution of equation (2.147). We have placed $y_1 = \sin^4 E_p$ (*red line*). The other curves (*blue* and *black lines*) correspond to the right side of equation (2.147) for various combinations of the parameters: $M = 0.5$ (*left*) and $M = 1.0$ (*right*), and $m = 0, \pi/2, \pi$

Incidentally, we observe that Eq. (2.139) contains *in nuce* the Lambert[37] theorem: if, at the epoch of observation, a body is found to be farther away from the Sun than the Earth, its apparent trajectory on the celestial sphere will be convex towards the Sun; in the opposite case, the apparent trajectory will be concave towards the Sun. From this statement we deduce that the singular case $D = 0$ occurs when, at the epoch t_2, the apparent trajectory of P has an inflection.

Solving the system (2.138) with respect to the unknown $\dot{\rho}$, we obtain:

$$\dot{\rho} = GM_\odot \left(\frac{1}{R_\oplus^3} - \frac{1}{r^3} \right) \frac{D_2}{D}, \qquad (2.149)$$

where the determinant D_2 is:

$$D_2 = \begin{vmatrix} \alpha_\circ & x_\oplus & \ddot{\alpha}_\circ + GM_\odot \dfrac{\alpha_\circ}{r^3} \\ \beta_\circ & y_\oplus & \ddot{\beta}_\circ + GM_\odot \dfrac{\beta_\circ}{r^3} \\ \gamma_\circ & 0 & \ddot{\gamma}_\circ + GM_\odot \dfrac{\gamma_\circ}{r^3} \end{vmatrix} = 2 \begin{vmatrix} \alpha_\circ & x_\oplus & \ddot{\alpha}_\circ \\ \beta_\circ & y_\oplus & \ddot{\beta}_\circ \\ \gamma_\circ & 0 & \ddot{\gamma}_\circ \end{vmatrix}. \qquad (2.150)$$

We are now ready to find the component of the heliocentric velocity of P. In fact, by deriving the (2.137) with respect to time, we have:

---

[37] A contemporary of Euler, Johann Heinrich Lambert (1728–1777) was a Swiss-French polymath who made important contributions in physics, mathematics, cartography, and philosophy. He was a pioneer of non-Euclidean geometry.

$$\dot{x} = \dot{x}_\oplus + \dot{\rho}\alpha_o + \rho\dot{\alpha}_o,$$
$$\dot{y} = \dot{y}_\oplus + \dot{\rho}\beta_o + \rho\dot{\beta}_o, \qquad (2.151)$$
$$\dot{z} = \dot{\rho}\gamma_o + \rho\dot{\gamma}_o,$$

where all quantities at the right hand side have been calculated for $t_2 = 0$ or can be found in almanacs. Recalling (2.68), from the (2.151) it is:

$$\dot{x}^2 + \dot{y}^2 + \dot{z}^2 = 2\,GM_\odot \left(\frac{1}{r} - \frac{1}{2\,a}\right). \qquad (2.152)$$

Since r is already known, this relation allows us determine the semi-major axis a. Remembering the considerations in Sect. 2.7, if a is finite and positive the orbit of the point P is elliptical, if $a = +\infty$ it is parabolic,[38] and finally, if a is finite and negative it is hyperbolic. Whether the orbit is elliptical, the Eq. (2.83):

$$\frac{1}{T} = \sqrt{\frac{GM_\odot}{4\pi^2\,a^3}}, \qquad (2.153)$$

provides the period T (a meaningless quantity for open orbits).

Consider now a generic plane in the heliocentric ecliptic system, imposing the condition that it contains the orbit of P. This implies that this plane contains also the origin of the coordinate system (where the Sun is); therefore its equation writes: $\lambda x + \mu y + z = 0$. The intersection with the ecliptical plane, $z = 0$, generates the line of nodes, $y = -(\lambda/\mu)\,x$. Its angular coefficient corresponds to the trigonometric tangent of the angle formed by the line of nodes with the x-axis, which is directed towards the point Υ. Therefore:

$$\tan \Omega = -\frac{\lambda}{\mu}. \qquad (2.154)$$

Moreover, the third of the direction cosines of the vector $\mathbf{N}$ normal to the plane (2.143):

---

[38] You understand from here that, in a random sample of orbits, the probability of a true parabolic orbit is just zero. This consideration also applies to the simple models of the universe developed in the third decade of 1900 by the Russian Alexandr Friedman (1888-1925) on the basis of a dynamical solution of the equations of general relativity. Three are the possible results, according to the sign of the curvature k, where $k = +1$ is valid for a closed universe with spherical geometry, $k = -1$ for an open universe with hyperbolic geometry. The open Euclidean universe corresponds to $k = 0$, making this solution very unlikely from the statistical point of view, while observations seem to prove that the space is actually flat. In the Newtonian case, this corresponds to negative, positive, or zero energy values.

$$\cos \widehat{Nx} = \frac{\lambda}{\sqrt{\lambda^2 + \mu^2 + 1}},$$

$$\cos \widehat{Ny} = \frac{\mu}{\sqrt{\lambda^2 + \mu^2 + 1}}, \qquad (2.155)$$

$$\cos \widehat{Nz} = \frac{1}{\sqrt{\lambda^2 + \mu^2 + 1}},$$

is obviously the cosine of the inclination angle i:

$$\cos i = \frac{1}{\sqrt{\lambda^2 + \mu^2 + 1}}. \qquad (2.156)$$

By determining λ and μ as functions of the orbital elements Ω and i through (2.154) and (2.156), the direction cosines of $\mathbf{N}$ are:

$$\begin{aligned}
\cos \widehat{Nx} &= \quad \sin i \, \sin \Omega, \\
\cos \widehat{Ny} &= - \sin i \, \cos \Omega, \qquad (2.157) \\
\cos \widehat{Nz} &= \quad \cos i.
\end{aligned}$$

Now observe that the orbital angular momentum per unit mass of P is constant and directed as $\mathbf{N}$. By indicating its module with h, the components along the three heliocentric ecliptic axes are:

$$\begin{aligned}
y\dot{z} - z\dot{y} &= h \, \sin i \, \sin \Omega, \\
z\dot{x} - x\dot{z} &= -h \, \sin i \, \cos \Omega, \qquad (2.158) \\
x\dot{y} - y\dot{x} &= h \, \cos i.
\end{aligned}$$

Squaring and summing, it is:

$$(y\dot{z} - z\dot{y})^2 + (z\dot{x} - x\dot{z})^2 + (x\dot{y} - y\dot{x})^2 = h^2, \qquad (2.159)$$

and therefore, knowing the components of the distance and of the heliocentric velocity of P at the epoch t_2, the module of the angular momentum per unit mass is found. But, from the relation (2.65), it is: $h^2 = GM_\odot a \, (1 - e^2)$, which provides the value of the eccentricity e, using the already calculated value of a. Furthermore, Eq. (2.158), being their left-hand sides known, allows us to calculate the longitude at the node Ω and the inclination i. Not surprisingly, the number of equations is larger than the number of unknowns. The overabundance is needed to remove the ambiguity on the sign of the main values. This is why, in a rotation of axes, the parameters are two (Euler) angles only, but the independent equations are three.

Now the argument of pericenter ω must be found. To this end, we indicate with θ_2 the true anomaly of P at the epoch of the second observation, referred to the focal axis. This is provided by (2.64) where we place $\theta_\circ = 0$. The ecliptic latitude of P

Fig. 2.20 The relation
between ecliptic latitude b
and true anomaly $\omega + \theta$

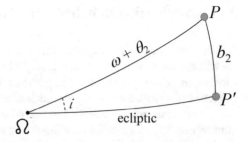

can be expressed by the equations (Fig. 2.20):

$$\sin b_2 = \sin i \; \sin(\omega + \theta_2),$$
$$\tan(\ell_2 - \Omega) = \cos i \; \tan(\omega + \theta_2), \tag{2.160}$$

which provide ω.

Finally, to obtain the epoch t_o of the passage through the pericenter, we just use the relation between true anomaly and time. For example, via (2.87) we calculate the eccentric anomaly E_2 at $t = t_2$ for the elliptical orbit, entering with the value θ_2 of the true anomaly. Then, t_o is readily given by (2.84).

At the end of this paragraph, it is worth emphasizing once more that the method suffers from two serious limitations: it does not account for the measurement errors, and disregards any possible perturbations to the pure Keplerian motion of the two massive points. The first limitation (which includes the uncertainty introduced in the truncation of the series (2.117)) is removed to a large extent by increasing the number of astrometric determinations. The system of equations (2.133) will then be over-abundant and must be solved by proper interpolation methods (for example, the least square method; see Appendix O.1). Actually, when all computations were made by hand, a variational approach to the determination of orbital elements was developed to add a new observation to an already found solution (called preliminary) without repeating from the beginning the long and tedious procedure on the complete set of data. The second uncertainty is removed by applying correctly the theory of planetary perturbations to the preliminary determined orbit.

As an application of what we learned in this chapter, a graphical technique to determine the orbital elements of binary stellar systems, widely used in the second half of the 19-th century, is illustrated in Appendix J. The results so far obtained are also useful for some simple considerations on space flight dynamics (for an in-depth study, *cf.* [10]).

2.11   Application to Ballistics and Space Flight

Consider an object with a negligible mass if compared to the Earth, $m_P \ll M_\oplus$, that we will call *probe* for convenience. It might be, for instance, a missile, or an artificial satellite, or a deep space ship. We also assume that the Earth is spherically symmetric (which implies that, for $r \geq R_\oplus$, its gravitational field is the same of a point-like mass $M_\oplus$ placed in its gravity center), and free of superficial roughness (smooth) and any atmosphere (to remain in the conditions of conservation of mechanical energy).

From (2.73), with $M_\oplus = 5.974 \times 10^{27}$ g and $R_\oplus = 6.373 \times 10^8$ cm, the escape velocity of *probe* is:

$$V_e = \sqrt{\frac{2GM_\oplus}{R_\oplus}} \simeq 11.2 \text{ km s}^{-1}, \tag{2.161}$$

and the velocity required to keep it in a hypothetical circular orbit of radius $R_\oplus$ (i.e., in an orbit shaving the surface of the Earth) is:

$$V_g = \sqrt{\frac{GM_\oplus}{R_\oplus}} = \frac{V_e}{\sqrt{2}} \simeq 7.9 \text{ km s}^{-1}. \tag{2.162}$$

Let us now determine the initial velocity V_i to be given (impulsively) to *probe* in order to insert it in an elliptical trajectory with apogee at an altitude h, i.e., at a distance $h + R_\oplus$ from the Earth center which, for the first Kepler law, is also the focus of the orbit. Since in (2.68) it is $r = R_\oplus$ and $a(1 + e) = R_\oplus + h$, i.e., $a = \dfrac{R_\oplus + h}{1 + e}$, then:

$$V_i = \sqrt{GM_\oplus \left(\frac{2}{R_\oplus} - \frac{1+e}{R_\oplus + h} \right)} = V_e \sqrt{1 - \frac{1+e}{2(1 + h/R_\oplus)}}. \tag{2.163}$$

Therefore, once the altitude h is assigned, the initial velocity of *probe* depends on the orbital eccentricity e.

Let us now ask what is the least expensive trajectory, in terms of work done (and therefore of fuel used for the propulsion), to transfer a *probe* from a circular Earth orbit of radius r_1 to a coplanar one of radius r_2. The answer was found in 1925 by Walter Hohmann.[39] It is, at least in the cases in which $r_2/r_1 \lesssim 12$, an elliptical orbit tangent to both circular orbits (Fig. 2.21).

Consider then the transfer orbit in the gravitational field of a body of mass M from the orbit with radius r_1 to the coplanar orbit r_2. The velocity of *probe* is equal

---

[39] The German engineer Walter Hohmann (1880-1945) was a pioneer of astronautics and a leading member in the *Verein für Raumschiffahrt*, an amateur Society for Space Travel born in Germany at the end of the 1920s, of which Wernher von Braun (1912-1977) was also an influential figure. With the advent of Nazism he decided to withdraw from rockets in order not to encourage their use as weapons, unlike von Braun who instead successfully developed the lethal V2 (Aggregat 4) at Peenemünde.

to the circular velocity at the radius r_1:

$$(V_1)_c = \sqrt{\frac{GM}{r_1}}. \tag{2.164}$$

Let us change impulsively this velocity to the value $(V_i)_e$ so that *probe* settles on an elliptical orbit with an apogee in r_2, with a major axis $a = (r_1 + r_2)/2$:

$$(V_1)_e = \sqrt{2GM\left(\frac{1}{r_1} - \frac{1}{r_1 + r_2}\right)} = \sqrt{\frac{GM}{r_1}}\sqrt{\frac{2r_2}{r_1 + r_2}}. \tag{2.165}$$

Thus, the impulse must produce a velocity variation:

$$\Delta V_1 = (V_1)_e - (V_1)_c = \sqrt{\frac{GM}{r_1}}\left(\sqrt{\frac{2r_2}{r_1 + r_2}} - 1\right). \tag{2.166}$$

Obviously, for the transfer from an outer to an inner orbit, the impulse must produce $\Delta V_1 = (V_1)_c - (V_1)_e$. When *probe* reaches its apogee, due to the conservation of momentum (Kepler's second law), the velocity on the elliptical orbit is reduced by a factor r_1/r_2. So:

$$(V_2)_e = (V_1)_e\frac{r_1}{r_2} = \sqrt{\frac{GM}{r_2}}\sqrt{\frac{2r_1}{r_1 + r_2}}. \tag{2.167}$$

The impulse to be given to *probe* in order to reach the orbit r_2 is $\Delta V_2 = (V_2)_c - (V_2)_e$:

$$\Delta V_2 = \sqrt{\frac{GM}{r_2}}\left(1 - \sqrt{\frac{2r_1}{r_1 + r_2}}\right). \tag{2.168}$$

For the transfer from outer to inner orbits, it must be: $\Delta V_2 = (V_2)_e - (V_2)_c$.

The time t needed to complete the transfer is half of the period P of the elliptical orbit. It is provided by the Eq. (2.89) or by the third Kepler law:

$$\Delta t = \pi\sqrt{\frac{1}{GM}\left(\frac{r_1 + r_2}{2}\right)^3}. \tag{2.169}$$

Consider the example of a Hohmann transfer from an altitude of $h = 300$ km above the Earth surface to a circular orbit at geosynchronous[40] altitude. In this case, $M = M_\oplus$, $r_1 = R_\oplus + h \simeq 6{,}670$ km, and $r_2 = r_{GSO}$. The radius of the (circular)

---

[40] The orbit of a satellite around Earth is said to be geosynchronous (GSO) if its period equals the sidereal day $P_s = 23^h\ 56^m\ 4^s = 86{,}164.1$ s, i.e., the 'average' angular velocity of the *probe* is the same as the diurnal rotation of the Earth.

Fig. 2.21 Hohmann orbit
(*black curve*) for a transfer
from Earth orbit to Mars
orbit. Open circles mark the
positions of the Earth (*blue*)
and Mars (*red*) at the time of
launch; solid circles show
the positions of two planets
at arrival time

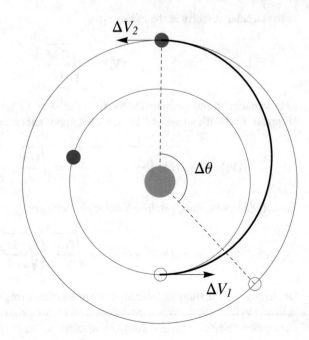

geosynchronous orbit is obtained from the third Kepler's law:

$$r_{GSO} = \left(\frac{P_s^2 \, G \, M_\oplus}{4\pi^2} \right)^{1/3} = 42,164 \, \text{km}, \qquad (2.170)$$

where P_s is the sidereal day. The circular velocities are $(V_1)_c = 7.73$ km s^{-1} and
$(V_2)_c = 3.07$ km s^{-1}. The impulse at the altitude h gives $\Delta V_1 = 2.43$ km s^{-1}. At
the geosynchronous orbit it is $\Delta V_2 = 1.47$ km s^{-1}, and the total velocity change is
$\Delta V = 3.9$ km s^{-1}.

Using the above formula we can calculate the change of velocities needed for a
transfer between two planets. As an example, we want to fly from the Earth to Mars.
In this case, we use $M = M_\odot$, $r_1 = r_\oplus$, and $r_2 = r_{\circ\!\!\!\!\nearrow}$ in (2.164) and (2.169). The
real travel to other planets can be fairly complicated. In general, the travel can be
subdivided in three stages:

1. the motion in the sphere of influence[41] (SOI) of the Earth, where the primary
   gravitational source is our planet;
2. from the SOI of the Earth to the SOI of Mars, where the primary gravitational
   source is the Sun;
3. the motion in the SOI of Mars.

We will consider the second stage of the flight by a Hohmann orbit, neglecting the
sizes of the SOIs of planets in comparison with the interplanetary distance to be

---

[41] A sphere of influence is a spheroidal region surrounding a celestial body where it exerts the
primary gravitational influence on an object with negligible mass; see also Appendix I.

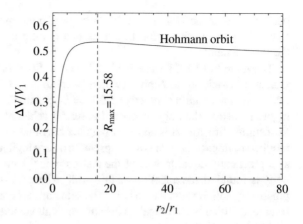

Fig. 2.22 Hohmann transfer orbit efficiency. Graphical representation of Eq. (2.173). Note that $R = r_2/r_1 \geq 1$

covered. To move from Earth to Mars we have to choose the proper configuration of the planets: the position of the Earth must coincide with a pericenter of the Hohmann orbit and the position of the Mars with the apocenter of the transfer orbit (Fig. 2.21). The period of opposition can be found from the difference of the angular velocities of the Earth ($\omega_\oplus$) and Mars (ω_σ): $\omega_{opp} = \omega_\oplus - \omega_\sigma$. So, we obtain:

$$T_{opp} = \frac{T_\oplus T_\sigma}{T_\sigma - T_\oplus}, \tag{2.171}$$

and $T_{opp} = 2.13$ year. The spacecraft has to be lunched when the angle between the positions of Mars at the start and at the end of the travel is $\Delta\theta = 135.66°$ (Fig. 2.21). The impulses given to leave the Earth orbit and at the Mars orbit are $\Delta V_1 = 2.94$ km s^{-1} and $\Delta V_2 = 2.65$ km s^{-1}, and the total change of velocity is $\Delta V = 5.59$ km s^{-1}. The flight time along the Hohmann orbit can be easily expressed through the orbital period of the Earth:

$$\Delta t = \frac{1}{4\sqrt{2}} \left(1 + \frac{r_\sigma}{1\,\text{AU}}\right)^{3/2} \text{yr}. \tag{2.172}$$

Since the distance to Mars is $r_\sigma = 1.5$ AU, we obtain $\Delta t \simeq 0.7$ years, that is a little less than 8.5 months.

The ratio $\Delta V/V_1$ as a function of relative distance $R = r_2/r_1$ gives us the efficiency of this transfer orbit (Fig. 2.22):

$$\frac{\Delta V}{V_1} = \sqrt{\frac{2R}{1+R}} - 1 + \sqrt{\frac{1}{R}}\left[1 - \sqrt{\frac{2}{1+R}}\right]. \tag{2.173}$$

The transfer is more effective for the nearest orbits and for the most distant. The function in (2.173) has the maximum at $R_{\text{max}} = 15.58$; in other words, the worst performance of the Hohmann orbit is to pass to an upper orbit that is 15.58 times larger than the starting orbit. In this case, a bi-elliptical orbit can be used to reach the large distance.

The relation (2.173), once again adapted to Solar orbits, shows that in order to launch a vehicle up to Jupiter (which is about 5 AU from the Sun), the impulse must generate an initial velocity of $\Delta V = 14.3 \ \mathrm{km \ s^{-1}}$, equal to 48% of $V_\oplus$. Much higher is instead the impulse needed to send a vehicle to the Sun, while the latter is the central attractor. This apparent paradox becomes clear if we think that the orbit required to impact the Sun is an ellipse with a very high flattening, with a semi-major axis practically equal to half of the Astronomical Unit. At the time of launch, the body is found at the aphelion of this orbit, where its (tangential) speed should be practically null. Instead the body has the same orbital speed of the Earth, which must be zeroed with a launch 'against the stream' with velocity equal and opposite to that of the planet: 29.8 $\mathrm{km \ s^{-1}}$.

References

1. G. Bertone, D. Hooper, *History of Dark Matter*, FERMILAB-PUB-16-157-A https://arxiv.org/pdf/1605.04909.pdf
2. M. Mayor, Nobel Lecture: Plurality of worlds in the cosmos: a dream of antiquity, a modern reality of astrophysics. Rev. Mod. Phys. **92**, 030502 (2020)
3. H. Goldstein, *Classical Mechanics* (Addison-Wesley, Boston, 1980)
4. L.D. Landau, E.M. Lifshitz, *Mechanics* (Elsevier, Amsterdam, 1982)
5. W.C. Nelson, E.E. Loft, *Space Mechanics* (Prentice-Hall Inc, Englewood Cliffs , N.J., 1962)
6. E. Harrison, *Cosmology: The Science of the Universe* (Cambridge University Press, Cambridge, 2000)
7. F.R. Moulton, *An Introduction to Celestial Mechanics* (The MacMillan Company, New York, 1960)
8. J. Dutka, On Gauss' priority in the discovery of the method of least squares. Arch. Hist. Exact Sci. **49**, 355 (1996)
9. A.D. Dubyago, *The Determination of Orbits* (The MacMillan Company, New York, 1961)
10. H.D. Curtis, *Orbital Mechanics for Engineering Students* (Elsevier, Amsterdam, 2005)

Chapter 3
The Three-Body Problem

Wolfgang Pauli

I don't mind your thinking slowly; I mind your publishing faster than you think.

The simplest generalization of the two-body problem is to introduce an additional third body. So doing generates the so-called three-body problem, which, while not generally solvable, is still a good tool for dealing with more realistic situations than the two-body approach alone. For instance, a geostationary satellite[1] turning in an Earth orbit feels also the gravitational attraction of the Moon, which cannot be overlooked if a better level of accuracy is pursued than one part on 10^4 on the acting force. Similarly, to analyze the simple motion of the small bodies in the asteroid belt (Fig. 3.1), it is inevitable to take into account, besides the Sun, the interaction with Jupiter as well. As we know, the general solution of the three-body problem requires the knowledge of 18 first integrals. The fact that we are limited by conservation laws to a smaller number requires us to follow paths other than the direct approach.

In this chapter we will address the problem of three bodies, focusing on attempts to solve it in particular cases as, for instance, when one of the three masses is negligible compared to the other two. In fact, not even this restrictive hypothesis is enough for us to solve the problem exactly. However, we can constrain the dynamical evolution of the system by identifying regions where the motion cannot take place, thus shedding light on some real astrophysical phenomena: for example, on the peculiar dynamical coupling among Sun, Jupiter, and the asteroid families of the Trojans and Greeks

---

[1] A geostationary orbit is a circular geosynchronous equatorial orbit; its geocentric radius (2.170) is such that the period of revolution, P_{gst}, of a massless particle is equal to a sidereal day P_s (see also Footnote 40). It is a special case of geosynchronous orbit, whose only constraint is $P_{gsy} = P_s$.

© The Author(s), under exclusive license to Springer Nature Switzerland AG 2022
E. Bannikova and M. Capaccioli, *Foundations of Celestial Mechanics*, Graduate Texts in Physics, https://doi.org/10.1007/978-3-031-04576-9_3

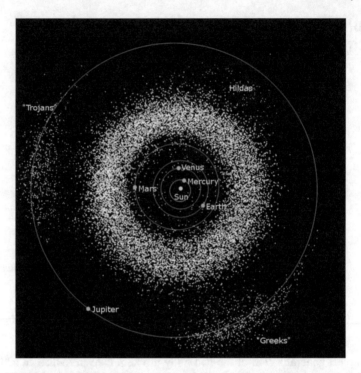

Fig. 3.1 Plot of the asteroid belt (*white*) superimposed on the planetary orbits out to that of Jupiter. Also shown (*green*) are the families of asteroids named Trojans and Greeks. Cf. Sect. 3.5.2 to understand why on average they form equilateral triangles with Jupiter and the Sun, clustering around the Lagrangian points L_4 and L_5 (Sect. 3.4). The asteroids of the Hilda family (*orange*) travel in a $3:2$ orbital resonance such that they reach aphelion (where Jupiter's perturbation is greatest) in a position diametrically opposite to the giant planet, or at an angular distance of Angle60 from it (i.e., again in L_3, L_4, and L_5). Picture from Wikipedia

(Fig. 3.1). The latter are small planets[2] preceding (Trojans) and following (Greeks) the giant planet in its orbit around the Sun. They form an angle of Angle60 with Jupiter and the Sun and resist well to planetary perturbations (in other words their positions are quasi-stable).

   In Sect. 3.1 we will consider a class of simple solutions of the three-body problem which are called stationary because the configuration assumed by the three massive

---

[2] No less than 2350 Trojans are known (this is the common name adopted for both families), equally distributed between the two fields of Greeks and Trojans. The tradition of calling these objects with the names of the Iliad heroes (up to availability) dates back to the German astronomer Max Wolf (1863–1932), pioneer of astrophotography. In 1906 he discovered an asteroid and called it Achilles; it was the first known member of a rich family. Before the rule of drawing on Homeric characters was clearly established, assigning Greek names to bodies following Jupiter and Trojan names to those preceding the planet, there was time to generate some historical mistakes. Thus the Greek Patroclus, discovered in 1906 by August Kopff (1882–1960), came to find itself in the Trojan camp, and the Trojan hero Hektor, also discovered by Kopff in 1907, in the Greek one. Today the term Trojans indicates also those asteroids anchored to the Lagrangian points L_4 and L_5 (cf. Sect. 3.5.3) pertaining to pairs of massive bodies such as Sun-Earth, Sun-Venus, Sun-Mars, and even Earth-Moon (dust, in this case).

points remains similar to itself during motion. In Sect. 3.1.1 the general problem will be restricted to the case, very useful in real astrophysical situations, in which one of the three bodies has a negligible mass. We will discover how it is possible to find a motion integral, named after Jacobi[3] and described in Sect. 3.1.2, from which the Hill[4] surfaces descend; they limit the motion of the massless body to certain regions of space. We will also derive the dynamical equilibrium points named after Lagrange, both under oversimplified assumptions and in the most general form of stationary solutions. Stability of these equilibrium points against perturbations will be assessed in Sect. 3.5. The remaining of this chapter will be devoted to study the variations of the orbital elements of a system of two bodies, one of which with negligible mass, under an external perturbation.

3.1 Stationary Solutions

We want to find those special configurations of three bodies that, under suitable initial conditions, are capable of producing stationary solutions. In other words, we pretend that the geometry of the system changes over time at most by similarity.[5] Found between 1765 and 1772 by Euler and Lagrange[6] (see also Sect. 3.1.3), these solutions, although apparently artificial, have proven to be of great practical importance.

---

[3] Carl Jacobi (1804–1851) was a German mathematician and lecturer. The son of a banker, he was the first Jew to be appointed a university professor in Germany. From 1829 he thought mathematics in Königsberg (the city of Immanuel Kant, now named Kaliningrad), where he had as a student Gustav Kirchhoff (1824–1887). He held this position until 1842 when, for physical collapse from overwork, he spent some time in Italy to rest. Returned to Germany, he moved to Berlin, where he remained until his death, due to a smallpox infection. His fundamental contributions include elliptic functions, dynamics, differential equations, determinants, and number theory.

[4] George William Hill (1838–1914): American mathematician and astronomer. Employee at the Nautical Almanac Office, he dealt with the problems of three and even four bodies in order to improve the calculations of the Lunar and planetary orbits.

[5] Note the difference between stationary and static; a configuration is stationary if there is a time-dependent reference frame where it looks static.

[6] Giuseppe Lodovico Lagrangia (1736–1813), known with his French name of Joseph-Louis Lagrange, was one of the most influential and productive mathematicians of the 18th century, certainly the most elegant. He was born in Turin to a family of French roots, impoverished by unfortunate speculation. The privations of his early life made the scientist say later that if he had been born rich, he probably would not have approached mathematics. Lagrange started his scientific career in Turin. In 1766, at the suggestion of Euler and d'Alembert, he was called by Frederick II of Prussia to succeed Euler as president of the Berlin Academy. In 1787 Louis XVI invited him to move to France to direct the mathematics section of the *Académie des Sciences*. He remained in Paris until his death, passing unscathed through the Revolution. Napoleon covered him with honors. Lagrange's contributions were in the fields of number theory, variation calculus, and differential equations. He was also a pioneer in group theory and, in astronomy, he worked on lunar librations and planetary motions, that is pure celestial mechanics. But his fundamental legacy is analytic mechanics.

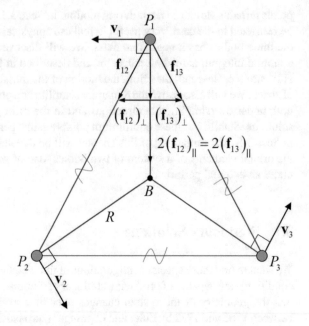

Fig. 3.2 Stationary solution for the motion of three bodies of equal mass placed at the vertexes of an equilateral triangle with properly-chosen identical initial velocities orthogonal to their vector radii

We start by proving (in an objectively unorthodox way) the existence of at least one stationary solution. For this purpose we place three massive points P_i, with equal masses $m_i = m$, at the vertexes of an equilateral triangle whose distances from the barycenter B are R (Fig. 3.2). As initial conditions, we choose three velocities $\mathbf{v}_i$ which are equal in modulus and perpendicular to the corresponding barycentric vector radii, orienting them so that the directions of the three angular momentum vectors coincide. It is easy to be convinced that, at the beginning of the motion, the point P_i feels the combined force $2\left(\mathbf{f}_{ij}\right)_\parallel$ exerted by the other two points j and k as if emanating from the barycenter B. Since:

$$2|\left(\mathbf{f}_{ij}\right)_\parallel| = \frac{G}{\sqrt{3}}\frac{m^2}{R^2},\tag{3.1}$$

if we make it that:

$$|\mathbf{v}| = \sqrt{\frac{G}{\sqrt{3}}\frac{m}{R}},\tag{3.2}$$

the point P_i is forced to move (instantaneously) around B on a circle with radius R. Since this behavior is the same for the other two points P_j and P_k, the instantaneous motion of the system is rigid and plane, with constant angular velocity:

$$\omega = \sqrt{\frac{G}{\sqrt{3}}\frac{m}{R^3}}.\tag{3.3}$$

In doing so, moment by moment a stationary movement is built.

Let us now ask what are the general conditions for the existence of stationary solutions for a system of three points P_i without applying any restriction to values of the masses m_i and to the shape of the initial configuration. Because these solutions concern only planar motions (cf. Appendix K for the proof), we introduce an inertial reference system with axes x and y on the invariable plane containing the three points and the origin on the center of gravity B. The equations of motion for the three points are:

$$\ddot{x}_i = -Gm_j \frac{x_i - x_j}{r_{ij}^3} - Gm_k \frac{x_i - x_k}{r_{ik}^3}$$

$$\hspace{6cm} (i \neq j \neq k = 1, 2, 3). \qquad (3.4)$$

$$\ddot{y}_i = -Gm_j \frac{y_i - y_j}{r_{ij}^3} - Gm_k \frac{y_i - y_k}{r_{ik}^3}$$

The condition of stationarity is expressed by imposing that the law of variation with time of the mutual distances $r_{ij} = |\overrightarrow{P_i P_j}|$ is the same for all pairs of points, i.e., it is independent of the indexes:

$$r_{ij}(t) = \rho(t) \, r_{ij}^\circ \qquad\qquad (i \neq j = 1, 2, 3), \qquad\qquad (3.5)$$

where r_{ij}° is the modulus of the vector $\overrightarrow{P_i P_j}$ at the conventional initial epoch $t = 0$. The condition expressed by (3.5) implies that a similar law also applies to the barycentric distances r_i of each of the three points:

$$r_i(t) = \rho(t) \, r_i^\circ \qquad\qquad (i = 1, 2, 3). \qquad\qquad (3.6)$$

The function $\rho(t)$ can be thought as a scaling factor. Finally, the self-similarity conditions require that the position angle $\phi_i(t)$ of each point P_i, which is the angle between the positive semi-axis x and the vector $\mathbf{r}_i$ counted in the direction of motion, varies over time with a law independent of the chosen point:

$$\varphi_i(t) = \varphi_i^\circ + \theta(t) \qquad\qquad (i = 1, 2, 3), \qquad\qquad (3.7)$$

where φ_i° is again the value of the position angle at $t = 0$. Using the conditions (3.5), (3.6), and (3.7), the coordinates of the point P_i can be written as:

$$x_i(t) = r_i(t) \cos\left[\varphi_i(t)\right] = r_i^\circ \rho(t) \cos\left[\varphi_i^\circ + \theta(t)\right] =$$
$$= \left\{ x_i^\circ \cos\left[\theta(t)\right] - y_i^\circ \sin\left[\theta(t)\right] \right\} \rho(t), \qquad (3.8)$$
$$y_i(t) = r_i(t) \sin\left[\varphi_i(t)\right] = r_i^\circ \rho(t) \sin\left[\varphi_i^\circ + \theta(t)\right] =$$
$$= \left\{ x_i^\circ \sin\left[\theta(t)\right] + y_i^\circ \cos\left[\theta(t)\right] \right\} \rho(t).$$

By replacing them in the equations of motion (3.4), we obtain a system of 6 second-order differential equations in $\rho(t)$ and $\theta(t)$, whose over-abundance provides the conditions for the existence of the stationary solutions. Since these are rather complicated equations, it is desirable to simplify them first. To this purpose we add to the first of (3.4), already modified by inserting (3.8) and multiplied by $\cos \theta$, the second multiplied by $\sin \theta$, and again we add to the first, multiplied by $-\sin \theta$, the second multiplied by $\cos \theta$ (i.e., we apply the usual rotation matrix). After some reductions, we obtain three pairs of condition equations for ($i \neq j \neq k = 1, 2, 3$):

$$
\begin{aligned}
x_i^\circ \ddot{\rho} - 2y_i^\circ \dot{\rho}\dot{\theta} - x_i^\circ \rho \dot{\theta}^2 - y_i^\circ \rho \ddot{\theta} &= -\left(m_j \frac{x_i^\circ - x_j^\circ}{(r_{ij}^\circ)^3} + m_k \frac{x_i^\circ - x_k^\circ}{(r_{ik}^\circ)^3} \right) \frac{G}{\rho^2}, \\
y_i^\circ \ddot{\rho} + 2x_i^\circ \dot{\rho}\dot{\theta} - y_i^\circ \rho \dot{\theta}^2 + x_i^\circ \rho \ddot{\theta} &= -\left(m_j \frac{y_i^\circ - y_j^\circ}{(r_{ij}^\circ)^3} + m_k \frac{y_i^\circ - y_k^\circ}{(r_{ik}^\circ)^3} \right) \frac{G}{\rho^2}.
\end{aligned}
\tag{3.9}
$$

By placing:

$$
\rho^2 \frac{d\theta}{dt} = \psi,
\tag{3.10}
$$

through time derivation and squaring, we have:

$$
\begin{aligned}
2\dot{\rho}\dot{\theta} + \rho\ddot{\theta} &= \frac{1}{\rho}\dot{\psi}, \\
\rho\dot{\theta}^2 &= \frac{\psi^2}{\rho^3},
\end{aligned}
\tag{3.11}
$$

with which the (3.9) assume the compact form:

$$
\begin{aligned}
\ddot{\rho} - \frac{y_i^\circ}{x_i^\circ}\frac{1}{\rho}\dot{\psi} - \frac{\psi^2}{\rho^3} &= -\frac{A_{ix}}{x_i^\circ}\frac{G}{\rho^2}, \\
\ddot{\rho} + \frac{x_i^\circ}{y_i^\circ}\frac{1}{\rho}\dot{\psi} - \frac{\psi^2}{\rho^3} &= -\frac{A_{iy}}{y_i^\circ}\frac{G}{\rho^2},
\end{aligned}
\tag{3.12}
$$

having placed:

$$
A_{ix} = m_j \frac{x_i^\circ - x_j^\circ}{(r_{ij}^\circ)^3} + m_k \frac{x_i^\circ - x_k^\circ}{(r_{ik}^\circ)^3}
$$

$$
(i \neq j \neq k = 1, 2, 3). \tag{3.13}
$$

$$
A_{iy} = m_j \frac{y_i^\circ - y_j^\circ}{(r_{ij}^\circ)^3} + m_k \frac{y_i^\circ - y_k^\circ}{(r_{ik}^\circ)^3}
$$

The constants A_{ix} and A_{iy} in (3.13) are proportional to the components of the forces $\mathbf{F}_{ij}^\circ$ and $\mathbf{F}_{ik}^\circ$ acting on the ith body at the initial epoch. In particular, it is:

$$A_{ix} = \frac{(F_{ij}^\circ)_x + (F_{ik}^\circ)_x}{G\, m_i}$$

$$(i \neq j \neq k = 1, 2, 3). \tag{3.14}$$

$$A_{iy} = \frac{(F_{ij}^\circ)_y + (F_{ik}^\circ)_y}{G\, m_i}$$

Note that it is always possible to choose the orientation of the axes in such a way that $x_i^\circ \neq 0$ and $y_i^\circ \neq 0$, unless the center of gravity falls at one of the bodies.

In short, a necessary (but not sufficient) condition for the existence of dynamical solutions preserving the relations of similitude among the mutual distances of the three bodies is that all the six equations (3.12) are satisfied. But they contain only two unknown functions, ρ and ψ, which respectively provide, through (3.10), size and orientation of the system of three points. Each pair of equations (3.12) is sufficient to define ρ and ψ when the initial conditions are specified. It is however necessary that all the three pairs provide the same result for the solution of the over-abundant system (3.12). Subtracting from the first equation of the system (3.12), with index i, the second equation with index j and doing the same for the second equation, and then repeating the operation with the equations having indexes i and k, two pairs of new equations are obtained:

$$\left(\frac{y_i^\circ}{x_i^\circ} - \frac{y_j^\circ}{x_j^\circ} \right) \dot{\psi} = \left(\frac{A_{ix}}{x_i^\circ} - \frac{A_{jx}}{x_j^\circ} \right) \frac{G}{\rho}$$

$$(j \neq i = 2, 3), \tag{3.15}$$

$$\left(\frac{x_j^\circ}{y_j^\circ} - \frac{x_i^\circ}{y_i^\circ} \right) \dot{\psi} = \left(\frac{A_{iy}}{y_i^\circ} - \frac{A_{jy}}{y_j^\circ} \right) \frac{G}{\rho}$$

which, if identically satisfied, reduce the system (3.12) to two equations in two unknown functions (i.e., the indexes reduce to one). Since $\dot{\psi}$ and ρ are independent, to identically satisfy (3.15) it is necessary that both terms vanish in each equation. The left-hand side term cancels if either:

$$\frac{y_1^\circ}{x_1^\circ} = \frac{y_2^\circ}{x_2^\circ} = \frac{y_3^\circ}{x_3^\circ}, \tag{3.16a}$$

or:

$$\frac{d\psi}{dt} = 0. \tag{3.16b}$$

We will see later that the first condition necessarily entails the second, but not vice versa. The second term cancels if:

$$\frac{A_{ix}}{x_i^\circ} = \frac{A_{jx}}{x_j^\circ} = \text{const}_1$$

$$(i \neq j = 1, 2), \tag{3.17}$$

$$\frac{A_{iy}}{y_i^\circ} = \frac{A_{jy}}{y_j^\circ} = \text{const}_2$$

that is, given the independence of x_i° and y_i° ($i = 1, 2, 3$), if the constants A_{ix}/x_i° and A_{iy}/y_i° are independent of the index. We can also see that these constants must be equal to each other observing how (3.12) are reduced when (3.16a) or (3.16b) are verified. So:

$$
\begin{aligned}
\frac{A_{ix}}{x_i^\circ} = n^2 \quad &\Rightarrow \quad m_j \frac{x_i^\circ - x_j^\circ}{(r_{ij}^\circ)^3} + m_k \frac{x_i^\circ - x_k^\circ}{(r_{ik}^\circ)^3} = n^2 x_i^\circ, \\
\frac{A_{iy}}{y_i^\circ} = n^2 \quad &\Rightarrow \quad m_j \frac{y_i^\circ - y_j^\circ}{(r_{ij}^\circ)^3} + m_k \frac{y_i^\circ - y_k^\circ}{(r_{ik}^\circ)^3} = n^2 y_i^\circ,
\end{aligned}
\tag{3.18}
$$

where n^2 is an arbitrary constant, necessarily positive.[7]

In conclusion, the necessary condition for the system (3.12) to give a solution is that, together with (3.18), either the (3.16a) or the (3.16b) are verified. We will now prove that these conditions provide all and alone the stationary solutions.

3.1.1   Collinear Solutions

Assume that condition (3.16a) is satisfied. It requires that, at the epoch $t = 0$, the three massive points are on the same line, which must obviously also contain the barycenter B. Since we constrained the system not to change its configuration over time, the alignment must remain indefinitely. In this case, the resulting of the forces acting on each point P_i is constantly directed towards the common center of mass, which is also a fixed point. Then, if the motion is possible, by the central-force theorem the specific angular momentum of each of the three points must be constant:

$$h_i = r_i^2 \dot{\theta} = c_i. \tag{3.19}$$

Using (3.6), (3.10), and (3.19), it follows that:

$$\rho^2 \dot{\theta} = \psi = \frac{c_i}{(r_i^\circ)^2} = c_\circ. \tag{3.20}$$

The fact that the three constants $c_i/(r_i^\circ)^2$ are equal implies that the condition $\dot{\psi} = 0$ is satisfied, as we anticipated above (3.16b).

---

[7] Using the (3.13) and assuming, for instance, that $x_i \leq x_j \leq x_k$ (or $y_i \leq y_j \leq y_k$), it is easy to prove for the index i that the constant must be positive.

Combined with the particular geometric configuration and the property expressed by (3.20), the condition (3.18) reduces the six equations (3.12) to one only:

$$\ddot{\rho} - \frac{c_{\circ}^2}{\rho^3} = -n^2 \frac{G}{\rho^2}, \qquad (3.21)$$

which expresses the law of variation for the function $\rho(t)$, scaling factor of the system. From (3.10) we have then the expression of the angular velocity:

$$\dot{\theta} = \frac{c_{\circ}}{\rho^2}, \qquad (3.22)$$

with which the system of points rotates rigidly around the center of gravity. Apart from the notation, (3.21) and (3.22) are identical to those obtained for the Keplerian motion in polar coordinates. The time derivative of (2.14) gives the first; as for the second, compare it with (2.17). Finally, the (3.18) prove that each component of the resulting force acting on the point P_i is proportional to the component, along the same axis, of the barycentric vector $\mathbf{r}_i = \overrightarrow{B P_i}$. Thus, the resulting force on each body of the system is proportional to the distance of that body from the center of gravity.

All that remains is to find the coordinates of the three bodies (at the time $t = 0$). To this end, we place the origin of the relative reference system in one of the masses; for instance, m_i. The equation of motion for the relative system can be immediately written using (1.38):

$$\ddot{\mathbf{r}}_i = -G(m_i + m_j)\frac{\mathbf{r}_i}{r_i^3} + Gm_k \left[\frac{\mathbf{r}_k - \mathbf{r}_i}{r_{ki}^3} - \frac{\mathbf{r}_k}{r_k^3} \right]. \qquad (3.23)$$

It is simple to verify that, in this case, we obtain the same (3.12) but with different expressions for A_{ix} and A_{iy}:

$$A_{ix} = (m_i + m_j)\frac{x_i^{\circ}}{(r_i^{\circ})^3} + m_k \left[\frac{x_k^{\circ}}{(r_k^{\circ})^3} - \frac{x_k^{\circ} - x_i^{\circ}}{(r_{ik}^{\circ})^3} \right]$$

$$\qquad\qquad\qquad\qquad\qquad (i \neq j \neq k = 1, 2, 3). \qquad (3.24)$$

$$A_{iy} = (m_i + m_j)\frac{y_i^{\circ}}{(r_i^{\circ})^3} + m_k \left[\frac{y_k^{\circ}}{(r_k^{\circ})^3} - \frac{y_k^{\circ} - y_i^{\circ}}{(r_{ik}^{\circ})^3} \right]$$

Since we are interested in collinear solutions, we choose the initial epoch such that the three points are aligned along the x-axis, so that $y_i^{\circ} = 0$ for all the three values of the index i (Fig. 3.3). Hence, we have to deal with the first of the (3.24) only. As found above, any stationary solution implies the condition (3.17):

$$\frac{A_{jx}}{x_j^{\circ}} = \frac{A_{kx}}{x_k^{\circ}}. \qquad (3.25)$$

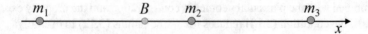

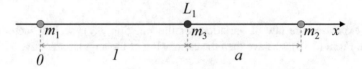

Fig. 3.3 Sketch of the configuration of three collinear points

Fig. 3.4 Configuration of the three masses for the Lagrangian point L_1 for the case in which $m_1 = 10\,m_3$ and $m_2 = 5\,m_3$

With the following choice of indexes: $i = 1$, $j = 2$, $k = 3$, it becomes:

$$\frac{A_{2x}}{x_2^\circ} = \frac{A_{3x}}{x_3^\circ}, \tag{3.26}$$

and the two surviving (3.24) rewrite as:

$$A_{2x} = (m_2 + m_1)\frac{x_2^\circ}{(r_2^\circ)^3} + m_3\left[\frac{x_3^\circ}{(r_3^\circ)^3} - \frac{x_3^\circ - x_2^\circ}{(r_{23}^\circ)^3}\right], \tag{3.27a}$$

$$A_{3x} = (m_3 + m_1)\frac{x_3^\circ}{(r_3^\circ)^3} + m_2\left[\frac{x_2^\circ}{(r_2^\circ)^3} - \frac{x_2^\circ - x_3^\circ}{(r_{23}^\circ)^3}\right]. \tag{3.27b}$$

It is apparent that there are three distinct solutions for the collinear configuration (that is, for the position of the third mass relative to the first two). They are called collinear Lagrangian or libration points, and are generally indicated with the symbols L_1, L_2, and L_3. For example, having fixed the positions of m_1 and m_2, the third mass m_3 can set either in L_1, the Lagrangian point located between m_1 and m_2 (Fig. 3.4), or outside the line joining m_1 to m_2, either in L_2, which is on the side of m_2 (Fig. 3.5), or in L_3, which places on the opposite side (Fig. 3.6).

Lagrangian Point L_1

Consider the case:

$$r_2^\circ = x_2^\circ = 1 + a, \tag{3.28}$$
$$r_3^\circ = x_3^\circ = 1, \tag{3.29}$$

such that the mass m_3 is situated between m_1 and m_2 (Fig. 3.4). We have:

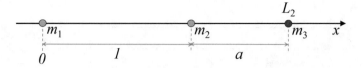

Fig. 3.5 Same as Fig. 3.4 for the Lagrangian point L_2

$$A_{2x} = \frac{m_2 + m_1}{(1 + a)^2} + m_3 \left[1 + \frac{1}{a^2} \right],$$

$$A_{3x} = m_3 + m_1 + m_2 \left[\frac{1}{(1 + a)^2} - \frac{1}{a^2} \right].$$

(3.30)

Together with (3.26), they give:

$$\frac{(m_2 + m_1)}{(1 + a)^3} + \frac{m_3}{1 + a} + \frac{m_3}{a^2(1 + a)^2} = m_3 + m_1 + \frac{m_2}{(1 + a)^2} - \frac{m_2}{a^2},$$

(3.31)

and, after simplifying:

$$(m_1 + m_3)\, a^5 + (3m_1 + 2m_3)\, a^4 + (3m_1 + m_3)\, a^3 +$$
$$-(3m_2 + m_3)\, a^2 - (3m_2 + 2m_3)\, a - (m_2 + m_3) = 0.$$

(3.32)

According to the Descartes rule of signs,[8] this 5th degree algebraic equation[9] has only one real and positive root (since the coefficients change their sign only once). Therefore the root of the (3.36) provides a unique value for the distance a between m_2 and m_3 in units of the distance between m_1 and m_3.

Lagrangian Point L_2

Consider now the case:

$$r_2^\circ = x_2^\circ = 1,$$
$$r_3^\circ = x_3^\circ = 1 + a,$$

(3.33)

with which the mass m_3 places on the right side of m_2 as in Fig. 3.5. It is:

---

[8] The maximum number of positive real roots of a ordered polynomial is given by the number of sign variations between consecutive coefficients, neglecting any null coefficients.

[9] Note that this equation is homogeneous with respect to the masses and can be always rewritten in terms of m_2/m_1 and m_3/m_1, for instance.

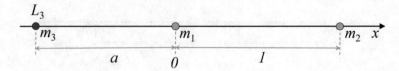

Fig. 3.6 Same as Fig. 3.4 for the Lagrangian point L_3. Note that now L_3 is on the negative side of the x-axis

$$A_{2x} = (m_2 + m_1) + m_3 \left[\frac{1}{(1+a)^2} - \frac{1}{a^2} \right],$$
$$A_{3x} = \frac{m_3 + m_1}{(1+a)^2} + m_2 \left[1 + \frac{1}{a^2} \right], \tag{3.34}$$

which, again through (3.26), give us:

$$(m_2 + m_1) - \frac{m_3}{a^2} + \frac{m_3}{(1+a)^2} = \frac{m_3 + m_1}{(1+a)^3} + \frac{m_2}{1+a} + \frac{m_2}{a^2(1+a)}, \tag{3.35}$$

and, by simplifying:

$$(m_1 + m_2)\, a^5 + (3m_1 + 2m_2)\, a^4 + (3m_1 + m_2)\, a^3 +$$
$$- (m_2 + 3m_3)\, a^2 - (2m_2 + 3m_3)\, a - (m_2 + m_3) = 0. \tag{3.36}$$

Lagrangian Point L_3

Finally, consider the case:

$$r_2^\circ = x_2^\circ = 1,$$
$$r_3^\circ = -x_3^\circ = a, \tag{3.37}$$

placing the mass m_3 on the left-hand side of m_1 (Fig. 3.6). We have:

$$A_{2x} = m_2 + m_1 + m_3 \left[-\frac{1}{a^2} + \frac{1}{(1+a)^2} \right],$$
$$A_{3x} = -\frac{m_3 + m_1}{a^2} + m_2 \left[1 - \frac{1}{(1+a)^2} \right], \tag{3.38}$$

and:

$$m_2 + m_1 - \frac{m_3}{a^2} + \frac{m_3}{(1+a)^2} = \frac{m_3 + m_1}{a^3} - \frac{m_2}{a} + \frac{m_2}{a(1+a)^2}, \tag{3.39}$$

from where:

$$(m_1 + m_2)a^5 + (2m_1 + 3m_2)\, a^4 + (m_1 + 3m_2)\, a^3 +$$
$$-(m_1 + 3m_3)\, a^2 - (2m_1 + 3m_3)\, a - (m_1 + m_3) = 0. \tag{3.40}$$

We can see that (3.32), (3.36), and (3.40) differ from each other only in the coefficients and are valid for any set of masses (but be careful in remembering how a is defined in each one of the three cases). They must be solved numerically. However, (3.32) readily shows that, for $m_1 = m_2 = m$ and $m_3 = 0$, the Lagrangian point L_1 is located in the baricenter of the system, as it is obvious for symmetry reasons. Also, the solutions for L_2 and L_3 are the same in this case; it means that these stationary points are located at the same distance $a = 0.698$ (in units of the distance between m_1 and m_2), L_2 from m_2 on its right (see Fig. 3.5) and L_3 from m_1 on its left (Fig. 3.6) respectively. We will consider the case $m_3 \simeq 0$ in more detain in Sect. 3.2. The reader can verify also that, if $m_2 \ll m_1$ and $m_3 = 0$, the Lagrangian points L_1 and L_2 are located very near the mass m_2 while L_3 places on the other side, almost on the circumference whose radius is the distance between m_1 and m_2.

3.1.2 Triangular Solutions

Let us see what happens if conditions (3.16b) and (3.18) are satisfied but (3.16a) is not. Since the configuration cannot be collinear, it must be triangular. We show that this triangle is necessarily equilateral. To this purpose, making use of the equations of the center of mass:

$$m_i x_i^\circ + m_j x_j^\circ + m_k x_k^\circ = 0,$$
$$m_i y_i^\circ + m_j y_j^\circ + m_k y_k^\circ = 0, \tag{3.41}$$

we eliminate x_k° and y_k° from (3.18) and obtain six equations of the type:

$$n^2 x_i^\circ = x_i^\circ \left[\frac{m_i + m_k}{\left(r_{ik}^\circ\right)^3} + \frac{m_j}{\left(r_{ij}^\circ\right)^3} \right] + m_j x_j^\circ \left[\frac{1}{\left(r_{ik}^\circ\right)^3} - \frac{1}{\left(r_{ij}^\circ\right)^3} \right],$$
$$n^2 y_i^\circ = y_i^\circ \left[\frac{m_i + m_k}{\left(r_{ik}^\circ\right)^3} + \frac{m_j}{\left(r_{ij}^\circ\right)^3} \right] + m_j y_j^\circ \left[\frac{1}{\left(r_{ik}^\circ\right)^3} - \frac{1}{\left(r_{ij}^\circ\right)^3} \right]. \tag{3.42}$$

In our hypotheses they are satisfied if $r_{ik}^\circ = r_{ij}^\circ$ (according to the independence of x_i from x_j, which is the same argument used for (3.18)), that is, by exchanging indexes, if:

$$r_{12}^\circ = r_{23}^\circ = r_{13}^\circ = a. \tag{3.43}$$

Therefore, at the time $t = 0$ the three massive points are located at the vertexes of an equilateral triangle, which must stay similar to itself during motion. Once two of

Fig. 3.7 Schematic
positions (5 Lagrangian
points) that a third mass can
take with respect to two
other massive points in order
to make a stationary
configuration

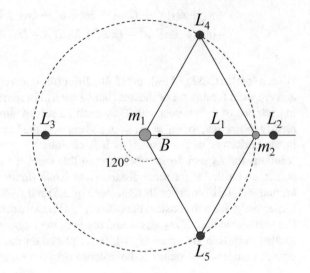

the three massive points have been set, the third has only two positions available on
the plane of motion,[10] the triangular Lagrangian points L_4 and L_5 (see Fig. 3.7 and
cf. Sect. 3.1.1).

Substituting (3.43) in (3.42), we have:

$$n^2 = \frac{m_1 + m_2 + m_3}{a^3} = \frac{M}{a^3}, \tag{3.44}$$

where $M = \sum_{i=1}^{3} m_i$ is the total mass. The resultant of the forces on each point is
derived from (3.14), (3.18), and (3.44). Passing to the vectors, we obtain:

$$\mathbf{f}_i^\circ = -G \, m_i \, \frac{M}{a^3} \, \mathbf{r}_i^\circ. \tag{3.45}$$

Since t_0 is arbitrary, then:

$$\mathbf{f}_i = -G \, m_i \, \frac{M}{\rho^3 a^3} \, \mathbf{r}_i. \tag{3.46}$$

As already seen in the previous paragraph about the collinear solution, $\mathbf{f}_i$ is directed
towards the center of mass and has a modulus proportional to the distance $\mathbf{r}_i = \overrightarrow{BP_i}$
of P_i from barycenter B. The latter can be derived from the relation defining the
center of mass with respect to the position of the point P_i. It is:

---

[10] It seems appropriate to emphasize the fact that, once given the initial conditions (positions and
velocities) of the two masses, stationary motion is allowed only in the plane orthogonal to the
angular momentum of these two masses.

$$M\,\mathbf{r}_i = m_j\overrightarrow{P_iP_j} + m_k\overrightarrow{P_iP_k}, \tag{3.47}$$

where, at $t = 0$, $|\mathbf{r}_i| = r_i^\circ$ and $|\overrightarrow{P_iP_j}| = |\overrightarrow{P_iP_k}| = r_{ij}^\circ = r_{ik}^\circ = a$. Remembering that the two vectors $\overrightarrow{P_iP_j}$ and $\overrightarrow{P_iP_k}$ form an angle of $60°$, from the modulus of (3.47) we obtain:

$$r_i^\circ = \frac{\sqrt{m_j^2 + m_j m_k + m_k^2}}{M}\,a. \tag{3.48}$$

Using this expression to replace the lenght a of the side of the triangle in (3.45), the force $\mathbf{f}_i$ takes the simple and familiar form:

$$\mathbf{f}_i = -G\frac{m_i M_i}{r_i^3}\,\mathbf{r}_i, \tag{3.49}$$

where the fictitious mass M_i is:

$$M_i = \frac{(m_j^2 + m_j m_k + m_k^2)^{3/2}}{(m_i + m_j + m_k)^2}. \tag{3.50}$$

Equation (3.49) shows that each point P_i behaves as if it only feels the effect of the mass M_i (specific to each point) placed at the barycenter. The same result can be obtained by other means remembering that (3.21) and (3.22), identifying a Keplerian motion, are valid for every solution of the stationary three-body problem.

3.1.3 Stationary Solutions: Summary

The results of this whole section can be summarized as follows.

1. Two stationary configurations of three bodies are possible, either collinear or triangular with all sides equal (equilateral).
2. The resultant of the forces on each point is proportional to the inverse of the square of its distance from the gravity center.
3. The motion of each point around the gravity center is Keplerian.

This set of stationary solutions of the three-body problem is named after Euler-Lagrange, as in 1765 the Swiss mathematician demonstrated the existence of the collinear solutions and 7 years later Lagrange added the triangular solutions. For two centuries they were also considered as the unique periodic solutions of the three-body problem. In the 1970s, with the aid of modern computer, the U.S. mathematician Roger Broucke (1932–2005) and the French astronomer Michel Henon (1931–2013) found other periodic cases for systems of equal masses, and later the rich family of weakly stable figure-8 periodic solutions began to appear [1].

3.2   The Circular Restricted Three-Body Problem

Since the three-body problem does not admit a general solution, we try to simplify it assuming that one of the points, henceforth indicated with P, has a negligible mass compared to the others two. In other words, we want to reach the condition in which the gravitational pull of P on P_1 and P_2 can be ignored. On the contrary, P will be always subject to the attraction by the two massive bodies, thus it can be considered as a test particle. This is possible as the equations of motion of P do not contain its mass; on the contrary, the latter is at factor in the expressions of the forces of P on P_1 and P_2, making them vanish. This simplified mathematical model is applicable to more real situations.

Due to the negligibility of the mass of P, the other two bodies P_1 and P_2 meet the conditions for a two-body problem; their mutual orbits are Keplerian. For simplicity, we further assume that:

(a) these orbits are circular, and that
(b) the sum of the masses of P_1 and P_2, their mutual distance $d = P_1 P_2$, and the universal gravitation constant G are units.

We also adopt the following notations/conventions:

(c) μ is the mass of P_2 and $1-\mu$ that of P_1, assuming, without loss of generality, that:
(d) $\mu \le 1-\mu$, that is: $\mu \le 0.5$.

Under these assumptions, only the motion of P remains to be determined; a three-body problem that takes the adjective 'restricted' to distinguish it from the general case where there are no constraints on masses [2], [3]. We use the inertial reference system $B\,[\xi, \eta, \zeta]$, with origin at the barycenter B of the masses P_1 and P_2 and axes ξ and η contained on the invariable plane of their motion. Having indicated with (ξ, η, ζ) the coordinates of the point P, using (3.4) the system of its equations of motion writes as:

$$\ddot{\xi} = -(1-\mu)\frac{\xi - \xi_1}{r_1^3} - \mu\frac{\xi - \xi_2}{r_2^3},$$

$$\ddot{\eta} = -(1-\mu)\frac{\eta - \eta_1}{r_1^3} - \mu\frac{\eta - \eta_2}{r_2^3}, \qquad (3.51)$$

$$\ddot{\zeta} = -(1-\mu)\frac{\zeta}{r_1^3} - \mu\frac{\zeta}{r_2^3},$$

where $(\xi_i, \eta_i, 0)$ are the coordinates of the massive point P_i $(i = 1, 2)$ and:

$$r_i = \sqrt{(\xi - \xi_i)^2 + (\eta - \eta_i)^2 + \zeta^2}, \qquad (3.52)$$

is the modulus of the vector $\overrightarrow{P_i P}$ (with $i = 1, 2$) (Fig. 3.8). We now introduce a new reference system $B[x,y,z]$, co-rotating with the massive points P_1 and P_2. To

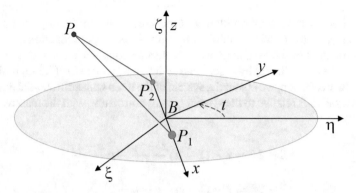

Fig. 3.8 Barycentric reference systems for the restricted three-body problem. The axes (x, y) and (ξ, η) lie on the plane of motion of the massive points P_1 and P_2

this purpose we set the new system in rotation about the $\zeta = z$ axis with the same angular velocity of the radius vector $\overrightarrow{P_1 P_2}$; the previous choice of the units implies that the mean motion is $n = 1$; cf. (2.83). We can also choose the x-axis in such a way that it contains P_1 and P_2, whose new coordinates are therefore $P_1(x_1, 0, 0)$ and $P_2(x_2, 0, 0)$. Remembering that $|\overrightarrow{P_1 P_2}| = 1$, from the system:

$$(1 - \mu) \, x_1 + \mu x_2 = 0,$$
$$|x_1 - x_2| = 1,$$
(3.53)

the values of x_1 and x_2 are derived.

The new coordinates of P are given by:

$$\xi = x \cos t - y \sin t,$$
$$\eta = x \sin t + y \cos t,$$
(3.54)
$$\zeta = z.$$

Replacing in (3.51) we obtain the new equations of the motion of P:

$$\left(\ddot{x} - 2\dot{y} - x\right) \cos t - \left(\ddot{y} + 2\dot{x} - y\right) \sin t =$$
$$= -\left[(1-\mu)\frac{x-x_1}{r_1^3} + \mu\frac{x-x_2}{r_2^3}\right] \cos t + \left[(1-\mu)\frac{y}{r_1^3} + \mu\frac{y}{r_2^3}\right] \sin t, \quad (3.55a)$$

$$\left(\ddot{x} - 2\dot{y} - x\right) \sin t + \left(\ddot{y} + 2\dot{x} - y\right) \cos t =$$
$$= -\left[(1-\mu)\frac{x-x_1}{r_1^3} + \mu\frac{x-x_2}{r_2^3}\right] \sin t - \left[(1-\mu)\frac{y}{r_1^3} + \mu\frac{y}{r_2^3}\right] \cos t, \quad (3.55b)$$

$$\ddot{z} = -(1 - \mu)\frac{z}{r_1^3} - \mu\frac{z}{r_2^3}. \quad (3.55c)$$

We can simplify them by the following procedure: equation (3.55a) multiplied by $\cos t$ is added to the (3.55b) multiplied by $\sin t$. After proper reductions, we obtain the first equation of the system (3.56). The second equation is derived in the same way by multiplying (3.55a) by $-\sin t$ and adding the result to the (3.55b) multiplied by $\cos t$. In conclusion, we obtain the system of the three scalar differential equations of the motion of P relative to the system of axes co-rotating with the massive points P_1 and P_2:

$$\ddot{x} - 2\dot{y} = x - (1 - \mu)\frac{x - x_1}{r_1^3} - \mu\frac{x - x_2}{r_2^3},$$

$$\ddot{y} + 2\dot{x} = y - (1 - \mu)\frac{y}{r_1^3} - \mu\frac{y}{r_2^3}, \tag{3.56}$$

$$\ddot{z} = -(1 - \mu)\frac{z}{r_1^3} - \mu\frac{z}{r_2^3},$$

with $r_i = \sqrt{(x - x_i)^2 + y^2 + z^2}$. The system (3.56) has an important property; its equations do not explicitly involve the independent variable t. This fact is due to the particular choice of the reference system which eliminates from equations (3.51) the coordinates ξ_i and η_i which depend explicitly on time.

The set of (3.56) is equivalent to a system of six scalar differential equations of the first order; thus the solution of the problem of the motion of P requires the knowledge of six motion integrals. One such integral, found by Jacobi, is obtained introducing the scalar function:

$$U = \frac{1}{2}(x^2 + y^2) + \frac{1 - \mu}{r_1} + \frac{\mu}{r_2}, \tag{3.57}$$

corresponding to the sum of the centrifugal and the gravitational potentials.[11] It is straightforwardly verified that:

---

[11] According the (D.8) of Appendix D, the centrifugal acceleration term to be added to the true acceleration (relative to the non rotating system) is $-\vec{\omega} \times (\vec{\omega} \times \mathbf{r}')$. Here we have: $\mathbf{r}' = x'\mathbf{i}' + y'\mathbf{j}' + z'\mathbf{k}'$ and $\vec{\omega} = \mathbf{k}'$, since $\omega = 1$. Therefore, the centrifugal potential is:

$$-\vec{\omega} \times (\vec{\omega} \times \mathbf{r}') = x'\mathbf{i}' + y'\mathbf{j}',$$

whose integral is:

$$U_c = \int (x'\mathbf{i}' + y'\mathbf{j}') \cdot d\mathbf{r}' = \int x' \, dx' + \int y' \, dy' = \frac{1}{2}\left(x'^2 + y'^2\right).$$

Moreover, since $|\mathbf{r}'| = |\mathbf{r}|$, then:

$$U_c = \frac{1}{2}\left(x^2 + y^2\right).$$

You may ask what happened to the Coriolis acceleration $2\vec{\omega} \times \dfrac{d\mathbf{r}'}{dt}$. Its components are already at the left-hand sides of the first two (3.56), as a consequence of the applied rotation.

$$\ddot{x} - 2\dot{y} = \frac{\partial U}{\partial x}, \qquad \ddot{y} + 2\dot{x} = \frac{\partial U}{\partial y}, \qquad \ddot{z} = \frac{\partial U}{\partial z}. \qquad (3.58)$$

Summing up the three (3.58) multiplied respectively by $\dot{x}$, $\dot{y}$ and $\dot{z}$, we have:

$$\dot{x}\ddot{x} + \dot{y}\ddot{y} + \dot{z}\ddot{z} = \frac{\partial U}{\partial x}\dot{x} + \frac{\partial U}{\partial y}\dot{y} + \frac{\partial U}{\partial z}\dot{z}, \qquad (3.59)$$

that, integrated with respect to time (remember that U does not depend explicitly on t), provides the Jacobi integral:

$$\dot{x}^2 + \dot{y}^2 + \dot{z}^2 = |\dot{\mathbf{r}}|^2 = 2U - C_J = x^2 + y^2 + 2\frac{1-\mu}{r_1} + 2\frac{\mu}{r_2} - C_J, \quad (3.60)$$

where $|\dot{\mathbf{r}}|$ is the modulus of the velocity of P in the rotating reference system, and C_J is an integration constant, named after Jacobi.

The Jacobi integral finds immediate use in the Tisserand[12] method for the identification of periodic comets. We know that the orbit of one comet is subject to planetary perturbations. For this reason, the same comet may exhibit quite different orbital elements in two successive passages. However, if non-gravitational effects are neglected (a hypothesis that for comets is not always acceptable[13]) and it is assumed that the perturbations on the orbital elements are caused by just one planet only (typically Jupiter), then the integral of Jacobi is conserved. Therefore, if two comets which appeared at different epochs with different orbital elements have the same value of the Jacobi integral, it is likely that they represent successive transits of a same object. In practice it can be demonstrated (see [3]) that, with good approximation considered the small mass of Jupiter with respect to the Sun, the constant C_J is given by the equation:

$$C_J = \frac{1}{a} + 2\sqrt{a\left(1 - e^2\right)}\,\cos i, \qquad (3.61)$$

where a, e and i are the Keplerian orbital elements of the comets.

3.3 The Zero-Velocity Curves

The Jacobi integral reduces by one unit the degree of the problem of the motion of the massless point in the gravitational field of P_1 and P_2, but it is not sufficient to solve it. Indeed, it has been shown that there are no other motion integrals, both in

---

[12] Félix Tisserand (1845–1896). French astronomer, author of the classical *Traitéde mécanique céleste*.

[13] Already Bessel in 1835 pointed out that comets could be subject to various types of non-gravitational forces, including transit in a resistant medium (hypothesis already considered by Johann Encke (1791–1865), pupil of Gauss and discoverer of the homonymous comet) and the effect of ejection of matter, both in impulsive and continuous form.

Cartesian coordinates and when the orbital elements are used [4–6]. Nonetheless, the study of function in (3.60):

$$f(x, y, z; C_J) = x^2 + y^2 + 2\frac{1-\mu}{r_1} + 2\frac{\mu}{r_2} - C_J, \tag{3.62}$$

is very instructive. In fact, being $|\dot{\mathbf{r}}|^2 \geq 0$, as a result of (3.60) it must also be: $f(x, y, z; C_J) \geq 0$. Therefore, even if it is not possible to determine the motion of P, the equation:

$$f(x, y, z; C_J) = 2U - C_J = 0, \tag{3.63}$$

allows us to define at least the zero-velocity boundaries of the space, called Hill surfaces, within which P can move, as a function of the value of the constant C_J. Because the rigorous solution of (3.63) involves considerable mathematical difficulties (once rationalized, the first term is equivalent to a polynomial of the 16-th grade), we limit ourselves to a qualitative study; the quantitative solution can always be obtained numerically. For convenience, we treat separately the three cases obtained by projecting the position of the point P on each of the principal planes of the coordinate system.

3.3.1   The (x, y) Plane

With the condition $z = 0$, (3.63) becomes:

$$f_1(x, y) + f_2(x, y) + f_3(x, y) = C_J, \tag{3.64}$$

where we have placed (see (3.60)):

$$f_1(x, y) = x^2 + y^2,$$
$$f_2(x, y) = \frac{2(1-\mu)}{\sqrt{(x - x_1)^2 + y^2}}, \tag{3.65}$$
$$f_3(x, y) = \frac{2\mu}{\sqrt{(x - x_2)^2 + y^2}}.$$

It is apparent that, as the projected distance $r = \sqrt{x^2 + y^2}$ of P from the barycenter B increases, f_1 also increases while f_2 and f_3 decrease. Consider the case in which the initial conditions assign the constant C_J a very high value ($r \gg d = 1$). It is possible that $f_1 \sim C_J$ since, if $x^2 + y^2 \sim C_J$, both $1/r_1$ and $1/r_2$ must be very small and so are the functions f_2 and f_3. The zero-velocity curve is therefore similar to a circumference with the large radius ($\simeq \sqrt{C_J}$), centered at the barycenter B of the system. It simply means that, at great distances compared to the separation between P_1 and P_2, the massless point P sees the two massive points under a very small angle.

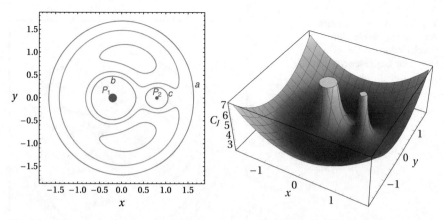

Fig. 3.9 *Left*: projection of the zero-velocity surfaces on the plane (x, y) of the coordinate system co-rotating with the two massive points P_1 and P_2, for different values of the constant C_J, marked by the change of color. *Right*: the surface plots (3.64) for an arbitrary choice of μ in (3.65)

Therefore the gravitational field appears increasingly isotropic. Since $f_1 \sim C_J \geq 0$, the region allowed to P (where the velocity is not imaginary) must be external to the pseudo-circumference (curve *(a)* in Fig. 3.9).

The solution just found is not unique. For a very large C_J, it is also possible that $f_2 \sim C_J$ if P is close to P_1; then $1/r_1 \gg 1$, while both $1/r_2$ and $(x^2 + y^2)$ will be reasonably small, reducing the weight of the functions f_1 and f_3. The zero-velocity curve would be once again a pseudo-circumference, now centered on P_1, whether f_1 and f_3 could be strictly null. Actually the curve looks like a small oval stretched in the direction of P_2 and symmetric with respect to the x-axis (curve *(b)*). With a similar reasoning it can be seen that there is a third closed curve similar to the previous one (but generally smaller since $\mu \leq 0.5$), which encircles the point P_2 (curve *(c)*). The condition $f \geq 0$ imposes P to remain confined within one of these two oval regions which, for C_J sufficiently large, have no points in common between them and with the outer circle. The region of the plane external to them (curves *(b)* and *(c)*) and inside the major quasi-circumference (curve *(a)*) is forbidden to P. It is worth stressing that, in absence of external forces able to modify the value of C_J (perturbations), P is forced to move perpetually within one of the three allowed regions, because the transit from one to another would involve crossing the region where the velocity is imaginary.

For smaller values of the constant C_J (which, in absence of perturbations, can vary only by changing the initial conditions), the forbidden surface shrinks since the size of the pseudo-circumference *(a)* decreases (stretching along the y-axis) while the contours of *(b)* and *(c)* grow. For suitable values of C_J, the curves *(b)* and *(c)* come in contact with each other and each of them with the curve *(a)*; for obvious reasons of symmetry, these contact points, which we call L_1, L_2 and L_3, must lie on the x-axis (see Fig. 3.10). They are also saddle points of the zero-velocity curves since:

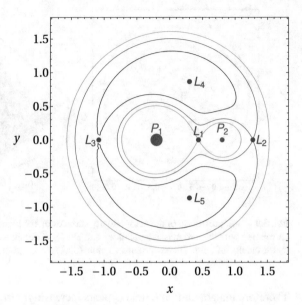

Fig. 3.10 Same as the left panel of Fig. 3.9 but for special values of C_J giving rise to the Lagrangian points

$$\frac{\partial}{\partial x} f(L_i) = \frac{\partial}{\partial y} f(L_i) = 0 \qquad (i = 1, 2, 3). \qquad (3.66)$$

The second condition is verified anywhere on the x-axis owing to the symmetry of f, while the first descends from the fact that $f(x, 0)$ has an extreme in L_i. The (3.66) shows that the partial derivatives of U (cf.Eq. (3.63)) are null in L_i, and therefore the accelerations in the co-rotating system also vanish.

In conclusion, L_1, L_2, and L_3 are stationary points since both the velocity and the acceleration of P (in the co-rotating system) vanish there. The stationarity with respect to z is ensured on the whole plane x, y by the fact that, for $z = 0$, the third of the (3.56) gives $\ddot{z} = 0$. This implies that in these three positions the massless point P co-rotates with P_1 and P_2 (rigidly as by hypothesis the motion of the massive points is circular).

Two other positions of equilibrium are obtained when, at a further decreasing of C_J, the zone forbidden to P shrinks down to two points (L_4 and L_5), symmetric with respect to the x-axis (Fig. 3.10). They form two equilateral triangles with P_1 and P_2. We demonstrate this statement by proving that the five stationary points we just found represent particular cases of the Lagrangian configurations discussed in Sect. 3.1, when one of the masses is null. We rewrite (3.18) defining the necessary and sufficient conditions for stationarity, using the conventions already adopted for this problem:

Fig. 3.11 Special choice of C_J which creates a communication channel between the two regions, now called Roche lobes, surrounding P_1 and P_2

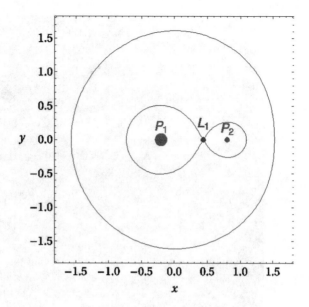

$$\mu = -n^2 x_1,$$
$$1 - \mu = n^2 x_2,$$
$$(1 - \mu)\frac{x - x_1}{r_1^3} + \mu\frac{x - x_2}{r_2^3} = n^2 x,$$
$$(1 - \mu)\frac{y - y_1}{r_1^3} + \mu\frac{y - y_2}{r_2^3} = n^2 y.$$

(3.67)

The first two equations are equivalent to the definition of the center of gravity. From them we derive that $n^2 = 1$, with which the last two come to coincide with the partial derivatives of (3.57).

An interesting astrophysical application of this analysis was made by the French astronomer and mathematician Édouard Albert Roche (1820–1883), author also of the famous Roche limit.[14] As shown in Fig. 3.11, with a particular choice of C_J the regions inside the two Hill curves surrounding P_1 and P_2 respectively can be put in communication through L_1, while keeping the matter confined. In this situation a test particle is free to transit from one region to the other, carrying matter and a part of the orbital angular momentum of the system. This happens, for instance, in some close stellar binary systems. Suppose that the region assigned to P_1 in the figure is occupied by a star in an evolutionary phase in which the outermost layers expand (this happens, for example, when a star runs out of central hydrogen and begins its ascent path of the so-called Giant Branch in the H-R diagram). When the situation represented in the figure occurs, the matter of the stellar envelope of P_1 begins to

---

[14] Minimum distance from the center of a celestial body to prevent a smaller object (orbiting it and holding itself together by self-gravity) from fragmenting due to tidal forces.

Fig. 3.12 An artistic conception of the accretion disk in a close binary system, drawn through the data of the real case of the WZ Sge system. Credit: P. Marenfeld and NOIRLab/NSF/AURA

transit in the region of competence of the star P_2 and, if the latter is a compact object (neutron star or black hole), this matter goes to place itself in a disk called accretion disk, from which it will then fall towards the degenerate star, creating production of X-rays and eventually the phenomenon of nova (Fig. 3.12). It goes without saying that the detailed study of these situations presents a high degree of complexity, if only for the fact that the bodies in play are finite, deformable, and very close to each other [7].

3.3.2   The (x, z) Plane

With the condition $y = 0$, (3.63) becomes:

$$f_1(x, z) + f_2(x, z) + f_3(x, z) = C_J, \tag{3.68}$$

where we placed (see (3.60)):

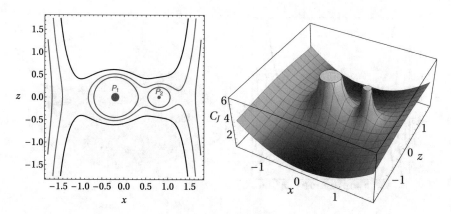

Fig. 3.13 Same as Fig. 3.9 for the projection of the zero-velocity curves on the x, z plane

$$f_1(x, z) = x^2,$$
$$f_2(x, z) = \frac{2(1 - \mu)}{\sqrt{(x - x_1)^2 + z^2}}, \qquad (3.69)$$
$$f_3(x, z) = \frac{2\mu}{\sqrt{(x - x_2)^2 + z^2}}.$$

This case differs from the previous one only in the form of f_1. Therefore, with respect to Fig. 3.9, only the curve *(a)* will be drastically modified, while the curves *(b)* and *(c)* remain almost unchanged. We leave to the reader to discuss the curves in Fig. 3.13 following the line of thought adopted above. On the x, z plane there are three saddle points, all contained on the x-axis, which identify three stationary positions. They coincide with the points L_1, L_2, and L_3 of the previous case, as shown by the fact that, for $z = 0$, the (3.69) coincide with the (3.65) when we set $y = 0$.

3.3.3 The (y, z) Plane

With the condition $x = 0$, (3.63) becomes:

$$f_1(y, z) + f_2(y, z) + f_3(y, z) = C_J, \qquad (3.70)$$

where (see (3.60)):

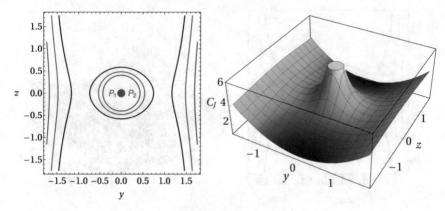

Fig. 3.14 Same as Fig. 3.9 for the projection of the zero-velocity curves on the y, z plane. In this projection, the two points P_1 and P_2, which are aligned along the x-axis, fall on each other

$$f_1(y, z) = y^2,$$

$$f_2(y, z) = \frac{2(1 - \mu)}{\sqrt{x_1^2 + y^2 + z^2}}, \tag{3.71}$$

$$f_3(y, z) = \frac{2\mu}{\sqrt{x_2^2 + y^2 + z^2}}.$$

It is relatively simple to understand the shape of the curves in Fig. 3.14. We start noticing that the functions f_i depend exclusively on y^2 and z^2; therefore, the zero-velocity curves must be symmetric with respect to both these axes. For a very large C_J, the condition (3.63) is verified if $f_1(y, z) \sim C_J$. This case produces curves that tend asymptotically to straight lines parallel to the z-axis. When P is near the origin of the coordinate system, all the three functions f_i give a non negligible contribution. The almost elliptical shape of the curves that are obtained is due to the presence of the term in y^2, without which they would be circles. The region included between the internal and the external curves, defined for the same value of C_J, is forbidden to the motion of P. As usual, its extension decreases with C_J.

The only saddle point is given by the origin of the axes, i.e., by the projection of the center of gravity B of the two massive points, P_1 and P_2. But this position is not necessarily of equilibrium. In fact, if P is in B with zero velocity, it will not remain indefinitely there because, as we see from the first of (3.56), its acceleration generally has not a null component along the x-axis. The equilibrium occurs only if $1 - \mu = \mu = 0.5$; in this case the barycenter B coincides with the position of the Lagrangian point L_1. The same condition turns into equilibrium points the other two double points contained in the y-axis and symmetric with respect to the origin, which thus coincide with L_4 and L_5.

3.4 Considerations on the Lagrangian Points

In the previous section we have seen that, although the restricted three-body problem does not admit a complete solution, the existence of the Jacobi integral (3.60) allows us to identify, according to the initial conditions, the regions of space within which the massless point P can move. If C_J has a very large value, the region allowed to P is either external to a quasi-cylinder (narrowed near to the x, y plane) which is symmetrical with respect to the z-axis, or it is internal to one of the two quasi-ellipsoids surrounding P_1 and P_2. These three regions are isolated and P cannot travel from one to the other. For smaller values of C_J they may come first in contact and then even overlap completely, so that P can move freely from one region to the other.

Finally, the Jacobi integral allows us to recognize five equilibrium positions on the x, y plane (i.e., with respect to the reference system co-rotating with P_1 and P_2), which coincide with the Lagrangian points[15] L_1, L_2, L_3, L_4, and L_5. Easy to understand in a purely mathematical sense, these stationary points are rather peculiar and improbable. We must realize that they are not traps in the motion of P but privileged positions which must be occupied at the beginning of the motion itself and then forever (in absence of perturbations). In fact, suppose that, at the epoch t_o, P transits though L_1 with the correct value of C_J, then stopping at that position forever. If it were possible, it would violate the reversibility of the equations of motion with respect to time. In fact, the same point P that is in L_1 at t_o, will have to stay there at any other time, including when $t < t_o$ (remember that in L_1 both velocity and acceleration vanish). But then two problems with the same initial conditions would come to have different solutions, which is against the deterministic model of classical mechanics. Does this imply that it is not possible to find initial conditions allowing P to transit L_1? Of course not! It just means that P cannot stop in L_1 in the absence of external forces.

It may be interesting to report here a practical use of the Lagrangian points L_1 and L_2 of the Earth-Sun systems. Although unstable (see the next section Sect. 3.5), they allow bodies of negligible mass (e.g. solar observatories as the NASA Solar and Heliospheric Observatory (Soho) and Advanced Composition Explorer (Ace) at L_1 and other satellites as the NASA Wilkinson Microwave Anisotropy Probe (WMAP), the ESA Herschel Space Observatory, Planck Surveyor, and Global Astrometric Interferometer for Astrophysics (Gaia), and the NASA James Webb Space Telescope (JWST) at L_2) to remain around these points on 'quasi-bound orbits', called halo orbits because of their shape, by means of few corrections. For a more detailed discussion, see [8].

---

[15] The numeration of Lagrangian points is usually that of decreasing C_J.

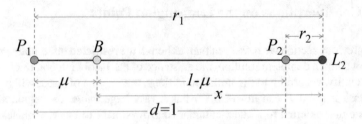

Fig. 3.15 Lagrangian point L_2: geometry for the determination of the coordinate x

3.4.1 Case of Dominant Mass

In the restricted three-body problem, the search for the barycentric coordinates (x, y) of the Lagrangian points L_1, L_2, and L_3 in the co-rotating reference system is straight-forward. Let us consider, as an example, the case of the collinear point L_2. We must search for the coordinate x only, since $y = 0$. From Fig. 3.15, which makes reference to the (3.53), we have:

$$
\begin{aligned}
r_1 - r_2 &= 1, \\
r_1 &= x + \mu, \\
r_2 &= x - (1 - \mu).
\end{aligned}
\tag{3.72}
$$

Now, the coordinates L_2, $(x, 0)$, must satisfy the the condition $\nabla U = 0$, where the Jacoby potential U is given by (3.57). Using (3.72), we have:

$$
\nabla U(L_2) = \frac{dU}{dx} = (1 - \mu)\left(-\frac{1}{(1+r_2)^2} + 1 + r_2\right) +
$$
$$
+ \mu\left(-\frac{1}{r_2^2} + r_2\right) = 0,
\tag{3.73}
$$

from where:

$$
\frac{\mu}{1 - \mu} = 3r_2^3 \frac{1 + r_2 + (r_2^2/3)}{(1 + r_2)^2(1 - r_2^3)}.
\tag{3.74}
$$

If $\mu \ll (1 - \mu)$, then $r_2 \ll 1$, and (3.74) reduces to:

$$
r_2 \approx \left(\frac{\mu}{3(1 - \mu)}\right)^{1/3},
\tag{3.75}
$$

and

$$
x \approx (1 - \mu) + \left(\frac{\mu}{3(1 - \mu)}\right)^{1/3} \simeq 1 + \left(\frac{\mu}{3}\right)^{1/3},
\tag{3.76}
$$

Fig. 3.16 Lagrangian points
for the case in which the
mass of P_1 dominates over
that of P_2

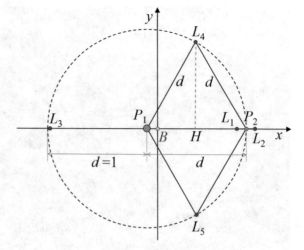

obviously, in units of the distance a between P_1 and P_2. A similar procedure allows
us to obtain the coordinates of L_1 and L_3.

The search for the coordinates of L_4 and L_5 is even simpler, whatever it is $\mu \leq 0.5$.
It is enough to prove, through the equation of the barycenter, that $P_1 B = -\mu$ and
observe that $P_1 H = 1/2$ (Fig. 3.16).

In summary, the coordinates (in units of the distance d between P_1 and P_2) of the
Lagrangian points for[16] $\mu \ll 1 - \mu$ are (Fig. 3.16):

$$
\begin{aligned}
\text{point } L_1 \quad & \left[\left(1 - \sqrt[3]{\frac{\mu}{3}}\right), 0\right], \\
\text{point } L_2 \quad & \left[\left(1 + \sqrt[3]{\frac{\mu}{3}}\right), 0\right], \\
\text{point } L_3 \quad & \left[-\left(1 + \frac{7\mu}{12}\right), 0\right], \\
\text{point } L_4 \quad & \left[\frac{(1 - 2\mu)}{2}, \frac{\sqrt{3}}{2}\right], \\
\text{point } L_5 \quad & \left[\frac{(1 - 2\mu)}{2}, -\frac{\sqrt{3}}{2}\right].
\end{aligned}
\tag{3.77}
$$

From the (3.77) we see that the Lagrangian points L_1 and L_2 in the system Sun-Earth
are $d\sqrt[3]{\dfrac{\mu}{3}} \simeq 1.5 \times 10^6$ km from the Earth, since $d = 1$ AU and $\mu = \dfrac{M_\oplus}{M_\odot + M_\oplus} \simeq$
3×10^{-6}.

---

[16] The limitation applies only to the collinear points L_1, L_2, and L_3.

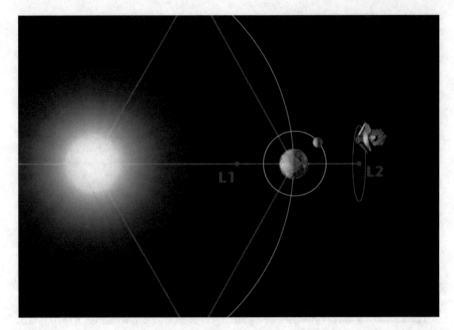

Fig. 3.17 Schematic view of the 'halo orbit' of James Webb Space Telescope around the Lagrangian point L_2. Clearly the distances of L_1 and L_2 from the Earth are not in scale with the distance Earth-Sun. Credit: NASA's Goddard Space Flight Center

This is the distance that NASA space telescope JWST, launched at Christmas 2021, had to travel to reach its final destination: the surroundings of the point L_2 where, by revolving with the Earth around the Sun, it is able to maintain contact with the terrestrial bases and at the same time the attitude of the shields that protect the infrared instrumentation from the solar radiation (Fig. 3.17).

3.5 Stability of the Lagrangian Points

The Lagrangian points are equilibrium positions. But, is this equilibrium stable? We will try to find it out by treating the issue through a first-order analysis.

To this end we suppose that a massless body P occupies an equilibrium position in the co-rotating system $B[x, y, z]$, i.e., that it sits on a Lagrangian point with zero relative velocity. We want to study its motion after applying a small impulsive perturbation. If P reacts in such a way that it remains indefinitely in the vicinity of the Lagrangian point, the equilibrium is stable. Otherwise, the position is unstable equilibrium.

3.5.1 Conditions for the Equilibrium

To study under which conditions a position of equilibrium is stable,[17] we expand the partial derivatives of the Jacobi potential at the right-hand side of the (3.58) in a Taylor series in the vicinity of the generic point Q of coordinates (x_o, y_o, z_o), retaining only the first order terms. This means that the result is valid only for small displacements (linear theory). We have:

$$\frac{\partial U}{\partial x} \approx \left(\frac{\partial U}{\partial x}\right)_o + (x - x_o)\left(\frac{\partial}{\partial x}\frac{\partial U}{\partial x}\right)_o +$$
$$+ (y - y_o)\left(\frac{\partial}{\partial y}\frac{\partial U}{\partial x}\right)_o + (z - z_o)\left(\frac{\partial}{\partial z}\frac{\partial U}{\partial x}\right)_o, \qquad (3.78)$$

and similar expressions for the derivatives with respect to y and z. Let us now identify the generic point Q with a Lagrangian point $L_i(x_i^o, y_i^o, z_i^o)$, remembering that (3.66) holds. This means that $z_i^o = 0$ and that the equations of motion (3.58) satisfy the conditions:

$$\ddot{x}_i^o - 2\dot{y}_i^o = \left(\frac{\partial U}{\partial x}\right)_o = 0,$$
$$\ddot{y}_i^o + 2\dot{x}_i^o = \left(\frac{\partial U}{\partial y}\right)_o = 0, \qquad (3.79)$$
$$\ddot{z}_i^o = \left(\frac{\partial U}{\partial z}\right)_o = 0,$$

since both velocity and acceleration are zero at a Lagrangian point in the co-moving system.

In conclusion, using the (3.79) and (3.78) and introducing the new coordinates (Fig. 3.18):

$$x' = x - x_i^o,$$
$$y' = y - y_i^o, \qquad (3.80)$$
$$z' = z,$$

which represent the displacement with respect to the libration point in the neighborhood of the equilibrium position, the equations of motion of P given by (3.58) translate into a system of linear differential equations with constant coefficients:

---

[17] A stability analysis of this kind becomes necessary when the Coriolis forces are introduced and the positions that are extremes of the Jacobi integral may become stable, as it happens just in the case of the Lagrangian points.

Fig. 3.18 System of coordinates relative to the Lagrangian point L_i, with $i = 1, \ldots, 5$

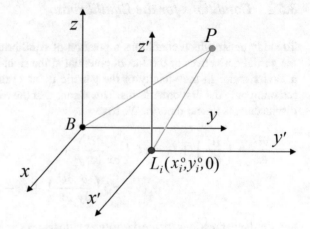

$$
\begin{cases}
\ddot{x}' - 2\dot{y}' = x' \left(\dfrac{\partial^2 U}{\partial x^2} \right)_{\!\!\circ} + y' \left(\dfrac{\partial^2 U}{\partial x \partial y} \right)_{\!\!\circ} + z' \left(\dfrac{\partial^2 U}{\partial x \partial z} \right)_{\!\!\circ}, \\[2mm]
\ddot{y}' + 2\dot{x}' = x' \left(\dfrac{\partial^2 U}{\partial x \partial y} \right)_{\!\!\circ} + y' \left(\dfrac{\partial^2 U}{\partial y^2} \right)_{\!\!\circ} + z' \left(\dfrac{\partial^2 U}{\partial y \partial z} \right)_{\!\!\circ}, \\[2mm]
\ddot{z}' = x' \left(\dfrac{\partial^2 U}{\partial x \partial z} \right)_{\!\!\circ} + y' \left(\dfrac{\partial^2 U}{\partial y \partial z} \right)_{\!\!\circ} + z' \left(\dfrac{\partial^2 U}{\partial z^2} \right)_{\!\!\circ},
\end{cases}
\qquad (3.81)
$$

where we made use of the Schwarz[18] theorem about the interchangeability of the differentiation order. Taking into account the expression for U, we obtain:

$$
\begin{aligned}
\frac{\partial^2 U}{\partial x^2} &= 1 - \frac{1 - \mu}{r_1^3} - \frac{\mu}{r_2^3} + 3(1 - \mu)\frac{(x + \mu)^2}{r_1^5} + 3\mu \frac{(x - 1 + \mu)^2}{r_2^5}, \\[2mm]
\frac{\partial^2 U}{\partial y^2} &= 1 - \frac{1 - \mu}{r_1^3} - \frac{\mu}{r_2^3} + 3(1 - \mu)\frac{y^2}{r_1^5} + 3\mu \frac{y^2}{r_2^5}, \\[2mm]
\frac{\partial^2 U}{\partial z^2} &= -\frac{(1 - \mu)}{r_1^3} - \frac{\mu}{r_2^3}, \\[2mm]
\frac{\partial^2 U}{\partial x \partial y} &= 3y \left[\frac{(1 - \mu)(x + \mu)}{r_1^5} + \frac{\mu(x - 1 + \mu)}{r_2^5} \right], \\[2mm]
\frac{\partial^2 U}{\partial x \partial z} &= \frac{\partial^2 U}{\partial y \partial z} = 0.
\end{aligned}
\qquad (3.82)
$$

Here we have taken into account the coordinates of the massive points: $x_1 = -\mu$, $x_2 = 1 - \mu$; see (3.53) to understand the reason for the signs. The equations in the last line of (3.82) were obtained using the condition $z_i^{\circ} = 0$. We will use these

---

[18] Hermann Schwarz (1843–1921) was a German mathematician, known for his work in complex analysis.

expression (3.82) to derive the equations for stability of Lagrangian points in the following sections.

From the theory of differential equations we know that the general solutions of systems of the type (3.81) are given by linear combinations of exponential functions. If the coefficients at the exponents are real, the solutions are monotonically growing functions[19] and, in our case, indicate instability; P tends to move indefinitely away from the position of equilibrium (although the solution found is no longer valid at large distances). If instead the coefficients are purely imaginary, the solutions are periodic (as we see using Euler formulas). In our case, this implies stability, in the sense that the massless body P is bound to oscillate around the equilibrium position. In order to be able to translate these general observations into quantitative terms, we again distinguish two cases.

3.5.2 Collinear Solutions: Points L_1, L_2 and L_3

Let $(x_{Li}^\circ, 0, 0)$ be the coordinates of L_i, one of the collinear libration points L_1, L_2, L_3. We calculate the second partial derivatives of the Jacobi potential and substitute them in (3.81), obtaining:

$$\begin{cases} \ddot{x}' - 2\dot{y}' = (1 + 2A_i)\,x', \\ \ddot{y}' + 2\dot{x}' = (1 - A_i)\,y', \\ \ddot{z}' = -A_i\,z', \end{cases} \tag{3.83}$$

where:

$$A_i = \frac{1 - \mu}{|x_{Li}^\circ - x_1|^3} + \frac{\mu}{|x_{Li}^\circ - x_2|^3} > 0, \tag{3.84}$$

being x_1 and x_2 the coordinates of the massive points P_1 and P_2. We note that the last of (3.83) is independent of the others. It is the differential equation characteristic of a harmonic oscillator with period $2\pi/\sqrt{A_i}$. Thus, in relation to the motion along the z-axis, the libration point is stable. The fact is not surprising because the potential Jacobi U is symmetric with respect to the plane xy.

We now solve the system of the first two (3.83). The solutions must be of the type:

$$\begin{aligned} x' &= \alpha\,e^{\lambda t}, \\ y' &= \beta\,e^{\lambda t}. \end{aligned} \tag{3.85}$$

Replacing in (3.83), we obtain a system of linear equations in the unknowns α and β:

$$\begin{cases} \left[\lambda^2 - (1 + 2A_i)\right]\alpha - 2\lambda\,\beta = 0, \\ 2\lambda\,\alpha + \left[\lambda^2 - (1 - A_i)\right]\beta = 0. \end{cases} \tag{3.86}$$

---

[19] The sign of the coefficient does not matter in how the solution must be indifferent to the direction of time.

This is a homogeneous system, but we are not interested to the trivial solutions: $x' = y' \equiv 0$. The Cramer condition for the existence of non-trivial solutions requires that the determinant of the system:

$$\begin{vmatrix} \lambda^2 - (1 + 2A_i) & -2\lambda \\ 2\lambda & \lambda^2 - (1 - A_i) \end{vmatrix}, \tag{3.87}$$

vanishes, which implies the equation:

$$\lambda^4 + (2 - A_i)\,\lambda^2 + (1 - A_i)(1 + 2A_i) = 0, \tag{3.88}$$

to be satisfied. The general solutions are then:

$$\begin{aligned} x' &= \alpha' e^{\lambda_1 t} + \alpha'' e^{\lambda_2 t} + \alpha''' e^{\lambda_3 t} + \alpha'''' e^{\lambda_4 t}, \\ y' &= \beta' e^{\lambda_1 t} + \beta'' e^{\lambda_2 t} + \beta''' e^{\lambda_3 t} + \beta'''' e^{\lambda_4 t}, \end{aligned} \tag{3.89}$$

where λ_1, λ_2, λ_3, and λ_4 are the roots of equation (3.88):

$$\lambda_{1,2,3,4} = \pm \sqrt{-1 + \frac{A_i \mp \sqrt{9A_i^2 - 8A_i}}{2}}. \tag{3.90}$$

The Lagrangian points are stable if the eigenvalues λ_i are imaginary. Indeed, the general solution can be represented in the form $\exp\left[\pm (l + ik) \right]$. It is periodic if $l = 0$ (in view of the Euler formula). For collinear points, from (3.88) and from the properties of polynomials, we have:

$$(\lambda_1 \lambda_2)(\lambda_3 \lambda_4) = (1 - A_i)(1 + 2A_i). \tag{3.91}$$

We can see from (3.90) that $\lambda_1 = -\lambda_2$ and $\lambda_3 = -\lambda_4$. The request that all roots are imaginary implies the following condition: $\lambda_1^2 = \lambda_2^2 < 0$ and $\lambda_3^2 = \lambda_4^2 < 0$ which, according to (3.91), corresponds to: $-1/2 < A_i < 1$. Since for collinear points it is always[20] $A_i > 1$, it means that all collinear solutions are unstable.

This is understandable from the physical points of view. Points L_2, L_3 are the result of equilibrium between centrifugal and gravitational forces which act in opposite directions. The gravitational forces are decreasing with the distance from the attractive masses but the centrifugal forces are increasing, and thus the equilibrium is unstable. The point L_1, which is located between two masses, is also unstable due to gravitational forces acting in opposite directions. The small perturbations will induce a motion towards one of the massive bodies.

---

[20] Cf. H.C. Plummer, *On periodic orbits in the neighbourhood of centres of libration*, Mont. Not. of Roy. astr. Soc., 62, 6, 2001. A numerical test up to $\mu \simeq 0.1$ is easily done by coupling the (3.13) with the coordinates of L_1, L_2, and L_3 in (3.105).

3.5.3 Triangular Solution: Points L_4 and L_5

The libration points are identified by the condition $|\overrightarrow{P_1 P}| = |\overrightarrow{P_2 P}| = 1$. Taking this into account, we calculate the second partial derivatives of the Jacobi potential using (3.82) and the coordinates for the triangular Lagrangian points: $x_\circ = 0.5 - \mu$, $y_\circ = \pm\sqrt{3}/2$, together with the condition $(r_1)_\circ = (r_2)_\circ$, that is:

$$
\begin{aligned}
\left(\frac{\partial^2 U}{\partial x^2} \right)_\circ &= \frac{3}{4}, \\[4pt]
\left(\frac{\partial^2 U}{\partial y^2} \right)_\circ &= \frac{9}{4}, \\[4pt]
\left(\frac{\partial^2 U}{\partial z^2} \right)_\circ &= -1, \\[4pt]
\left(\frac{\partial^2 U}{\partial x \partial y} \right)_\circ &= \frac{3\sqrt{3}}{4}(1 - 2\mu), \\[4pt]
\left(\frac{\partial^2 U}{\partial x \partial z} \right)_\circ &= \left(\frac{\partial^2 U}{\partial y \partial z} \right)_\circ = 0.
\end{aligned}
\tag{3.92}
$$

Replacing them in (3.81), we obtain:

$$
\begin{aligned}
\ddot{x}' - 2\dot{y}' &= \frac{3}{4} x' + \frac{\sqrt{27}}{4}(1 - 2\mu)\, y', \\[4pt]
\ddot{y}' + 2\dot{x}' &= \frac{\sqrt{27}}{4}(1 - 2\mu)\, x' + \frac{9}{4} y', \\[4pt]
\ddot{z}' &= -z'.
\end{aligned}
\tag{3.93}
$$

Again, the third equation can be solved separately; the solution is a periodic motion with period 2π. This proves that any libration point is stable with respect to z. To solve the system of the first two (3.93), we proceed as in the previous case. The characteristic equation:

$$
\lambda^4 + \lambda^2 + \frac{27}{4}\mu\,(1 - \mu) = 0,
\tag{3.94}
$$

has 4 roots:

$$
\lambda_{1,2,3,4} = \pm\sqrt{\frac{-1 \pm \sqrt{1 - 27\mu\,(1 - \mu)}}{2}}.
\tag{3.95}
$$

If:

$$
1 - 27\mu\,(1 - \mu) \geq 0,
\tag{3.96}
$$

all the roots of (3.94) are purely imaginary and thus the solutions of (3.93) are periodic, i.e., stable. If instead the condition (3.96) is not verified, the roots are complex; they contain a real part that generally makes the solutions unstable. The inequality (3.96) is verified by the values of the mass of P_2 meeting the conditions:

$$\mu \geq \frac{1}{2} + \sqrt{\frac{23}{108}}, \qquad \mu \leq \frac{1}{2} - \sqrt{\frac{23}{108}}. \qquad (3.97)$$

Since we have assumed $0 < \mu \leq 0.5$, only the second condition is meaningful. In conclusion, the linear theory shows that the Lagrangian points L_4 and L_5 are stable (within the limits of the restricted problem) if and only if the condition:

$$\mu \lesssim 0.0385, \qquad \text{that is:} \quad \frac{m_2}{m_1} = \frac{\mu}{1 - \mu} \lesssim 0.0401, \qquad (3.98)$$

is satisfied. This is largely verified by the Sun-Jupiter system, which justifies the existence of the families of the Greek and Trojan small planets mentioned at the beginning of this chapter (see Fig. 3.1).

3.6 The Variation of the Elements

As an introduction of the perturbation theory, here we consider the effects of an external force acting on an isolated system of two points P_1 and P of masses m_1 and m respectively, assuming that m is negligible compared to m_1. This means that:

1. the center of gravity B of the system coincides with the position of P_1, and
2. the reduced mass μ coincides the mass m of P.

With these simplifications the two-body problem is reduced to that of a single body subjected to a central force exerted by a fixed point, which allows us to apply a perturbing force only to P, neglecting the effects on P_1 [9].

The idea behind the method devised by Lagrange (see also Sect. 4.12) is straightforward: reduce a problem that cannot be solved (or is too complicated) to another one with known solution, exploiting this solution in order to simplify the equations of motion of the original problem. To better understand the Lagrangian strategy, we can reverse the reasoning starting from a problem with the known solution, which we will call unperturbed problem; classically it will be the two-body problem. The presence of additional forces perturbs the simple problem, complicating it or even making it unsolvable.

The originality of the Lagrangian approach lies especially in the choice of variables that are constant of the motion in the unperturbed problem. In this way, in fact, we are reduced to studying the variation of these constants (in the unperturbed motion) under the action of the perturbing terms. Therefore, any simplifications introduced for solving these equations will not affect the accuracy of the solution of the unperturbed problem, which stays as the zero-th order approximation of the solution of the perturbed problem.

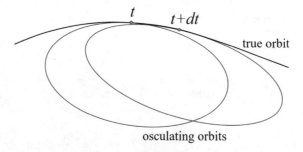

Fig. 3.19 Sketch showing the concept of osculating orbit. At each time t, the position and the velocity in the true orbit allow us to define a Keplerian orbit tangent to the true one. It is the orbit that the point P would follow if at t the perturbation would cease abruptly. At $t + dt$, the osculating orbit has typically different elements. In conclusion, the true orbit can be traced through the points of osculation of a Keplerian orbit with variable elements

We must introduce the concept of osculating orbit.[21] To this end, let us consider a perturbed Keplerian motion, i.e., the motion of a point P subject to a central Newtonian force and to an additional force that perturbs the motion otherwise conical. If, for some reason, at the generic instant of time t the disturbance ceases suddenly, P will meet the conditions to start a purely Keplerian motion with initial conditions given by the position and velocity at the time t. So, at any t there is a conical orbit which is tangent to the perturbed one, corresponding to the trajectory that the point P would take if the disturbance ceased (Fig. 3.19). In other words, we can interpret the perturbed motion of P as a succession of infinitesimal elements of conical motion corresponding to initial conditions which are continuous functions of time.

The elements of the Keplerian orbit of P relative to P_1, indicated with the symbol f, depend on the position and velocity coordinates of P, i.e., $f = f(x, y, z, \dot{x}, \dot{y}, \dot{z})$. The corresponding variation is given by:

$$\frac{df}{dt} = \frac{\partial f}{\partial x}\dot{x} + \frac{\partial f}{\partial y}\dot{y} + \frac{\partial f}{\partial z}\dot{z} + \frac{\partial f}{\partial \dot{x}}\ddot{x} + \frac{\partial f}{\partial \dot{y}}\ddot{y} + \frac{\partial f}{\partial \dot{z}}\ddot{z}. \tag{3.99}$$

At the epoch $\bar{t}$, the coordinates of position and velocity of P on the true and osculating orbits coincide while, in the presence of a perturbation $\mathbf{J}$ with components F_x, F_y and F_z, the accelerations assume different values:

$$\begin{aligned} \ddot{x} &= \ddot{x}_\mathrm{o} + F_x, \\ \ddot{y} &= \ddot{y}_\mathrm{o} + F_y, \\ \ddot{z} &= \ddot{z}_\mathrm{o} + F_z. \end{aligned} \tag{3.100}$$

Here we have made the convention, valid also in the following, that the variables with the subscript represent quantities relative to the osculating orbit. From (3.99) to

---

[21] The etymology is from the Latin verb '*osculari*', meaning to kiss.

Fig. 3.20 Relation between true and osculating orbits. The latter contains the plane (x, y)

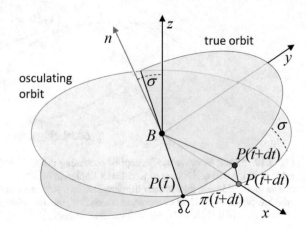

(3.100), we obtain the fundamental variational equation:

$$\left(\frac{df}{dt}\right)_* = \frac{df}{dt} - \left(\frac{df}{dt}\right)_o = \frac{\partial f}{\partial \dot{x}} F_x + \frac{\partial f}{\partial \dot{y}} F_y + \frac{\partial f}{\partial \dot{z}} F_z, \qquad (3.101)$$

remembering that $(df/dt)_o \equiv 0$.

To make explicit the previous equation for the six orbital elements, we choose a convenient inertial reference system, placing the origin in P_1 (which is also the center of gravity of the system) and the xy plane coincident with that of the osculating orbit at the time $\bar{t}$. We also orient the x-axis as $\overrightarrow{BP(\bar{t})}$, so that:

$$\begin{aligned} x(\bar{t}) &= r(\bar{t}), \\ y(\bar{t}) &= z(\bar{t}) = 0. \end{aligned} \qquad (3.102)$$

Moreover, let σ and π be the inclination and the longitude of the node of the true orbit on the osculating one (Fig. 3.20). Obviously:

$$\sigma(\bar{t}) = \pi(\bar{t}) = 0. \qquad (3.103)$$

In other words, $\sigma(t)$ can be thought of as the inclination of the osculating orbit corresponding to an epoch $t \neq \bar{t}$, counted with respect to the osculating orbit at $t = \bar{t}$; π is then the node identified by the two orbits counted starting from the position of P at $\bar{t}$.

In an infinitesimal interval of $\bar{t}$, the angular momentum per unit mass of P is the same on the true and the osculating orbit; from (2.65) it is:

$$h' = \sqrt{m_1 \, G \, a \, (1 - e^2)}, \qquad (3.104)$$

with components:

$$h'_x = y\dot{z} - z\dot{y} = \quad h' \sin\sigma \, \sin\pi,$$
$$h'_y = z\dot{x} - x\dot{z} = -h' \sin\sigma \, \cos\pi, \qquad (3.105)$$
$$h'_z = x\dot{y} - y\dot{x} = \quad h' \cos\sigma.$$

Here we have used the definition of angular momentum with respect to the coordinates (x, y, z). Moreover, in the second terms the three components have been written using the angular coordinates σ and π. Their derivatives with respect to time, calculated at $\bar{t}$ in which the (3.102) and (3.103) hold, are:

$$\left(\frac{dh'_x}{dt}\right)_o = 0,$$
$$\left(\frac{dh'_y}{dt}\right)_o = -h' \left(\frac{d\sigma}{dt}\right)_o, \qquad (3.106)$$
$$\left(\frac{dh'_z}{dt}\right)_o = \left(\frac{dh'}{dt}\right)_o.$$

They coincide with the components of the momentum of the external force, $\mathbf{M} = \mathbf{r} \times \mathbf{F}$, evaluated at the same time $\bar{t}$ through (3.102) and (3.103). Since, according to (3.102), $\mathbf{r} = x\mathbf{i}$, the momentum of the force can be written as:

$$\mathbf{M} = \mathbf{r} \times \mathbf{F} = (x\mathbf{i}) \times (F_x\mathbf{i} + F_y\mathbf{j} + F_z\mathbf{k}) = -r\,F_z\mathbf{j} + r\,F_y\,\mathbf{k}. \qquad (3.107)$$

It is apparent that:

$$\left(\frac{dh'_x}{dt}\right)_o = 0,$$
$$\left(\frac{dh'_y}{dt}\right)_o = -h'\left(\frac{d\sigma}{dt}\right)_o = -r F_z, \qquad (3.108)$$
$$\left(\frac{dh'_z}{dt}\right)_o = \left(\frac{dh'}{dt}\right)_o = r F_y.$$

from where:

$$\left(\frac{1}{h'}\frac{dh'_y}{dt}\right)_o = -\left(\frac{d\sigma}{dt}\right)_o = -r\,\frac{F_z}{h'} = -r\,W,$$
$$\left(\frac{1}{h'}\frac{dh'_z}{dt}\right)_o = \left(\frac{1}{h'}\frac{dh'}{dt}\right)_o = r\,\frac{F_y}{h'} = r\,T, \qquad (3.109)$$

having used (3.108) and placing, as it is common practice in the literature:

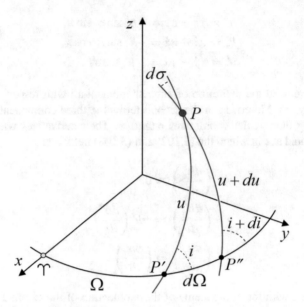

Fig. 3.21 Osculating orbit at the epochs $\bar{t}$ and $\bar{t} + dt$

$$
\begin{aligned}
S &= F_x/h' & \text{radial component,} \\
T &= F_y/h' & \text{transverse component,} \\
W &= F_z/h' & \text{normal component.}
\end{aligned}
\tag{3.110}
$$

3.6.1 Variation of the Orientation Elements

Now consider Fig. 3.21, where we have represented the osculating orbits at the epochs $\bar{t}$ and $\bar{t} + dt$. The fixed reference frame makes it possible to evaluate the orientation elements i, Ω and ω. Applying the law of sines to the spherical triangle $P P' P''$ (see Appendix A), indicating with u the true anomaly counted from the ascending node on the fixed reference system at $t = \bar{t}$, $u = \theta + \omega$, we have (see the law of sines in (A.15)):

$$
\sin(i + di)\ \sin(d\Omega) = \sin u\ \sin(d\sigma),
\tag{3.111}
$$

or, remembering that, if dx is infinitesimal, $\sin(dx) = dx$ and $\cos(dx) = 1$ are exact relations at the first order:

$$
\sin(i + di)\ d\Omega = \sin u\ d\sigma,
\tag{3.112}
$$

from which, using the first of (3.109):

$$\frac{d\Omega}{dt} = \sin u \ \csc i \ \frac{d\sigma}{dt} = r \ W \ \sin u \ \csc i. \tag{3.113}$$

From the same triangle we obtain also:

$$\cos(i + di) = \cos i \ \cos(d\sigma) - \sin i \ \sin(d\sigma) \ \cos u, \tag{3.114}$$

which simplifies as follows:

$$\cos(i + di) - \cos i = -\sin i \ \cos u \ d\sigma,$$
$$\cos i \ \cos(di) - \sin i \ \sin(di) - \cos i = -\sin i \ \cos u \ d\sigma, \tag{3.115}$$
$$di = \cos u \ d\sigma.$$

Therefore, in terms of time derivatives and using the first one of the (3.109), we have:

$$\frac{di}{dt} = \cos u \ \frac{d\sigma}{dt} = r \ W \ \cos u. \tag{3.116}$$

From Fig. 3.21 or using the (A.9), it is:

$$\cos(u + du) = \cos u \ \cos(d\Omega) + \sin u \ \sin(d\Omega) \ \cos i, \tag{3.117}$$

from which, with the same procedure previously detailed:

$$\frac{du}{dt} = \frac{d\theta}{dt} + \frac{d\omega}{dt} = -r \ W \ \sin u \ \cot i. \tag{3.118}$$

3.6.2 Variation of the Geometric Elements

The variation of the semi-major axis is obtained by differentiating the (2.68) with respect to time (using (2.67) to make explicit the expression of energy) and remembering that $m/m_1 \simeq 0$:

$$\frac{1}{a^2} \frac{da}{dt} = \frac{2}{r^2} \frac{dr}{dt} + \frac{2}{G m_1} \left(\frac{dx}{dt} \frac{d^2 x}{dt^2} + \frac{dy}{dt} \frac{d^2 y}{dt^2} + \frac{dz}{dt} \frac{d^2 z}{dt^2} \right). \tag{3.119}$$

At the time of osculation, the coordinates of P are $x = r$ and $y = z = 0$, and therefore the velocity components are immediately provided by (3.105):

$$\frac{dx}{dt} = \frac{dr}{dt},$$

$$\frac{dy}{dt} = \frac{h'}{r}, \qquad (3.120)$$

$$\frac{dz}{dt} = 0.$$

Moreover, through (3.100), it is:

$$\frac{d^2x}{dt^2} = -G\,m_1\frac{x}{r^3} + F_x = -\frac{G\,m_1}{r^2} + F_x,$$

$$\frac{d^2y}{dt^2} = -G\,m_1\frac{y}{r^3} + F_y = F_y, \qquad (3.121)$$

$$\frac{d^2z}{dt^2} = -G\,m_1\frac{z}{r^3} + F_z = F_z,$$

and replacing these expressions in (3.119):

$$\frac{1}{a^2}\frac{da}{dt} = \frac{2}{G\,m_1}\left(\frac{dr}{dt}F_x + \frac{h'}{r}F_y\right). \qquad (3.122)$$

Since at the epoch $\bar{t}$ the velocity has the same value, dr/dt, on the true and the osculating orbits, we can evaluate it on the latter by differentiating with respect to time the reciprocal of the radius vector (see (2.64) and (2.65)):

$$r = \frac{a\,(1-e^2)}{1+e\cos\theta} = \frac{h'^2}{G\,m_1\,(1+e\cos\theta)}. \qquad (3.123)$$

It is:

$$\frac{1}{r^2}\frac{dr}{dt} = \frac{G\,m_1 e\sin\theta}{h'^2}\frac{d\theta}{dt}. \qquad (3.124)$$

Since $r^2\dfrac{d\theta}{dt} = h'$, we finally have:

$$\frac{dr}{dt} = G\,m_1\frac{e\sin\theta}{h'}, \qquad (3.125)$$

which, replaced in (3.122), gives:

$$\frac{da}{dt} = \frac{2a^2}{G\,m_1}\left[\left(G\,m_1\,e\sin\theta\right)S + \left(\frac{h'^2}{r}\right)T\right]. \qquad (3.126)$$

The variation of the mean motion n is calculated starting from the expression of the third Kepler law (2.83) which, in our hypotheses, becomes:

$$\frac{4\pi^2}{P^2} = \frac{G\,m_1}{a^3}, \tag{3.127}$$

that is:

$$n = \sqrt{\frac{G\,m_1}{a^3}}. \tag{3.128}$$

Differentiating with respect to time and using (3.126), it is:

$$\frac{dn}{dt} = -\frac{3}{\sqrt{G\,m_1}\,a}\left[\left(G\,m_1\,e\,\sin\theta\right)S + \left(\frac{h'^2}{r}\right)T\right]. \tag{3.129}$$

In order to obtain the variation of the eccentricity, we operate the replacement:

$$e = \sin\psi \qquad\qquad (0 \le \psi < \pi/2), \tag{3.130}$$

legitimated only if the orbit is elliptical or parabolic, since then $0 \le e \le 1$. With the new variable, the derivative of the angular momentum per unit mass with respect to time (3.104) becomes:

$$\frac{dh'}{dt} = \frac{G\,m_1}{2h'}\left[\cos^2\psi\,\frac{da}{dt} - 2a\,\sin\psi\,\cos\psi\,\frac{d\psi}{dt}\right], \tag{3.131}$$

from where:

$$\frac{d\psi}{dt} = \frac{1}{2a\,\sin\psi\,\cos\psi}\left[\cos^2\psi\,\frac{da}{dt} - \frac{2h'}{G\,m_1}\,\frac{dh'}{dt}\right], \tag{3.132}$$

or, using the second of the (3.109) and the (3.126):

$$\frac{d\psi}{dt} = \frac{1}{a\,\sin\psi\,\cos\psi}\left[\left(a^2\,\sin\psi\,\cos^2\psi\,\sin\theta\right)S + \right.$$
$$\left. + \frac{1}{G\,m_1}\left(a^2\cos^2\psi\,\frac{h'^2}{r} - h'^2r\right)T\right], \tag{3.133}$$

and again, using the (2.48) and (2.52):

$$\frac{d\psi}{dt} = a\,\cos\psi\left[\left(\sin\theta\right)S + \left(\cos\theta + \cos E\right)T\right]. \tag{3.134}$$

From (3.130) it is: $\dfrac{de}{dt} = \cos\psi\,\dfrac{d\psi}{dt}$, so:

$$\frac{de}{dt} = a \cos^2 \psi \left[(\sin \theta) \, S + (\cos \theta + \cos E) \, T \right] =$$
$$= a(1 - e^2) \left[(\sin \theta) \, S + (\cos \theta + \cos E) \, T \right]. \qquad (3.135)$$

Now we calculate the variation of the true anomaly with respect to the osculating orbit at $\bar{t}$, remembering that, also in the latter, θ is a function of time. For our purposes it is sufficient to differentiate with respect to time the relation (2.48) so modified:

$$\frac{h'^2}{G\,m_1 r} = 1 + \sin \psi \, \cos \theta, \qquad (3.136)$$

keeping the radius vector r constant, as its variation is the same in the true and the osculating orbit:

$$\frac{d\theta}{dt} = \frac{1}{\sin \psi \, \sin \theta} \left[\cos \psi \, \cos \theta \, \frac{d\psi}{dt} - \frac{2h'}{G\,m_1 r} \frac{dh'}{dt} \right]. \qquad (3.137)$$

From this latter, using the (3.131) and (3.134), after some manipulation we obtain:

$$\frac{d\theta}{dt} = \frac{1}{\sin \psi} \left[\left(\frac{h'^2}{G\,m_1} \cos \theta \right) S - \left(r + \frac{h'^2}{Gm_1} \right) \sin \theta \, T \right]. \qquad (3.138)$$

Through (3.118) we finally have:

$$\frac{d\omega}{dt} = \left[-\frac{h'^2}{G\,m_1} \cos \theta \, \csc \psi \right] S +$$
$$+ \left[\frac{G\,m_1 r + h'^2}{G\,m_1} \sin \theta \, \csc \psi \right] T - \left[r \sin(\theta + \omega) \, \cot i \right] W. \qquad (3.139)$$

In order to calculate the variation of the time of passage at the periapsis t_o, we refer the mean anomaly to the epoch of osculation, $\bar{t}$:

$$M = n(t - t_o) = n \left(t - \bar{t} \right) + n \left(\bar{t} - t_o \right) = n \left(t - \bar{t} \right) + M_o. \qquad (3.140)$$

The total derivative of this expression with respect to time (including the variations of the orbital elements) is:

$$\frac{dM}{dt} = \frac{dM_o}{dt} + \frac{dn}{dt} \left(t - \bar{t} \right) + n. \qquad (3.141)$$

Since, on the osculating orbit, it is $(dM/dt)_o = n$, the expression:

$$\left(\frac{dM}{dt}\right)_* = \frac{dM_o}{dt} + \frac{dn}{dt}(t - \bar{t}), \tag{3.142}$$

represents the variation of the mean anomaly due to the variation of the orbital elements only and not due to the progression of time; remember that M changes on the osculating orbit too. The expression (3.142) consists of two terms, which give the variation of the zero point (dM_o/dt) and of the velocity (dn/dt) respectively. To compute the variation of M with respect to time, we differentiate the Kepler equation (2.54), allowing only the variations of the orbital elements (a fact that we remark by putting the subscript $*$ at the time derivatives):

$$\left(\frac{dM}{dt}\right)_* = (1 - \sin\psi \cos E)\left(\frac{dE}{dt}\right)_* - \sin E \cos\psi \frac{d\psi}{dt}, \tag{3.143}$$

which, through (2.52), becomes:

$$\left(\frac{dM}{dt}\right)_* = \frac{r}{a}\left(\frac{dE}{dt}\right)_* - \sin E \cos\psi \frac{d\psi}{dt}. \tag{3.144}$$

In order to obtain the variation of the eccentric anomaly, we differentiate the relation (2.52) with respect to time:

$$\left(\frac{dr}{dt}\right)_* = (1 - \sin\psi \cos E)\frac{da}{dt} + a \sin\psi \sin E \left(\frac{dE}{dt}\right)_* +$$
$$- a \cos E \cos\psi \frac{d\psi}{dt}. \tag{3.145}$$

Since the variation of r is null when passing from the osculating orbit to the true one $((dr/dt)_* = 0)$, from (3.145) we have:

$$\left(\frac{dE}{dt}\right)_* = \frac{1}{a \sin\psi \sin E}\left[a \cos\psi \cos E \frac{d\psi}{dt} - \frac{r}{a}\frac{da}{dt}\right]. \tag{3.146}$$

Substituting in (3.144) and using (3.126) and (3.134), after some simplifications we obtain:

$$\left(\frac{dM}{dt}\right)_* = -\left[2r \cos\psi - \frac{h'^2}{G m_1} \cot\psi \cos\theta\right] S +$$
$$- \left[\left(\frac{h'^2}{G m_1} + r\right)\cot\psi \sin\theta\right] T. \tag{3.147}$$

Table 3.1 Variations of the orbital elements for a perturbed two-body problem with $m \ll m_1$. S, T, W are the components of the perturbing force expressed as accelerations per unit angular momentum and $p = a(1 - e^2)$ is the focal parameter

	radial component S	transverse component T	normal component W
$\dfrac{da}{dt}$	$2a^2 e \sin \theta$	$2p\dfrac{a^2}{r}$	0
$\dfrac{de}{dt}$	$p \sin \theta$	$p\left(\cos \theta + \cos E\right)$	0
$\dfrac{di}{dt}$	0	0	$r \cos(\theta + \omega)$
$\dfrac{d\Omega}{dt}$	0	0	$r \sin(\theta + \omega) \csc i$
$\dfrac{d\omega}{dt}$	$-\dfrac{p}{e} \cos \theta$	$\dfrac{(r + p)}{e} \sin \theta$	$-r \sin(\theta + \omega) \cot i$
$\dfrac{d\theta}{dt}$	$\dfrac{p}{e} \cos \theta$	$-\dfrac{(r + p)}{e} \sin \theta$	0

We again remember that (3.147) represents the variation of the mean anomaly due to the variation of the elements of the osculating orbit. The total variation the true anomaly is instead given by the integral of (3.141):

$$M = M_\circ + n_\circ(t - \bar{t}) + \int_{\bar{t}}^{t} n \, dt + \iint_{\bar{t}}^{t} \frac{dn}{dt} \, dt^2, \qquad (3.148)$$

where the integration constants result from the condition $M = M_\circ$ for $t = \bar{t}$.

In conclusion, we have calculated the variation of the orbital elements of a two-body problem $(m \ll m_1)$ in the condition in which the perturbation acts on the body of negligible mass only. The formulas for the infinitesimal variations as a function of the instantaneous components of the perturbing force per units mass[22] are summarized in Table 3.1.

---

[22] These planetary equations were presented by Gauss in his *Theoria motus Corporum Coelestium in Sectionibus Conicis Solem Ambientium*, appeared in 1809.

References

1. R. Broucke, A. Elipe, A. Riaguas, On the figure-8 periodic solutions in the three-body problem. Chaos, Solitons Fractals **30**, 513–520 (2006)
2. E. Finlay-Freundlich, *Celestial Mechanics* (Pergamon Press, New York, 1958)
3. F.R. Moulton, *An Introduction to Celestial Mechanics* (The MacMillan Company, New York, 1960)
4. J. Barrow-Green, *J. Poincaré and the Three Body Problem* (American Mathematical Society, 1996)
5. A.R. Forsyth, *Theory of Differential Equations*, vol. 3 (Dover, New York, 1959)
6. E.T. Whittaker, *A Treatise on the Analytical Dynamics of Particles and Rigid Bodies: with an Introduction to the Problem of Three Bodies* (Dover, New York, 1944)
7. B. Paczynski, Evolutionary processes in close binary systems. Ann. Rev. Astron. Astrophys. **9**, 183–208 (1971)
8. G. Gómez et al., *Dynamics and Mission Design Near Libration Points* (World Scientific, Singapore, 2001)
9. A.D. Dubyago, *The Determination of Orbits* (The MacMillan Company, New York, 1961)

Chapter 4
Analytical Mechanics

Pierre-Simon Laplace
*All the effects of Nature are only the mathematical consequences
of a small number of immutable laws.*

In Chap. 1 we mentioned the fundamental fact that the N-body problem cannot be solved for $N \geq 3$, even in the simplified case of massive points subject to internal gravitational forces only, since it is not possible to predict the evolution of the system over an arbitrarily long time interval starting from assigned initial conditions. A similar difficulty is normally encountered also in the two-body problem if the forces in play are different from just (1.1). This case occurs, for example, when:

1. the two bodies have finite dimensions, unless they are spherically symmetric;
2. the forces, while conservative, are not only gravitational;
3. the two bodies move in a resistant medium;
4. the system is not isolated.

In each one of the circumstances listed above, the equations of motion cannot be generally integrated in a rigorous way and must be treated by techniques that approximate the solutions. It is therefore convenient to abandon the Newtonian formalism in favor of a more advanced physical-mathematical language. The reason is that usually the solutions of the equations of motion are sought by series expansions of the variables, suitably truncated at some level of approximation (which normally represents a compromise between precision and degree of difficulty). The new formalism reaches its maximum effectiveness when it is possible to treat the assigned problem as a perturbed case of a simpler one, which has already an analytical solution. It is then convenient that the expansions applies only to the perturbing terms, so that the truncation does not also affect the solution of the unperturbed problem. To this end we need to introduce new variables such that, when the perturbing forces vanish (that is, when the approximation is reduced to the zero-th order), they act as constants (in the sense that they do not explicitly depend on time). In this chapter we will consider

E. Bannikova and M. Capaccioli, *Foundations of Celestial Mechanics*, Graduate Texts
in Physics, https://doi.org/10.1007/978-3-031-04576-9_4

the methods of analytic mechanics with which it is possible to obtain this important result, limiting ourselves to the aspects pertinent to celestial mechanics [1, 2].

We approach the Hamilton-Jacobi formalism by first considering the principle of virtual work.[1] It expresses the most general condition for the static equilibrium of a mechanical system using the concept of virtual displacement, $\delta \mathbf{r}$. This change in the configuration of a system must obey the following rules. It must be:

1. instantaneous ($dt = 0$),
2. infinitesimal ($\delta \mathbf{r}^2$ must be negligible), and
3. arbitrary, though respecting the constraints, which must be considered fixed at the time of the virtual displacement, since $dt = 0$.

The principle states that the work done by the forces for virtual displacements from an equilibrium position, $\delta W = \sum_i \mathbf{f}_i \, \delta \mathbf{r}_i$, is non-positive. In particular, if the constraints are smooth and reversible, the necessary and sufficient condition for the equilibrium is that $\delta W = 0$: the virtual work of the acting forces must be zero.

The principle of virtual work encompasses the whole statics. It allows to assert that, if the system of the acting forces is conservative, at equilibrium the potential energy $W = -\mathcal{U}$ must be stationary (either maximum or minimum), and minimum if the equilibrium is stable (theorem of Lagrange-Dirichlet); [3]. Note the extreme generality of this principle, which can be applied without knowing the structures of the bodies and without specifying the reference system. Appendices Q and R provide a detailed and complementary description, including a quick presentation of Maupertuis[2] and Fermat[3] principles, emphasizing once again the link between conservation laws and symmetries (Noether[4] theorem).

In conclusion, we can stipulate that the static behaviour of a conservative system is entirely described by the potential function, which is indeed the unique tool to identify the system itself. To achieve such unification in dynamics, it is necessary to

---

[1] The principle of virtual work, which has an ancient origin, was formulated in the language of calculus by Johann Bernoulli in 1717 and in a more effective form by Lagrange in 1811.

[2] Pierre Louis Moreau de Maupertuis (1698–1759): French physicist, mathematician, and philosopher who introduced Newtonianism in France. He was the first to formulate, in contrast to Fermat and Leibniz, the mechanical principle of least action, bringing finalism in physics.

[3] Pierre de Fermat (1601–1665): together with Descartes, one of the two greatest mathematical minds of the first mid of the 17-th century, was among the founders of analytic geometry and of the probability theory, precursor of differential calculus and of modern number theory, famous for the his conjecture, also known as the last Fermat theorem: for any integer $n > 2$, the equation $a^n + b^n = c^n$ has no solution in the domain of positive integer numbers but the trivial one; '*It is impossible to separate a cube into two cubes, or a fourth power into two fourth powers, or in general, any power higher than the second, into two like powers. I have discovered a truly marvelous proof of this, which this margin is too narrow to contain*', he annotated in Latin at the margin of Diophantus' *Arithmetica*. The theorem remained unproven for over 300 years until 1994. Fermat authored also the principle according to which, in order to move from the point A to B, light chooses the trajectory requiring the least amount of time.

[4] Amalie Emmy Noether (1882–1935) was a German mathematician. She worked in mathematical physics and abstract algebra. Her name is linked to the 1915 theorem which, in the field of theoretical physics, highlights a deep connection between symmetries and conservation laws. Cf. note at Sect. 4.8.

find an appropriate state function (equivalent to the potential in the static problem for conservative systems), which describes the behavior of each mechanical system, remembering that the dynamic state of a system depends not only on coordinates but also on velocities. The function sought has the name of a great mathematician of Italian origin, Joseph Louis Lagrange.

4.1   The Function of Lagrange

Consider a system of N massive points P_i of mass m_i. Making use of the d'Alembert[5] principle, at any time t and for each point P_i we consider the link between the force $\mathbf{f}_i$ (assuming that the virtual work of internal forces is zero) and the variation of the momentum $\mathbf{p}_i = m_i \dot{\mathbf{r}}_i$:

$$\mathbf{f}_i - \dot{\mathbf{p}}_i = 0 \qquad\qquad (i = 1, \ldots, N). \qquad\qquad (4.1)$$

Formally, $\dot{\mathbf{p}}_i$ is a force (of inertia) which balances the active force. At any moment, the condition:

$$\sum_{i=1}^{N} (\mathbf{f}_i \, \delta \mathbf{r}_i - \dot{\mathbf{p}}_i \, \delta \mathbf{r}_i) = 0, \qquad\qquad (4.2)$$

is verified on the true trajectory. Therefore it is also:

$$\int_{t_1}^{t_2} \sum_{i=1}^{N} (\mathbf{f}_i \, \delta \mathbf{r}_i - \dot{\mathbf{p}}_i \, \delta \mathbf{r}_i) dt = 0. \qquad\qquad (4.3)$$

Integrating by parts the second term of (4.3) and assuming that, at the limits of the integration interval, the virtual displacements are null (since all the virtual trajectories share the same extremes), we have (see also Appendix R):

---

[5] This principle is equivalent to second principle of Newton's dynamics. The French mathematician, physicist, philosopher, and astronomer, Jean-Baptiste Le Rond d'Alembert (1717–1783), was one of the major protagonists of Enlightenment. Together with Denis Diderot, he conceived and directed the famous *Encyclopédie*, the first volume of which appeared in 1751. D'Alembert treated the sections of mathematics and sciences until he abandoned the project because of disagreements with Diderot. A friend of Lagrange and historical opponent of Alexis Claude Clairaut, he was a member and then perpetual secretary of the *Académie française*.

$$\int_{t_1}^{t_2} \left[\sum_{i=1}^{N} \left(\mathbf{f}_i \delta \mathbf{r}_i + \mathbf{p}_i \frac{d}{dt}(\delta \mathbf{r}_i) \right) - \frac{d}{dt} \left(\sum_{i=1}^{N} \mathbf{p}_i \delta \mathbf{r}_i \right) \right] dt =$$

$$= \int_{t_1}^{t_2} \sum_{i=1}^{N} \left(\mathbf{f}_i \delta \mathbf{r}_i + \mathbf{p}_i \frac{d}{dt}(\delta \mathbf{r}_i) \right) dt - \sum_{i=1}^{N} \mathbf{p}_i \delta \mathbf{r}_i \Big|_{t_1}^{t_2} =$$

$$= \int_{t_1}^{t_2} \sum_{i=1}^{N} \left(\mathbf{f}_i \delta \mathbf{r}_i + \mathbf{p}_i \frac{d}{dt}(\delta \mathbf{r}_i) \right) dt = 0. \qquad (4.4)$$

Let us verify that, at the first order, it is:

$$\delta \left(\frac{d}{dt} \mathbf{r}_i \right) = \frac{d}{dt}(\delta \mathbf{r}_i), \qquad (4.5)$$

by comparing, at the same time t, not only the position $\mathbf{r}_i$ of P_i on the true trajectory with the varied position $\mathbf{r}_i + \delta \mathbf{r}_i$, but also the true velocity $\dot{\mathbf{r}}_i$ with that on the virtual trajectory:

$$(\dot{\mathbf{r}}_i)_{\text{virt}} = \dot{\mathbf{r}}_i + \delta \dot{\mathbf{r}}_i = \dot{\mathbf{r}}_i + \delta \left(\frac{d \mathbf{r}_i}{dt} \right). \qquad (4.6)$$

Since the velocity on the virtual trajectory can be written also as:

$$(\dot{\mathbf{r}}_i)_{\text{virt}} = \frac{d}{dt}(\mathbf{r}_i + \delta \mathbf{r}_i) = \dot{\mathbf{r}}_i + \frac{d}{dt}(\delta \mathbf{r}_i), \qquad (4.7)$$

the comparison with (4.6) proves the identity (4.5). Note that the above considerations are made possible only because, by hypothesis, the virtual displacements occur at a constant time and they hold only at the first order.

That said, the variation of the kinetic energy, $T_i = \frac{1}{2} m_i \dot{\mathbf{r}}_i^2$, is:

$$\delta T_i = m_i \dot{\mathbf{r}}_i \delta \left(\frac{dr}{dt} \right) = \mathbf{p}_i \delta \dot{\mathbf{r}}_i = \mathbf{p}_i \frac{d}{dt}(\delta \mathbf{r}_i) \qquad (i = 1, \ldots, N). \qquad (4.8)$$

Through this relation, noting that $\delta W_i = \mathbf{f}_i \delta \mathbf{r}_i$ gives the virtual work of the external forces on P_i, the variational integral (4.4) takes the compact form:

$$\int_{t_1}^{t_2} \sum_{i=1}^{N} \left(\delta W_i + \delta T_i \right) dt = \int_{t_1}^{t_2} \left(\delta W + \delta T \right) dt = 0, \qquad (4.9)$$

where T and W represent the total kinetics energy and the work respectively.

Equation (4.9) contains the generalized Hamilton[6] principle. It can be rewritten in the following way:

$$\delta \int_{t_1}^{t_2} T dt + \int_{t_1}^{t_2} \delta W dt = 0, \qquad (4.10)$$

with the purpose of enlightening that, in general, the identity:

$$\int_{t_1}^{t_2} \delta W dt = \delta \int_{t_1}^{t_2} W dt, \qquad (4.11)$$

is not verified. Remember that δW is not the variation of W but the infinitesimal work accomplished by the active forces in the transition from the true to the varied trajectory. In particular, if the system is conservative, it results: $\delta W = \delta \mathcal{U}$, where $\mathcal{U}$ is the potential, and (4.9) becomes simply:

$$\int_{t_1}^{t_2} (\delta T + \delta \mathcal{U}) dt = \delta \int_{t_1}^{t_2} (T + \mathcal{U}) dt = \delta \int_{t_1}^{t_2} \mathcal{L} dt = 0. \qquad (4.12)$$

The function:

$$\mathcal{L}(\mathbf{r}_1, \mathbf{r}_2, \ldots, \mathbf{r}_N, \dot{\mathbf{r}}_1, \dot{\mathbf{r}}_2, \ldots, \dot{\mathbf{r}}_N, t) = \mathcal{L}(\mathbf{r}, \dot{\mathbf{r}}, t) = T + \mathcal{U} = T - \mathcal{W}, \qquad (4.13)$$

is called Lagrange function or simply Lagrangian. The (4.9) is summarized by saying that the first variation of the integral of the Lagrange function for a conservative system is zero on the true trajectory, i.e.:

$$\delta \int_{t_1}^{t_2} \mathcal{L}(\mathbf{r}, \dot{\mathbf{r}}, t) \, dt = 0. \qquad (4.14)$$

4.2 Lagrange Function in Generalized Coordinates

The configuration of a mechanical system made of N massive points P_j is described at any time t by $3N$ coordinates x_i, with an appropriate rule for the index i to establish the association between coordinates and points; for instance the Cartesian coordinates of P_j could be $(x_{3j-2}, x_{3j-1}, x_{3j})$, with $j = 1, \ldots, N$. They need not be necessarily Cartesian; for instance, they could be spherical or cylindrical. Obviously, not all the $3N$ coordinates are required if the system is constrained. In the presence of a number

---

[6] Sir William Rowan Hamilton (1805–1865) was an Irish mathematician, physicist, and astronomer. Child prodigy, his greatest contribution is perhaps the reformulation of Newtonian mechanics in a new form (Hamiltonian mechanics), which is part of rational mechanics and has proven central to the modern study of classical field theories such as electromagnetism.

m of time-independent holonomic constraints,[7] given by $m < 3N$ equations of the type:

$$g_j(\xi_1, \xi_2, \ldots, \xi_{3N}) = 0 \qquad (j = 1, \ldots, m), \qquad (4.15)$$

we say that the system has $n = (3N - m)$ degrees of freedom. In this case, the knowledge of the configuration of the system at any time t requires only n generalized coordinates, q_k $(k = 1, \ldots, n)$. They are given by the old coordinates ξ_i through:

$$q_k = g_k(\xi_1, \xi_2, \ldots, \xi_{3N}) \qquad (k = 1, \ldots, m). \qquad (4.16)$$

The functions (4.16) can be completely arbitrary, provided they are independent of each other.

Consider the set of $3N$ functions g given by (4.15) and (4.16), and assume that their first partial derivatives are continuous, so that we can apply the Schwarz[8] theorem. They form a system of equations:

$$g_h(\xi_1, \xi_2, \ldots, \xi_{3N}) = \begin{cases} 0 & \text{for} \quad h = 1, \ldots, m, \\ q_{h-m} & \text{for} \quad h = m+1, \ldots, 3N, \end{cases} \qquad (4.17)$$

corresponding to a point transformation with Jacobian[9] matrix:

$$J = \|a_{ij}\| = \left\| \frac{\partial g_i}{\partial \xi_j} \right\|, \qquad (4.18)$$

with characteristic n. Thus, the coordinates ξ_i can be all expressed as functions of n generalized coordinates:

$$\xi_i = \xi_i(q_1, q_2, \ldots, q_n) \qquad (i = 1, \ldots, 3N). \qquad (4.19)$$

By differentiating the equations of the system (4.17) with respect to time, a new set of equations is obtained:

$$\frac{dg_h}{dt} = \sum_{i=1}^{3N} \frac{\partial g_h}{\partial \xi_i} \dot{\xi}_i = \begin{cases} 0 & \text{for} \quad h = 1, \ldots, m, \\ \dot{q}_{h-m} & \text{for} \quad h = m+1, \ldots, 3N, \end{cases} \qquad (4.20)$$

---

[7] A system constraint is said to be holonomic if it can be expressed by an equation involving the coordinates q_i and the time t only (that is, not the momenta p_i or higher derivatives of time).

[8] Herman Schwarz (1843–1921): German mathematician, known for his contributions to functional analysis, differential geometry, and calculus of variations. Student of Karl Weierstrass (1815–1897), among his students he had Ernst Zermelo (1871–1953), then famous for his work on the foundations of mathematics.

[9] The matrix of all the first partial derivatives of a function defined in an Euclidean space, linearly approximates a differentiable function around a given point, extending the notion of derivation to multivariate functions.

which link linearly the time derivatives of the original coordinates to those of the generalized coordinates through coefficients depending on $\xi_1, \ldots, \xi_{3N}$. Therefore, since the kinetic energy is a positive defined[10] quadratic form in $\dot{\xi}_i$, it remains so also in $\dot{q}_k$, with coefficients depending on q. Finally, if the force field depends on a general potential $\mathcal{U}$ which is only a function of the generalized coordinates q and of their derivatives $\dot{q}$, we obtain the following important result: the Lagrangian function, expressed in generalized coordinates, depends only on the coordinates themselves and their first derivatives with respect to time. Explicit dependence on time is also possible, if this is required by the constraints or by the potential function; but this is not the case for us here.

In conclusion, the Lagrangian function, $\mathcal{L}(q, \dot{q}, t)$, represents the state function we were looking for. Besides a possible explicit dependence on time, it depends on $2n$ coordinates, which are exactly twice the spatial coordinates, those which identify the system in space at any moment. No surprise, though, if we think that the Lagrangian function describes the dynamic state of the system. Even in Newtonian formalism, the description of motion requires both positions (coordinates in numbers equal to the degrees of freedom of the system) and the corresponding velocities.

4.3 The Equations of Lagrange

We now want to apply the Hamilton principle (4.14) to the Lagrangian function in generalized coordinates.[11] To this purpose, we must calculate the variation of $\mathcal{L}(q, \dot{q}, t)$, remembering that it occurs at a constant time. It is:

$$\delta\mathcal{L} = \sum_{k=1}^{n} \frac{\partial\mathcal{L}}{\partial q_k}\delta q_k + \sum_{k=1}^{n} \frac{\partial\mathcal{L}}{\partial\dot{q}_k}\delta\dot{q}_k, \qquad (4.21)$$

or, according to (4.14):

$$\int_{t_1}^{t_2} \delta\mathcal{L}\, dt = -\int_{t_1}^{t_2} \sum_{k=1}^{n} \left(\frac{d}{dt}\frac{\partial\mathcal{L}}{\partial\dot{q}_k} - \frac{\partial\mathcal{L}}{\partial q_k}\right)\delta q_k\, dt = 0. \qquad (4.22)$$

In fact, through an integration by parts and recalling that the variations are assumed to vanish at the extremes of integration, it is readily verified that:

---

[10] Second degree homogeneous integer rational function, as easily verified in Cartesian and spherical coordinates. For more details see Sect. 5.5.

[11] For a discussion on variational principles and their applications to mechanics, see Appendix Q.

$$\int_{t_1}^{t_2} \frac{\partial \mathcal{L}}{\partial \dot{q}_k} \delta \dot{q}_k \, dt = \int_{t_1}^{t_2} \frac{\partial \mathcal{L}}{\partial \dot{q}_k} \frac{d}{dt} (\delta q_k) \, dt =$$

$$= -\int_{t_1}^{t_2} \frac{d}{dt} \frac{\partial \mathcal{L}}{\partial \dot{q}_k} \delta q_k \, dt + \frac{\partial \mathcal{L}}{\partial \dot{q}_k} \delta q_k \Big|_{t_1}^{t_2} = -\int_{t_1}^{t_2} \frac{d}{dt} \frac{\partial \mathcal{L}}{\partial \dot{q}_k} \delta q_k \, dt. \qquad (4.23)$$

Since the virtual displacements, δq_k, are completely arbitrary and independent, (4.22) can only be satisfied if the n equations:

$$\frac{d}{dt} \frac{\partial \mathcal{L}}{\partial \dot{q}_k} - \frac{\partial \mathcal{L}}{\partial q_k} = 0 \qquad (k = 1, \ldots, n), \qquad (4.24)$$

hold at any time t. They are called second kind Lagrange equations[12] and exist for systems (both continue and made of discrete elements) subject to holonomic constraints and to forces depending on a potential.

We want now put ourselves in the condition where the Lagrangian function does not depends explicitly on time, $(\partial \mathcal{L}/\partial t \equiv 0)$, so that the total derivative is:

$$\frac{d}{dt} \mathcal{L}(q, \dot{q}) = \sum_{k=1}^{n} \dot{q}_k \frac{\partial \mathcal{L}}{\partial q_k} + \sum_{k=1}^{n} \ddot{q}_k \frac{\partial \mathcal{L}}{\partial \dot{q}_k}. \qquad (4.25)$$

Since the kinetic energy T is a second degree homogeneous function in $\dot{q}_k$, the Euler theorem (cf. Sect. 5.5.1) assures that:

$$2T = \sum_{k=1}^{n} \dot{q}_k \frac{\partial T}{\partial \dot{q}_k}. \qquad (4.26)$$

Therefore, if the potential $\mathcal{U}$ is assumed to be independent of the generalized velocity (as it is the gravitational potential), then:

$$2T = \sum_{k=1}^{n} \dot{q}_k \frac{\partial T}{\partial \dot{q}_k} = \sum_{k=1}^{n} \dot{q}_k \frac{\partial}{\partial \dot{q}_k} (T + \mathcal{U}) = \sum_{k=1}^{n} \dot{q}_k \frac{\partial \mathcal{L}}{\partial \dot{q}_k}. \qquad (4.27)$$

Differentiating the (4.27) with respect to time:

$$2\frac{dT}{dt} = \sum_{k=1}^{n} \dot{q}_k \frac{d}{dt} \frac{\partial \mathcal{L}}{\partial \dot{q}_k} + \sum_{k=1}^{n} \ddot{q}_k \frac{\partial \mathcal{L}}{\partial \dot{q}_k}, \qquad (4.28)$$

---

[12] These equations are often called Euler-Lagrange equations because both mathematicians worked around Lagrange's original idea.

and using the second kind Lagrange equations (4.24), we have:

$$2\frac{dT}{dt} = \sum_{k=1}^{n} \dot{q}_k \frac{\partial \mathcal{L}}{\partial q_k} + \sum_{k=1}^{n} \ddot{q}_k \frac{\partial \mathcal{L}}{\partial \dot{q}_k}. \tag{4.29}$$

The latter, subtracted to (4.25), gives:

$$\frac{d\mathcal{L}}{dt} - 2\frac{dT}{dt} = \frac{d}{dt}(T + \mathcal{U}) - 2\frac{dT}{dt} = -\left(\frac{dT}{dt} - \frac{d\mathcal{U}}{dt}\right) = 0, \tag{4.30}$$

and, by integration with respect to time:

$$T - \mathcal{U} = \mathcal{E} = \text{const.} \tag{4.31}$$

Therefore, the principle of energy conservation is a direct consequence of the Lagrange equations. Finally, by rewriting the kinetic energy through the relation (4.26), we observe that the total energy of the system:

$$\mathcal{E} = T - \mathcal{U} = 2T - \mathcal{L} = \text{const}, \tag{4.32}$$

can be expressed by the Lagrangian function:

$$\mathcal{E} = T - \mathcal{U} = \sum_{k=1}^{n} \left(\dot{q}_k \frac{\partial \mathcal{L}}{\partial \dot{q}_k}\right) - \mathcal{L}(q, \dot{q}), \tag{4.33}$$

where we used the fact that $\mathcal{U}$ does not depend on $\dot{q}_k$.

4.4 The Hamilton Function

In the previous section we have shown that the Lagrangian function depends on n generalized coordinates q_k and on the corresponding n generalized velocities $\dot{q}_k$. However, these n pairs of dynamic coordinates are not completely independent; in fact, the virtual displacement δq_k affects the virtual variation $\delta \dot{q}_k$ of the corresponding velocity coordinate. In order to make up for the need of expressing the Lagrangian function by $2n$ coordinates that are totally independent, we reconsider (4.14) using the new notation now introduced:

$$\int_{t_1}^{t_2} \delta \mathcal{L} \, dt = \int_{t_1}^{t_2} \left[\delta \mathcal{L} - \frac{d}{dt}\left(\sum_{k=1}^{n} p_k \delta q_k\right)\right] dt = 0. \tag{4.34}$$

The generalized moments, p_k, so called for analogy with the Newtonian formalism (they have the dimensions of moments if q_k are lengths), can be arbitrarily chosen, provided that the product $p_k q_k$ has the dimension of an action, in what the integral:

$$\int_{t_1}^{t_2} \frac{d}{dt} \left(\sum_{k=1}^{n} p_k \delta q_k \right) dt = \left. \sum_{k=1}^{n} p_k \delta q_k \right|_{t_1}^{t_2},$$ (4.35)

is in any case vanishing due to the cancellation of δq_k at the extremes of the integration interval. Through the relation (4.21), once the time derivative is made explicit, the variational integral (4.34) becomes:

$$\int_{t_1}^{t_2} \sum_{k=1}^{n} \left[\left(\frac{\partial \mathcal{L}}{\partial \dot{q}_k} - p_k \right) \delta \dot{q}_k + \left(\frac{\partial \mathcal{L}}{\partial q_k} - \dot{p}_k \right) \delta q_k \right] dt = 0.$$ (4.36)

If we place:

$$p_k = \frac{\partial \mathcal{L}}{\partial \dot{q}_k},$$ (4.37)

the coefficient of $\delta \dot{q}_k$ in the integrand function is canceled, and therefore it no longer matters what the variation of $\dot{q}_k$ is for the purpose of the variational integral. Furthermore, when we have chosen the definition of p_k given by (4.37), for (4.36) to be satisfied it is necessary that:

$$\dot{p}_k = \frac{\partial \mathcal{L}}{\partial q_k}.$$ (4.38)

This last result can be derived as a consequence of the application of the generalized coordinates and of (4.24). In fact, (4.37) is the natural extension in Lagrangian coordinates of the Newtonian expression:

$$m_i v_i = p_i = \frac{1}{2} \frac{\partial}{\partial v_i} \left(\sum_{k=1}^{n} m_k v_k^2 \right) = \frac{\partial}{\partial v_i} (T + \mathcal{U}) = \frac{\partial \mathcal{L}}{\partial v_i},$$ (4.39)

under the assumption that $\partial \mathcal{U}/\partial v_i \equiv 0$. By inserting into (4.24) the definition of generalized moment given by (4.37), we immediately have its derivative with respect time (4.38).

Through (4.37) and (4.38), which are called first kind equations, the total differential of the Lagrangian function (4.21) becomes:

$$d\mathcal{L}(q, \dot{q}, t) = \sum_{k=1}^{n} \frac{\partial \mathcal{L}}{\partial q_k} dq_k + \sum_{k=1}^{n} \frac{\partial \mathcal{L}}{\partial \dot{q}_k} d\dot{q}_k + \frac{\partial \mathcal{L}}{\partial t} dt =$$

$$= \sum_{k=1}^{n} \dot{p}_k dq_k + \sum_{k=1}^{n} p_k d\dot{q}_k + \frac{\partial \mathcal{L}}{\partial t} dt, \qquad (4.40)$$

where the derivative with respect to time is included. Replacing the second term in the right-hand side of (4.40) with the expression:

$$\sum_{k=1}^{n} p_k \, d\dot{q}_k = d \left(\sum_{k=1}^{n} p_k \, \dot{q}_k \right) - \sum_{k=1}^{n} \dot{q}_k dp_k, \qquad (4.41)$$

it is:

$$d \left(\sum_{k=1}^{n} p_k \, \dot{q}_k - \mathcal{L} \right) = - \sum_{k=1}^{n} \dot{p}_k \, dq_k + \sum_{k=1}^{n} \dot{q}_k \, dp_k - \frac{\partial \mathcal{L}}{\partial t} dt. \qquad (4.42)$$

The function:

$$\mathcal{H}(p, q, t) = \sum_{k=1}^{n} p_k \, \dot{q}_k - \mathcal{L}(q, \dot{q}, t), \qquad (4.43)$$

of which the expression (4.42) is the total differential, takes the name of Hamiltonian function. The comparison with (4.33), modified using (4.37), immediately proves that, if the Hamiltonian function does not depend explicitly on time (property that must be shared by the associated Lagrangian function), it expresses the total energy of the system.

Formally, the total differential of $\mathcal{H}$ is given by:

$$d\mathcal{H}(p, q, t) = \sum_{k=1}^{n} \frac{\partial \mathcal{H}}{\partial q_k} dq_k + \sum_{k=1}^{n} \frac{\partial \mathcal{H}}{\partial p_k} dp_k + \frac{\partial \mathcal{H}}{\partial t} dt. \qquad (4.44)$$

By comparing with (4.42), it follows that:

$$\left. \begin{aligned} \dot{p}_k &= -\frac{\partial \mathcal{H}}{\partial q_k} \\[2mm] \dot{q}_k &= \frac{\partial \mathcal{H}}{\partial p_k} \end{aligned} \right\} \qquad (k = 1, \ldots, n), \qquad (4.45a)$$

$$\frac{\partial \mathcal{H}}{\partial t} = -\frac{\partial \mathcal{L}}{\partial t}. \qquad (4.45b)$$

The $2n$ differential equations of the first order (4.45a) are called Hamilton or canonical equations to emphasize their extraordinary symmetry and formal simplicity. Through these equations it is immediate to prove that:

$$\frac{d\mathcal{H}}{dt} = \frac{\partial \mathcal{H}}{\partial t}.$$ (4.46)

In fact:

$$\frac{d\mathcal{H}}{dt} = \frac{\partial \mathcal{H}}{\partial t} + \sum_{k=1}^{n}\left(\frac{\partial \mathcal{H}}{\partial q_k}\dot{q}_k + \frac{\partial \mathcal{H}}{\partial p_k}\dot{p}_k\right) =$$

$$= \frac{\partial \mathcal{H}}{\partial t} + \sum_{k=1}^{n}(-\dot{p}_k\dot{q}_k + \dot{p}_k\dot{q}_k) = \frac{\partial \mathcal{H}}{\partial t}.$$ (4.47)

Hence, if $\mathcal{H}$ does not explicitly depend on time ($\partial \mathcal{H}/dt = 0$), it remains constant during motion. The comparison of (4.33) with (4.43) shows that, under the assumption that the potential does not depend on velocity, $\mathcal{H}$ is the total energy of the system. The canonical equations can be deduced directly from the Hamilton principle (4.14), rewritten as:

$$\delta\int_{t_1}^{t_2} \mathcal{L}\, dt = \delta\int_{t_1}^{t_2}\left(\sum_{k=1}^{n} p_k\dot{q}_k - \mathcal{H}(p,q,t)\right) dt = 0,$$ (4.48)

using the (4.43). By expanding the variation at the right-hand side:

$$\int_{t_1}^{t_2}\left(\sum_{k=1}^{n} p_k\delta\dot{q}_k + \sum_{k=1}^{n}\dot{q}_k\delta p_k - \sum_{k=1}^{n}\frac{\partial \mathcal{H}}{\partial q_k}\delta q_k - \sum_{k=1}^{n}\frac{\partial \mathcal{H}}{\partial p_k}\delta p_k\right) dt = 0,$$ (4.49)

and integration by parts, we obtain:

$$\int_{t_1}^{t_2}\sum_{k=1}^{n} p_k\delta\dot{q}_k dt = \int_{t_1}^{t_2}\sum_{k=1}^{n} p_k\frac{d}{dt}(\delta q_k)dt =$$

$$= \sum_{k=1}^{n} p_k\delta q_k\Big|_{t_1}^{t_2} - \int_{t_1}^{t_2}\sum_{k=1}^{n}\dot{p}_k\delta q_k dt = -\int_{t_1}^{t_2}\sum_{k=1}^{n}\dot{p}_k\delta q_k dt,$$ (4.50)

with which the variational integral becomes:

$$\int_{t_1}^{t_2}\left[\sum_{k=1}^{n}\left(-\dot{p}_k - \frac{\partial \mathcal{H}}{\partial q_k}\right)\delta q_k + \sum_{k=1}^{n}\left(\dot{q}_k - \frac{\partial \mathcal{H}}{\partial p_k}\right)\delta p_k\right] dt = 0.$$ (4.51)

Given the independence of the virtual variations δq_k and δp_k of the $2n$ dynamical coordinates, this equation is identically satisfied if and only if all the coefficients are zero, that is, if the $2n$ (4.45a) are verified, which is what we wanted to prove.

Finally, we note that the Hamilton function is always defined but for an additive term corresponding to the total derivative with respect to time of an arbitrary function of coordinates, conjugate moments, and time, $f(q, p, t)$. In fact, let it be:

$$\mathcal{H}' = \mathcal{H} + \frac{df}{dt}, \tag{4.52}$$

where f is any function. It results:

$$\delta \int_{t_1}^{t_2} \left(\sum_{k=1}^{n} p_k \dot{q}_k - \mathcal{H}' \right) dt = \delta \int_{t_1}^{t_2} \left(\sum_{k=1}^{n} p_k \dot{q}_k - \mathcal{H} \right) dt - \delta \int_{t_1}^{t_2} \frac{df}{dt} dt =$$

$$= \delta \int_{t_1}^{t_2} \left(\sum_{k=1}^{n} p_k \dot{q}_k - \mathcal{H} \right) dt - \delta(f \Big|_{t_1}^{t_2}) =$$

$$= \delta \int_{t_1}^{t_2} \left(\sum_{k=1}^{n} p_k \dot{q}_k - \mathcal{H} \right) dt, \tag{4.53}$$

since the variations of f at the extremes of the integration interval are null: $\delta f_1 = f[q(t_1), p(t_1), t_1] = \delta f_2 = f[q(t_2), p(t_2), t_2] = 0$. Therefore, both $\mathcal{H}$ and $\mathcal{H}'$ satisfy the same variational integral.

4.5 Some Considerations on the Canonical Equations

Let $\mathcal{H}(p, q, t)$ be the Hamiltonian function of a mechanical system with n degrees of freedom. The system of $2n$ first order differential equations:

$$\left. \begin{array}{l} \dot{q}_k = \dfrac{\partial \mathcal{H}}{\partial p_k} \\[2mm] \dot{p}_k = -\dfrac{\partial \mathcal{H}}{\partial q_k} \end{array} \right\} \qquad (k = 1, \ldots, n), \tag{4.54}$$

forms the set of equations of motion which replace the n Lagrange equations (4.24). The solutions are $2n$ functions of the type:

$$\left. \begin{array}{l} q_k = q_k(t, a_1, a_2, \ldots, a_{2n}) \\[2mm] p_k = p_k(t, a_1, a_2, \ldots, a_{2n}) \end{array} \right\} \qquad (k = 1, \ldots, n), \tag{4.55}$$

which provide position and velocity of each point of the mechanical system at any time t. Obviously, the integration introduces $2n$ arbitrary constants, a_i, which are quantified by specifying the initial conditions.

At this point it is legitimate to wonder what the advantages of the new formulation could be over the ordinary Newtonian treatment. Indeed, apart from the greater formal elegance, very often the system (4.54) is more difficult to handle than the classical equations of motion. We must wait a few paragraphs to appreciate the power of this new formalism. For the moment we restrict our interest to the relations (4.55). The original idea of considering the $2n$ constant a_i appearing there as independent analytic quantities is due to Joseph-Louis Lagrange. Indeed, by solving the system of equations (4.55) with respect to a_i and keeping time in the role of a parameter, we obtain $2n$ solutions in the form[13]:

$$ a_i = a_i(t, p_1, p_2, \ldots, p_n, q_1, q_2, \ldots, q_n) \qquad (i = 1, \ldots, 2n). \qquad (4.56) $$

Equations (4.56) are the inverse of (4.55) and vice versa. In fact, by solving the system (4.56) with respect to p_k and q_k, we obtain again the equations of the system (4.55). Moreover, the arbitrariness of the constants a_i is evident, as they can all be replaced by an equal number of functions of $a_1, a_2, \ldots, a_{2n}$. We will see later the importance of this property.

The above considerations show that every Hamiltonian $\mathcal{H}(p_i, q_i, t)$ can be associated with $2n$ arbitrary functions of generalized coordinates, corresponding conjugate moments, and possibly time, $a_i(q_i, p_i, t)$, which, by (4.54), have the property of being constant with time and independent of each other. Each of such functions, which are called first integrals or constants of motion, satisfies a linear partial differential equation:

$$ \frac{da_i}{dt} = \frac{\partial a_i}{\partial t} + \sum_{k=1}^{n} \frac{\partial a_i}{\partial p_k} \dot{p}_k + \sum_{k=1}^{n} \frac{\partial a_i}{\partial q_k} \dot{q}_k = 0, \qquad (4.57) $$

which, through the Hamilton equations (4.54), becomes:

$$ \frac{\partial a_i}{\partial t} - \sum_{k=1}^{n} \frac{\partial a_i}{\partial p_k} \frac{\partial \mathcal{H}}{\partial q_k} + \sum_{k=1}^{n} \frac{\partial a_i}{\partial q_k} \frac{\partial \mathcal{H}}{\partial p_k} = 0. \qquad (4.58) $$

This is the equation that must be satisfied by each first integral, since it does not contain any arbitrary constant. Conversely, if the function a_i is solution (4.58), it also satisfies the (4.57) and therefore is a first integral.

In conclusion, the determination of a system of $2n$ first integrals (4.54) coincides with the determination of $2n$ particular solutions (4.58). The problem is therefore reduced to the consideration of this partial differential equation.

---

[13] Recall that the solution exists because the $2n$ functions q_k and p_k are independent; hence the determinant of their Jacobian matrix is different from zero.

4.6 Constants of Motion, Poisson and Lagrange Brackets

The importance (4.58) is such that we want to rewrite it in a more compact and convenient way through an operator named Poisson[14] brackets. Given two functions, $f(q, p, t)$ and $g(q, p, t)$, of n coordinates, relative moments, and time, the Poisson brackets operator is:

$$(f, g) = \sum_{k=1}^{n} \frac{\partial f}{\partial p_k} \frac{\partial g}{\partial q_k} - \sum_{k=1}^{n} \frac{\partial f}{\partial q_k} \frac{\partial g}{\partial p_k}. \tag{4.59}$$

Another very useful operator, named Lagrange brackets, is given by the expression:

$$[f, g] = \sum_{k=1}^{n} \frac{\partial p_k}{\partial f} \frac{\partial q_k}{\partial g} - \sum_{k=1}^{n} \frac{\partial q_k}{\partial f} \frac{\partial p_k}{\partial g}. \tag{4.60}$$

For more details on these two operators and theirs properties, see Appendix M or consult the specialized literature,[15] e.g. [4, 5].

With the notation now introduced, (4.58) rewrites as:

$$\frac{\partial a_i}{\partial t} + (\mathcal{H}, a_i) = 0. \tag{4.61}$$

Therefore, the necessary and sufficient condition for a function $a_i(q, p, t)$ to be a first integral for the Hamiltonian $\mathcal{H}$ is that it satisfies (4.61).

Consider the Poisson bracket between two first integrals, (a_i, a_j). It is easy to verify that it too satisfies the condition (4.61). Indeed, through the first of the (M.8) we have:

$$\frac{\partial}{\partial t}(a_i, a_j) = (\frac{\partial a_i}{\partial t}, a_j) + (a_i, \frac{\partial a_j}{\partial t}), \tag{4.62}$$

and, through the Jacobi identity (M.3):

$$\begin{aligned}(\mathcal{H}, (a_i, a_j)) &= ((a_j, \mathcal{H}), a_i) + ((\mathcal{H}, a_i), a_j) = \\ &= -((\mathcal{H}, a_j), a_i) + ((\mathcal{H}, a_i), a_j),\end{aligned} \tag{4.63}$$

---

[14] Siméon Denis Poisson (1781–1840): French mathematician and physicist, protected by Lagrange and especially by Laplace, who considered him almost like a son. He investigated electrostatics, magnetism, dynamics of solids, and analytical mechanics. His name is connected to the generalization of the famous Laplace equation (see 5.21) and to the homonym law of (discrete) probability distribution, $P(x)$, describing the occurrence (in a given time interval) of x random events in a process with a mean λ: $P(x) = \dfrac{e^{-\lambda}\lambda^x}{x!}$. We recall that the standard deviation, which expresses the deviation of the measures from the mean value, is equal to $\sqrt{\lambda}$.

[15] It is worth noting that rules other than the one adopted here for the signs are present in the literature; a fact that may generate some minor difficulties and some doubts when comparing different texts.

where we have used the second of the (M.2). Summing up, we obtain:

$$\frac{\partial}{\partial t}(a_i, a_j) + (\mathcal{H}, (a_i, a_j)) =$$

$$= (\frac{\partial a_i}{\partial t}, a_j) + ((\mathcal{H}, a_i), a_j) - (\frac{\partial a_j}{\partial t}, a_i) - ((\mathcal{H}, a_j), a_i) =$$

$$= (\frac{\partial a_i}{\partial t} + (\mathcal{H}, a_i), a_j) - (\frac{\partial a_j}{\partial t} + (\mathcal{H}, a_j), a_i) = 0, \qquad (4.64)$$

since both a_i and a_j satisfy (4.61). This proves the so-called Poisson theorem: every Poisson bracket between two first integrals is in turn a first integral.

It is useful to clarify right away the range of applicability of this theorem. Obviously it cannot always give us a new first integral, since the total number is limited to $2n$. Therefore, the result can sometimes be of no relevance, being a constant or just a function of the integrals generating it. In some cases, however, the Poisson theorem actually provides genuinely new first integrals. As an example, consider two of the three area integrals (1.20):

$$a_1 = \sum_{k=1}^{n} m_k (y_k \dot{z}_k - z_k \dot{y}_k),$$

$$a_2 = \sum_{k=1}^{n} m_k (z_k \dot{x}_k - x_k \dot{z}_k). \qquad (4.65a)$$

By applying the definition of Poisson bracket, we have:

$$(a_1, a_2) = \sum_{k=1}^{n} m_k (x_k \dot{y}_k - y_k \dot{x}_k), \qquad (4.65b)$$

which is precisely the third area integral. With that the cycle closes; the application of the Poisson theorem to any pair of the three previous integrals (4.65) does not provide any new one.

We now state the following theorem: the Lagrange bracket between two constants of motion a_i and a_j:

$$[a_i, a_j] = \sum_{k=1}^{n} \frac{\partial p_k}{\partial a_i} \frac{\partial q_k}{\partial a_j} - \sum_{k=1}^{n} \frac{\partial q_k}{\partial a_i} \frac{\partial p_k}{\partial a_j}, \qquad (4.66)$$

is independent of time and function of the constants only. We prove it in Appendix M, showing that the theorem has a general validity.

Concerning the integrals of motion, unlike the Poisson brackets, the Lagrange brackets do not provide any new results under any condition. In fact, in order to form the function $[a_i, a_j]$ it is required the knowledge of all the (4.55), that is, the knowledge of the solution to the problem. However, the two operators are so strictly related to be considered one the inverse of the other. This relation between Poisson and Lagrange brackets can be interpreted in terms of matrices, as we show in Appendix M.

4.7 Lagrange and Poisson Brackets for the Elliptical Orbit

As a useful example, we now calculate the Lagrange brackets for the elements of an elliptical orbit in the framework of the two-body problem, and from them the Poisson brackets (see also [2, 6]). Meanwhile we note that the two-body problem, as placed at Sect. 2.1, has 3 degrees of freedom (as many as they are the independent coordinates of one of the two points only, since the other point is considered to be fixed). Therefore, the solution requires $2 \times 3 = 6$ integrals of motion. They can be, for example, the six orbital elements introduced in Sect. 2.7:

- the semi-major axis a,
- the eccentricity e,
- the mean anomaly $\delta_\circ = -nt_\circ$ at the epoch $t = 0$ (used, for convenience, in place of the time of passage at the periapsis $t_\circ$; n is the mean motion),
- the longitude of the ascending node Ω,
- the distance of the periapsis from the ascending node ω, and
- the inclination of the orbital plane i.

The first three elements are dimensional in that they characterize the orbit on its plane. In the following we will indicate them generically with the symbol D_j ($j = 1, \ldots, 3$). The other three elements orient the orbit in space. We will represent them by the symbol O_j ($j = 1, \ldots, 3$). We now recall that the Lagrange brackets are anticommutative operators (see Appendix M); therefore, we will have to calculate only a total[16] of 15 (and not 6×6), three of which are of the type:

$$[O_i, O_j] = \sum_{k=1}^{3} \left(\frac{\partial p_k}{\partial O_i} \frac{\partial q_k}{\partial O_j} - \frac{\partial p_k}{\partial O_j} \frac{\partial q_k}{\partial O_i} \right) \qquad (i \neq j = 1, \ldots, 3), \qquad (4.67)$$

other three of the type:

$$[D_i, D_j] = \sum_{k=1}^{3} \left(\frac{\partial p_k}{\partial D_i} \frac{\partial q_k}{\partial D_j} - \frac{\partial p_k}{\partial D_j} \frac{\partial q_k}{\partial D_i} \right) \qquad (i \neq j = 1, \ldots, 3), \qquad (4.68)$$

and finally nine of the mixed type:

---

[16] To compute the number of independent brackets it is enough to evaluate the number of possible combinations in which $n = 6$ elements are paired in groups of $m = 2$, that is, to calculate the binomial coefficient $\dfrac{n!}{m!(n-m)!}$.

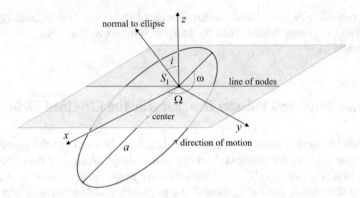

Fig. 4.1 Graphycal references to calculate the Lagrange and Poisson brackets for an elliptical orbit

$$[O_i, D_j] = \sum_{k=1}^{3} \left(\frac{\partial p_k}{\partial O_i} \frac{\partial q_k}{\partial D_j} - \frac{\partial p_k}{\partial D_j} \frac{\partial q_k}{\partial O_i} \right) \qquad (i, j = 1, \ldots, 3). \qquad (4.69)$$

As generalized coordinates we will use directly the Cartesian coordinates, x, y, and z, in a generic orthogonal reference system $S_1 [x, y, z]$, centered on the focus of the elliptical orbit where the second massive body is located. We therefore put:

$$q_1 = x,$$
$$q_2 = y, \qquad\qquad (4.70)$$
$$q_3 = z.$$

The Lagrangian function is:

$$\mathcal{L} = T + \mathcal{U} = \frac{1}{2}(\dot{x}^2 + \dot{y}^2 + \dot{z}^2) + G\frac{m_1 + m_2}{\sqrt{x^2 + y^2 + z^2}}, \qquad (4.71)$$

from where, through (4.37), we obtain the conjugated moments:

$$p_1 = \dot{x},$$
$$p_2 = \dot{y}, \qquad\qquad (4.72)$$
$$p_3 = \dot{z}.$$

We now introduce a second orthogonal reference system, $S_2 [x', y', z']$, concentric to S_1 and such that the plane $x'y'$ contains the orbit and the x'-axis coincides with the focal axis. This is the orbital reference system. For each point $P(x', y')$ of the orbit, the following transformation equations hold:

$$q_1 = x = a_{11} x' + a_{21} y',$$
$$q_2 = y = a_{12} x' + a_{22} y', \qquad (4.73)$$
$$q_3 = z = a_{13} x' + a_{23} y',$$

which, after differentiation with respect to time, turn into:

$$p_1 = \dot{x} = (a_{11}\,\dot{x}' + a_{21}\,\dot{y}'),$$
$$p_2 = \dot{y} = (a_{12}\,\dot{x}' + a_{22}\,\dot{y}'), \tag{4.74}$$
$$p_3 = \dot{z} = (a_{13}\,\dot{x}' + a_{23}\,\dot{y}').$$

The direction cosines, a_{ij}, are given by the relations (B.10a) of Appendix B, having placed $\delta = \widehat{x''x'} = \omega$:

$$a_{11} = \cos\Omega\cos\omega - \sin\Omega\sin\omega\cos i,$$
$$a_{12} = \sin\Omega\cos\omega + \cos\Omega\sin\omega\cos i,$$
$$a_{13} = \sin\omega\sin i,$$

$$a_{21} = -\cos\Omega\sin\omega - \sin\Omega\cos\omega\cos i,$$
$$a_{22} = -\sin\Omega\sin\omega + \cos\Omega\cos\omega\cos i, \tag{4.75}$$
$$a_{23} = \cos\omega\sin i,$$

$$a_{31} = \sin\Omega\sin i,$$
$$a_{32} = -\cos\Omega\sin i,$$
$$a_{33} = \cos i.$$

They were also given in Chap. 2 for the two-body problem (equations (2.123)) and are repeated here for convenience. Through the (4.73) and (4.74), the partial derivatives with respect to the orbital elements appearing in the bracket (4.67), (4.68), and (4.69), may be written as:

$$\frac{\partial q_k}{\partial O_i} = x'\frac{\partial a_{1k}}{\partial O_i} + y'\frac{\partial a_{2k}}{\partial O_i}, \tag{4.76a}$$

$$\frac{\partial p_k}{\partial O_i} = \dot{x}'\frac{\partial a_{1k}}{\partial O_i} + \dot{y}'\frac{\partial a_{2k}}{\partial O_i}, \tag{4.76b}$$

$$\frac{\partial q_k}{\partial D_i} = a_{1k}\frac{\partial x'}{\partial D_i} + a_{2k}\frac{\partial y'}{\partial D_i}, \tag{4.76c}$$

$$\frac{\partial p_k}{\partial D_i} = a_{1k}\frac{\partial \dot{x}'}{\partial D_i} + a_{2k}\frac{\partial \dot{y}'}{\partial D_i}, \tag{4.76d}$$

where the index k varies from 1 to 3. In fact, x' and y' are independent from Ω, ω, and i, while the direction cosines, a_{ij}, are independent of a, e, and δ_o.

Let us now calculated the bracket of the type (4.67) between the orientation elements O_i ($i = 1, \ldots, 3$). It is:

$$[O_i, O_j] = \sum_{k=1}^{3} \left[\left(\dot{x}' \frac{\partial a_{1k}}{\partial O_i} + \dot{y}' \frac{\partial a_{2k}}{\partial O_i} \right) \left(x' \frac{\partial a_{1k}}{\partial O_j} + y' \frac{\partial a_{2k}}{\partial O_j} \right) - \right.$$

$$\left. - \left(\dot{x}' \frac{\partial a_{1k}}{\partial O_j} + \dot{y}' \frac{\partial a_{2k}}{\partial O_j} \right) \left(x' \frac{\partial a_{1k}}{\partial O_i} + y' \frac{\partial a_{2k}}{\partial O_i} \right) \right] =$$

$$= (x' \dot{y}' - \dot{x}' y') \sum_{k=1}^{3} \left(\frac{\partial a_{2k}}{\partial O_i} \frac{\partial a_{1k}}{\partial O_j} - \frac{\partial a_{2k}}{\partial O_j} \frac{\partial a_{1k}}{\partial O_i} \right). \tag{4.77}$$

The quantity: $h = (x' \dot{y}' - \dot{x}' y')$, is the angular momentum per unit mass or, through the (2.65) and (2.83):

$$x' \dot{y}' - \dot{x}' y' = \sqrt{G (m_1 + m_2) a (1 - e^2)} = n a^2 \sqrt{1 - e^2}. \tag{4.78}$$

Replacing this latter in (4.77) and expanding the summation, we obtain the three Lagrange brackets for the orientation elements. The partial derivatives of the direction cosines relative to the elements O_i can be calculated directly from the relations (4.75). We have three groups of 9 derivatives each:

$$\begin{array}{lll}
\frac{\partial a_{11}}{\partial \Omega} = -a_{12}, & \frac{\partial a_{12}}{\partial \Omega} = a_{11}, & \frac{\partial a_{13}}{\partial \Omega} = 0, \\[2mm]
\frac{\partial a_{21}}{\partial \Omega} = -a_{22}, & \frac{\partial a_{22}}{\partial \Omega} = a_{21}, & \frac{\partial a_{23}}{\partial \Omega} = 0, \\[2mm]
\frac{\partial a_{31}}{\partial \Omega} = -a_{32}, & \frac{\partial a_{32}}{\partial \Omega} = a_{31}, & \frac{\partial a_{33}}{\partial \Omega} = 0,
\end{array} \tag{4.79a}$$

$$\begin{array}{lll}
\frac{\partial a_{11}}{\partial \omega} = a_{21}, & \frac{\partial a_{12}}{\partial \omega} = a_{22}, & \frac{\partial a_{13}}{\partial \omega} = a_{23}, \\[2mm]
\frac{\partial a_{21}}{\partial \omega} = -a_{11}, & \frac{\partial a_{22}}{\partial \omega} = -a_{12}, & \frac{\partial a_{23}}{\partial \omega} = -a_{13}, \\[2mm]
\frac{\partial a_{31}}{\partial \omega} = 0, & \frac{\partial a_{32}}{\partial \omega} = 0, & \frac{\partial a_{33}}{\partial \omega} = 0,
\end{array} \tag{4.79b}$$

$$\begin{array}{lll}
\frac{\partial a_{11}}{\partial i} = a_{31} \sin \omega, & \frac{\partial a_{12}}{\partial i} = a_{32} \sin \omega, & \frac{\partial a_{13}}{\partial i} = a_{33} \sin \omega, \\[2mm]
\frac{\partial a_{21}}{\partial i} = a_{31} \cos \omega, & \frac{\partial a_{22}}{\partial i} = a_{32} \cos \omega, & \frac{\partial a_{23}}{\partial i} = a_{33} \cos \omega, \\[2mm]
\frac{\partial a_{31}}{\partial i} = \sin \Omega \cos i, & \frac{\partial a_{32}}{\partial i} = - \cos \Omega \cos i, & \frac{\partial a_{33}}{\partial i} = - \sin i.
\end{array} \tag{4.79c}$$

At this point, all that remains is to complete the somewhat laborious but conceptually simple calculation. For example, if we set $O_i = \Omega$ and $O_j = i$, using the (4.77), (4.79a), and (4.79c), we readily have:

$$[\Omega, i] = n a^2 \sqrt{1 - e^2} \left[(a_{12} a_{31} - a_{11} a_{32}) \cos \omega + (a_{21} a_{32} - a_{22} a_{31}) \sin \omega \right] =$$

$$= n a^2 \sqrt{1 - e^2} \left[a_{23} \cos \omega + a_{13} \sin \omega \right] = n a^2 \sqrt{1 - e^2} \sin i. \tag{4.80}$$

Similarly, it is immediate to verify that:

$$[\omega, \Omega] = [i, \omega] = 0. \tag{4.81}$$

We now calculate the brackets of the type (4.68). In consequence of (4.76c) and (4.76d), it is:

$$
\begin{aligned}
[D_i, D_j] &= \sum_{k=1}^{3} \left[\left(a_{1k} \frac{\partial \dot{x}'}{\partial D_i} + a_{2k} \frac{\partial \dot{y}'}{\partial D_i} \right) \left(a_{1k} \frac{\partial x'}{\partial D_j} + a_{2k} \frac{\partial y'}{\partial D_j} \right) - \right. \\
&\quad \left. - \left(a_{1k} \frac{\partial \dot{x}'}{\partial D_j} + a_{2k} \frac{\partial \dot{y}'}{\partial D_j} \right) \left(a_{1k} \frac{\partial x'}{\partial D_i} + a_{2k} \frac{\partial y'}{\partial D_i} \right) \right] = \\
&= (a_{11}^2 + a_{12}^2 + a_{13}^2) \left(\frac{\partial \dot{x}'}{\partial D_i} \frac{\partial x'}{\partial D_j} - \frac{\partial \dot{x}'}{\partial D_j} \frac{\partial x'}{\partial D_i} \right) + \\
&\quad + (a_{21}^2 + a_{22}^2 + a_{23}^2) \left(\frac{\partial \dot{y}'}{\partial D_i} \frac{\partial y'}{\partial D_j} - \frac{\partial \dot{y}'}{\partial D_j} \frac{\partial y'}{\partial D_i} \right) + \\
&\quad + (a_{11}a_{21} + a_{12}a_{22} + a_{13}a_{23}) \times \\
&\quad \times \left(\frac{\partial \dot{y}'}{\partial D_i} \frac{\partial x'}{\partial D_j} - \frac{\partial \dot{x}'}{\partial D_j} \frac{\partial y'}{\partial D_i} + \frac{\partial \dot{x}'}{\partial D_i} \frac{\partial y'}{\partial D_j} - \frac{\partial \dot{y}'}{\partial D_j} \frac{\partial x'}{\partial D_i} \right),
\end{aligned} \tag{4.82}
$$

or, through the relations (B.12) of Appendix B due to the orthogonality of the reference system:

$$[D_i, D_j] = \left(\frac{\partial \dot{x}'}{\partial D_i} \frac{\partial x'}{\partial D_j} - \frac{\partial \dot{x}'}{\partial D_j} \frac{\partial x'}{\partial D_i} + \frac{\partial \dot{y}'}{\partial D_i} \frac{\partial y'}{\partial D_j} - \frac{\partial \dot{y}'}{\partial D_j} \frac{\partial y'}{\partial D_i} \right). \tag{4.83}$$

To make this expression explicit, recall that, according to the (2.51a):

$$
\begin{aligned}
x' &= r \cos \theta = a (\cos E - e), \\
y' &= r \sin \theta = a \sqrt{1 - e^2} \sin E,
\end{aligned} \tag{4.84}
$$

where θ and E are the true and the eccentric anomalies. Moreover, since from (2.84) we have:

$$\frac{dE}{dt} = \frac{n}{1 - e \cos E}, \tag{4.85}$$

it is also:

$$
\begin{aligned}
\dot{x}' &= -\sqrt{\frac{G(m_1 + m_2)}{a}} \frac{\sin E}{1 - e \cos E}, \\
\dot{y}' &= \sqrt{\frac{G(m_1 + m_2)}{a}} \frac{\sqrt{1 - e^2} \cos E}{1 - e \cos E}.
\end{aligned} \tag{4.86}
$$

Given the above result, it is immediate to calculate the partial derivatives which appear in (4.83). Moreover, since the Lagrange brackets do not depend explicitly on time, we can evaluate them at an arbitrary epoch. We choose the time of passage at the periapsis $t = t_o$, since in this way: $E = 0$, and $r = a(1 - e)$. Concerning the derivative with respect to δ_o, we must remember that:

$$M = n\left(t - t_o\right) = nt + \delta_o = E - e \sin E, \tag{4.87}$$

so:

$$\frac{\partial E}{\partial \delta_o} = \frac{1}{1 - e \cos E}. \tag{4.88}$$

We have:

$$
\begin{aligned}
&\frac{\partial x'}{\partial a} = 1 - e, && \frac{\partial x'}{\partial e} = -a, && \frac{\partial x'}{\partial \delta_o} = 0, \\[2mm]
&\frac{\partial y'}{\partial a} = 0, && \frac{\partial y'}{\partial e} = 0, && \frac{\partial y'}{\partial \delta_o} = a\sqrt{\frac{1+e}{1-e}}, \\[2mm]
&\frac{\partial \dot{x}'}{\partial a} = 0, && \frac{\partial \dot{x}'}{\partial e} = 0, && \frac{\partial \dot{x}'}{\partial \delta_o} = -\frac{na}{(1-e)^2}, \\[2mm]
&\frac{\partial \dot{y}'}{\partial a} = -\frac{n}{2}\sqrt{\frac{1+e}{1-e}}, && \frac{\partial \dot{y}'}{\partial e} = \frac{na}{(1-e)\sqrt{1-e^2}}, && \frac{\partial \dot{y}'}{\partial \delta_o} = 0,
\end{aligned}
\tag{4.89}
$$

and thus, replacing in (4.83):

$$[a, e] = [e, \delta_o] = 0, \qquad [\delta_o, a] = -\frac{na}{2}. \tag{4.90}$$

Let us now calculate the 9 brackets of the type (4.69). Through the relations (4.76a), (4.76b), (4.76c), and (4.76d), it is:

$$
\begin{aligned}
[O_i, D_j] &= \sum_{k=1}^{3}\left[\left(\dot{x}'\frac{\partial a_{1k}}{\partial O_i} + \dot{y}'\frac{\partial a_{2k}}{\partial O_i}\right)\left(a_{1k}\frac{\partial x'}{\partial D_j} + a_{2k}\frac{\partial y'}{\partial D_j}\right) + \right. \\[2mm]
&\quad \left. -\left(a_{1k}\frac{\partial \dot{x}'}{\partial D_j} + a_{2k}\frac{\partial \dot{y}'}{\partial D_j}\right)\left(x'\frac{\partial a_{1k}}{\partial O_i} + y'\frac{\partial a_{2k}}{\partial O_i}\right)\right] = \\[2mm]
&= \left(a_{11}\frac{\partial a_{11}}{\partial O_i} + a_{12}\frac{\partial a_{12}}{\partial O_i} + a_{13}\frac{\partial a_{13}}{\partial O_i}\right)\left(\dot{x}'\frac{\partial x'}{\partial D_j} - x'\frac{\partial \dot{x}'}{\partial D_j}\right) + \\[2mm]
&\quad + \left(a_{21}\frac{\partial a_{21}}{\partial O_i} + a_{22}\frac{\partial a_{22}}{\partial O_i} + a_{23}\frac{\partial a_{23}}{\partial O_i}\right)\left(\dot{y}'\frac{\partial y'}{\partial D_j} - y'\frac{\partial \dot{y}'}{\partial D_j}\right) + \\[2mm]
&\quad + \left(a_{11}\frac{\partial a_{21}}{\partial O_i} + a_{12}\frac{\partial a_{22}}{\partial O_i} + a_{13}\frac{\partial a_{23}}{\partial O_i}\right)\left(\dot{y}'\frac{\partial x'}{\partial D_j} - y'\frac{\partial \dot{x}'}{\partial D_j}\right) + \\[2mm]
&\quad + \left(a_{21}\frac{\partial a_{11}}{\partial O_i} + a_{22}\frac{\partial a_{12}}{\partial O_i} + a_{23}\frac{\partial a_{13}}{\partial O_i}\right)\left(\dot{x}'\frac{\partial y'}{\partial D_j} - x'\frac{\partial \dot{y}'}{\partial D_j}\right).
\end{aligned}
\tag{4.91}
$$

From the (B.12) of Appendix B it follows that:

$$\sum_{k=1}^{3} a_{1k} \frac{\partial a_{1k}}{\partial O_i} = \sum_{k=1}^{3} a_{2k} \frac{\partial a_{2k}}{\partial O_i} = 0, \tag{4.92}$$

while:

$$\sum_{k=1}^{3} a_{1k} \frac{\partial a_{2k}}{\partial O_i} = -\sum_{k=1}^{3} a_{2k} \frac{\partial a_{1k}}{\partial O_i}. \tag{4.93}$$

Thus, the (4.91) reduces to:

$$[O_i, D_j] = -\sum_{k=1}^{3} a_{1k} \left(\frac{\partial a_{2k}}{\partial O_i} \right) \left(\dot{x}' \frac{\partial y'}{\partial D_j} + y' \frac{\partial \dot{x}'}{\partial D_j} - x' \frac{\partial \dot{y}'}{\partial D_j} - \dot{y}' \frac{\partial x'}{\partial D_j} \right) =$$

$$= -\sum_{k=1}^{3} a_{1k} \left(\frac{\partial a_{2k}}{\partial O_i} \right) \frac{\partial}{\partial D_j} (\dot{x}' y' - x' \dot{y}'). \tag{4.94}$$

The derivatives with respect to D_j appearing at the second member of (4.94) become:

$$\frac{\partial}{\partial a} (\dot{x}' y' - x' \dot{y}') = -\frac{n\,a}{2} \sqrt{1 - e^2},$$

$$\frac{\partial}{\partial e} (\dot{x}' y' - x' \dot{y}') = +\frac{n\,a^2\,e}{\sqrt{1 - e^2}}, \tag{4.95}$$

$$\frac{\partial}{\partial \delta_o} (\dot{x}' y' - x' \dot{y}') = 0.$$

The expressions containing the direction cosines are instead made explicit using the (4.75) and the (4.79a), (4.79b), (4.79c), together with the (B.9) of Appendix B:

$$\sum_{k=1}^{3} a_{2k} \frac{\partial a_{1k}}{\partial \Omega} = -a_{21}\,a_{12} + a_{22}\,a_{11} = a_{33} = \cos i,$$

$$\sum_{k=1}^{3} a_{2k} \frac{\partial a_{1k}}{\partial i} = \left(a_{21}\,a_{31} + a_{22}\,a_{32} + a_{23}\,a_{33} \right) \sin \omega = 0, \tag{4.96}$$

$$\sum_{k=1}^{3} a_{2k} \frac{\partial a_{1k}}{\partial \omega} = a_{21}\,a_{21} + a_{22}\,a_{22} + a_{23}\,a_{23} = 1.$$

Combining the previous results in (4.94) it is:

$$[\Omega, a] = -\frac{n\,a}{2} \sqrt{1 - e^2} \, \cos i, \tag{4.97a}$$

$$[\Omega, e] = \frac{n\,a^2 e}{\sqrt{1 - e^2}} \, \cos i, \tag{4.97b}$$

Table 4.1 Non-null
Lagrange brackets

$$
\begin{aligned}
[a, \delta_o] &= -[\delta_o, a] = \frac{n\,a}{2} \\[2mm]
[a, \Omega] &= -[\Omega, a] = \frac{n\,a\sqrt{1-e^2}}{2}\cos i \\[2mm]
[a, \omega] &= -[\omega, a] = \frac{n\,a\sqrt{1-e^2}}{2} \\[2mm]
[e, \Omega] &= -[\Omega, e] = \frac{-\,n\,a^2 e}{\sqrt{1-e^2}}\cos i \\[2mm]
[e, \omega] &= -[\omega, e] = \frac{-\,n\,a^2 e}{\sqrt{1-e^2}} \\[2mm]
[\Omega, i] &= -[i, \Omega] = n\,a^2\sqrt{1-e^2}\,\sin i
\end{aligned}
$$

$$[\Omega, \delta_o] = 0, \tag{4.97c}$$

$$[i, a] = 0, \tag{4.97d}$$

$$[i, e] = 0, \tag{4.97e}$$

$$[i, \delta_o] = 0, \tag{4.97f}$$

$$[\omega, a] = -\frac{n\,a}{2}\sqrt{1-e^2}, \tag{4.97g}$$

$$[\omega, e] = \frac{n\,a^2 e}{\sqrt{1-e^2}}, \tag{4.97h}$$

$$[\omega, \delta_o] = 0. \tag{4.97i}$$

The Table 4.1 collects all the Lagrange brackets different from zero; all the other brackets that can build with the elements of orbit are null.

By applying the property that the matrices of Lagrange and Poisson brackets are one the reciprocal of the other, it is easy to calculate the Poisson brackets for the same elements of the elliptical orbit. First we calculate the determinant of the square matrix whose elements are the Lagrange bracket:

$$
\|A'\| = \begin{Vmatrix}
0 & 0 & 0 & [a, \Omega] & [a, \delta_o] & [a, \omega] \\
0 & 0 & 0 & [e, \Omega] & 0 & [e, \omega] \\
0 & 0 & 0 & [i, \Omega] & 0 & 0 \\
-[a, \Omega] & 0 & -[i, \Omega] & 0 & 0 & 0 \\
-[a, \delta_o] & 0 & 0 & 0 & 0 & 0 \\
-[a, \omega] & -[e, \omega] & 0 & 0 & 0 & 0
\end{Vmatrix}. \tag{4.98}
$$

It results:

$$|A'| = \big([a, \delta_o][e, \omega][i, \Omega]\big)^2, \tag{4.99}$$

Table 4.2 Non-null Poisson brackets

$$
\begin{aligned}
(a, \delta_o) &= -[a, \delta_o]^{-1} &&= -\frac{2}{n\,a} \\[2ex]
(e, \delta_o) &= [a, \omega]\,\{[a, \delta_o][e, \omega]\}^{-1} &&= -\frac{1-e^2}{n\,a^2 e} \\[2ex]
(e, \omega) &= -[e, \omega]^{-1} &&= \frac{\sqrt{1-e^2}}{n\,a^2 e} \\[2ex]
(\Omega, i) &= [\Omega, i]^{-1} &&= -\frac{1}{n\,a^2\sqrt{1-e^2}\,\sin i} \\[2ex]
(i, \omega) &= [e, \Omega]\{[e, \omega][\Omega, i]\}^{-1} &&= -\frac{\cot i}{n\,a^2\sqrt{1-e^2}}
\end{aligned}
$$

and therefore, by applying the relation (M.26), all the Poisson brackets are obtained. Those that are non-null, listed in Table 4.2, are five in total and, analogous to the Lagrange brackets, depend only on a, e, and i. Note, however, that the bracket:

$$(\delta_o, i) = [e, \Omega][a, \omega]\{[a, \delta_o][e, \omega][\Omega, i]\}^{-1} - [a, \Omega]\{[a, \delta_o][i, \Omega]\}^{-1} = 0, \qquad (4.100)$$

vanishes only after we have made explicit the Lagrange brackets which it is formed from.

Finally, we observe that, in consequence of the third relation (M.2), from Table 4.2 it is immediate to compute the Poisson brackets for a different set of elements of an elliptical orbit that are obtained as linear combinations of a, e, δ_o, Ω, ω, and i.

4.8 Canonical Transformations

A situation in which the solution of the Hamilton equations (4.45a) is particularly simple occurs when the Hamiltonian function is a constant of motion:

$$\frac{d\mathcal{H}}{dt} = \frac{\partial \mathcal{H}}{\partial t} = 0, \qquad (4.101)$$

and it is also independent of the generalized coordinates q_k ($k=1, \ldots, n$), which for this reason are said to be cyclic.[17] In this case, in fact, it results:

---

[17] In this context, we often refer to Noether's symmetries, through a theorem applied in many fields of physics. Demonstrated in 1915 and published three years later, it states that to every differentiable symmetry of the action of a physical system it corresponds a conservation law. In other words, to each continuous transformation which leaves the Lagrangian function unchanged (symmetric transformation), it corresponds a function that does not varies during the time evolution of the system. We have already made an indirect use of this property in the Chap. 1 when, in relation

$$\dot{p}_k = -\frac{\partial \mathcal{H}}{\partial q_k} = 0, \tag{4.102}$$

that is:

$$p_k = \beta_k = \text{const}, \tag{4.103}$$

for any value of the index k. Moreover, being: $\mathcal{H} = \mathcal{H}(p_1, p_2, \ldots, p_n) = \mathcal{H}(\beta_1, \beta_2, \ldots, \beta_n)$, it must be also:

$$\dot{q}_k = \frac{\partial \mathcal{H}}{\partial \beta_k} = \omega_k = \text{const}, \tag{4.104}$$

or:

$$q_k = \omega_k t + \alpha_k. \tag{4.105}$$

Equations (4.103) and (4.105) solve completely the problem of the motion represented by the Hamiltonian $\mathcal{H}$ indicated above.

At this point it is appropriate to ask whether such a condition has only a mere academic value. We might indeed guess that, in real situations, the conditions necessary for the generalized coordinates to be all cyclic are never realized. We shall see, instead, that the suspicion is not founded and that it is always possible, through a suitable transformation of coordinates, to find a reference system whose coordinates are all cyclic. Meanwhile, it is worth to remember that a mechanical system can be described by an infinite number of coordinate systems, defined by arbitrary transformation equations of the type:

$$q_k = q_k(Q_1, Q_2, \ldots, Q_n, P_1, P_2, \ldots, P_n, t)$$
$$(k = 1, \ldots, n), \tag{4.106}$$
$$p_k = p_k(Q_1, Q_2, \ldots, Q_n, P_1, P_2, \ldots, P_n, t)$$

which require only that the Jacobian determinant is different from zero so that the system (4.106) is invertible, that is, it can be solved with respect to Q_i and P_i:

$$Q_i = Q_i(q_1, q_2, \ldots, q_n, p_1, p_2, \ldots, p_n, t)$$
$$(i = 1, \ldots, n). \tag{4.107}$$
$$P_i = P_i(q_1, q_2, \ldots, q_n, p_1, p_2, \ldots, p_n, t)$$

Through the (4.107) it is possible to replace the original dynamical coordinates q_k and p_k with a new set of coordinates Q_i and P_i, that transform the Hamilton function $\mathcal{H}(p, q, t)$ into $\mathcal{H}'(P, Q, t)$. The latter can possibly have an analytical form that is more convenient to our purposes. We, for example, are not interested in transformations whatever, but only in the subset of those capable of keeping, in the new coordinates, the canonical form (4.45a) of the Hamilton equations:

---

to the N-body problem, we proved the existence of 10 first integrals through the system invariances to translations, rotations, and changes of the time origin.

$$\dot{Q}_i = \frac{\partial \mathcal{H}'}{\partial P_i}$$

$$(i = 1, \ldots, n). \qquad (4.108)$$

$$\dot{P}_i = -\frac{\partial \mathcal{H}'}{\partial Q_i}$$

Transformations meeting this requirement are called canonical or contact; all the others are useless even if capable of making the coordinates cyclic, since they do not allow us to exploit the advantages of canonicity.

4.8.1 Characteristic Function

At this point we need to remember that the Hamilton equations are a consequence of the variational principle (4.48). Using the most general form for the integrand function (4.53), it writes as:

$$\int_{t_1}^{t_2} \delta \left[\sum_{k=1}^{n} (p_k \dot{q}_k) - \mathcal{H}(p, q, t) + \frac{d}{dt} f(q, p, t) \right] dt = 0. \qquad (4.109)$$

In fact, as shown at the end of Sect. 4.4, the integrand of (4.48), i.e., the Lagrangian function, depends on an addictive constant which is the total time derivative of an arbitrary function $f(q, p, t)$. Therefore, if the transformation given by the system (4.106) or (4.107) is canonical, the new Hamiltonian $\mathcal{H}'(P, Q, t)$ must satisfy the general condition:

$$\int_{t_1}^{t_2} \delta \left[\sum_{k=1}^{n} (P_k \dot{Q}_k) - \mathcal{H}'(P, Q, t) + \frac{d}{dt} f'(Q, P, t) \right] dt = 0. \qquad (4.110)$$

This proves that the arguments of the integrals (4.109) and (4.110) must coincide. In other words, if the transformation is canonical, then:

$$\frac{d}{dt} f'(Q, P, t) - \frac{d}{dt} f(q, p, t) =$$

$$= \sum_{k=1}^{n} p_k \dot{q}_k - \sum_{k=1}^{n} P_k \dot{Q}_k - \mathcal{H}(p, q, t) + \mathcal{H}'(P, Q, t). \qquad (4.111)$$

Let us place:

$$F = f'(Q, P, t) - f(q, p, t), \qquad (4.112)$$

from where, through (4.111):

$$\frac{d}{dt} F = \sum_{k=1}^{n} p_k \dot{q}_k - \sum_{k=1}^{n} P_k \dot{Q}_k - \mathcal{H}(p, q, t) + \mathcal{H}'(P, Q, t). \qquad (4.113)$$

Contrary to what the position (4.112) seems to indicate, the function F cannot depend on all the $4n$ dynamic coordinates q_k, p_k, Q_k, P_k $(k = 1, \ldots, n)$; the transformation equations (4.107) prove that only $2n$ of these are independent of each other. Also, the function F must keep track of both the old and the new coordinates or, which is the same, of the two functions f and f'. As suggested by the form of the second member of (4.113), we assume for the moment that: $F = F(q, Q, t)$; then $\partial^2 F/\partial q_i \partial Q_k \neq 0$. Expanding the total derivative, we have:

$$\frac{d}{dt} F(q, Q, t) = \sum_{k=1}^{n} \frac{\partial F}{\partial q_k} \dot{q}_k + \sum_{k=1}^{n} \frac{\partial F}{\partial Q_k} \dot{Q}_k + \frac{\partial F}{\partial t}, \qquad (4.114)$$

and, by comparison with (4.113):

$$p_k = \frac{\partial}{\partial q_k} F(q, Q, t) \qquad (4.115a)$$

$$(k = 1, \ldots, n),$$

$$P_k = -\frac{\partial}{\partial Q_k} F(q, Q, t) \qquad (4.115b)$$

$$\mathcal{H}'(P, Q, t) - \mathcal{H}(p, q, t) = \frac{\partial F(q, Q, t)}{\partial t}. \qquad (4.115c)$$

These show that the function $F(q, Q, t)$ characterizes completely the canonical transformation. For this reason it is called characteristic function of the transformation.

Let us try to elaborate the meaning of the result just obtained. Given a Hamiltonian function $\mathcal{H}(p, q, t)$, we build a function F depending on the n generalized coordinates q_k, the new coordinates Q_k (whose form for the moment is not known), and possibly the time. We can impose that the function $F(q, Q, t)$ is the generator of a canonical transformation by asking that its partial derivatives with respect to q_k equal the corresponding conjugate moments p_k, as required by (4.115a). This is equivalent to a constraint on the choice of the new coordinates Q_k. If this condition is verified, the new moments P_k conjugated to the coordinates Q_k are given by (4.115b), and the new form of the Hamiltonian is linked to the old one by (4.115c). In this way it is certain that the transformation is canonical and, therefore, that the Hamilton equations in the form of (4.108) hold.

As a further clarification, we consider the following example by introducing the function:

$$F(q, Q) = \sum_{k=1}^{n} q_k Q_k \qquad (k = 1, \ldots, n), \qquad (4.116)$$

which does not explicitly depend on time. For it to be characteristic of a canonical transformation, the n conditions set by replacing the function in (4.115a) must be verified:

$$p_k = \frac{\partial F}{\partial q_k} = Q_k \qquad (k = 1, \ldots, n). \qquad (4.117)$$

Moreover, the new moments conjugated to Q_k will be given by the (4.115b), that is:

$$P_k = -\frac{\partial F}{\partial Q_k} = -q_k \qquad (k = 1, \ldots, n). \qquad (4.118)$$

Finally, since F does not depend on time, the new Hamiltonian will be identical to the old one:

$$\mathcal{H}'(P, Q) = \mathcal{H}(p, q). \qquad (4.119)$$

This canonical transformation is particular as it procures just to the inversion of the role of the generalized coordinates with that of the moments conjugated to them. A fact that proves once again the total independence of the dynamical coordinates from the traditional Newtonian interpretation of terms of space coordinates and moments.

4.8.2 The Various Forms of the Characteristic Function

The transformation by means of the function (4.116) is useful to further clarify the meaning of the statement (Sect. 4.8) that, besides time, the characteristic function of a canonical transformation depends on $2n$ generalized coordinates q_k and Q_k. Since the generalized coordinates are interchangeable with their conjugate moments without affecting the Hamiltonian function and the form of the Hamilton equations, it is legitimate to expect that the characteristic function of a canonical transformation can also occur in the forms:

$$F_2 = F_2(q, P, t), \qquad (4.120a)$$
$$F_3 = F_3(p, Q, t), \qquad (4.120b)$$
$$F_4 = F_4(p, P, t). \qquad (4.120c)$$

In fact, noted that:

$$\sum_{k=1}^{n} p_k \dot{q}_k = \frac{d}{dt}\left(\sum_{k=1}^{n} p_k q_k\right) - \sum_{k=1}^{n} q_k \dot{p}_k, \qquad (4.121a)$$

$$\sum_{k=1}^{n} P_k \dot{Q}_k = \frac{d}{dt}\left(\sum_{k=1}^{n} P_k Q_k\right) - \sum_{k=1}^{n} Q_k \dot{P}_k, \qquad (4.121b)$$

by replacing the expression $\sum_{k=1}^{n} P_k \dot{Q}_k$ at the right-hand side of the (4.113) with that given by the (4.121b), the characteristic functions becomes (4.120a):

$$\frac{d}{dt}F_2(q, P, t) = \frac{d}{dt}F(q, Q, t) + \frac{d}{dt}\left(\sum_{k=1}^{n} P_k Q_k\right) =$$

$$= \sum_{k=1}^{n} p_k \dot{q}_k + \sum_{k=1}^{n} Q_k \dot{P}_k - \mathcal{H} + \mathcal{H}', \quad (4.122)$$

$$p_k = \frac{\partial}{\partial q_k}F_2(q, P, t)$$
$$Q_k = \frac{\partial}{\partial P_k}F_2(q, P, t)$$
$$(4.123a)$$

$$\mathcal{H}'(P, Q, t) - \mathcal{H}(p, q, t) = \frac{\partial F_2(q, P, t)}{\partial t}. \quad (4.123b)$$

Note that the difference between $F_2(q, P, t)$ and $F(q, Q, t)$ is given by a function of the type (4.116), which is characteristic of the inversion between Q_k and P_k.

Proceeding in a quite similar way, we rewrite the expression of the total derivative of the characteristic function (4.113) via (4.121a):

$$\frac{d}{dt}F_3(p, Q, t) = \frac{d}{dt}F(q, Q, t) - \frac{d}{dt}\left(\sum_{k=1}^{n} p_k q_k\right) =$$

$$= -\sum_{k=1}^{n} q_k \dot{p}_k - \sum_{k=1}^{n} P_k \dot{Q}_k - \mathcal{H} + \mathcal{H}', \quad (4.124)$$

and thus:

$$q_k = -\frac{\partial}{\partial p_k}F_3(p, Q, t)$$
$$P_k = -\frac{\partial}{\partial Q_k}F_3(p, Q, t)$$
$$(k = 1, \ldots, n), \quad (4.125a)$$

$$\mathcal{H}'(P, Q, t) - \mathcal{H}(p, q, t) = \frac{\partial F_3(p, Q, t)}{\partial t}. \quad (4.125b)$$

Finally, through both the relations (4.121a) and (4.121b):

$$\frac{d}{dt}F_4(p, P, t) = \frac{d}{dt}F(q, Q, t) - \frac{d}{dt}\left(\sum_{k=1}^{n} p_k q_k\right) + \frac{d}{dt}\left(\sum_{k=1}^{n} P_k Q_k\right) =$$

$$= -\sum_{k=1}^{n} q_k \dot{p}_k + \sum_{k=1}^{n} Q_k \dot{P}_k - \mathcal{H} + \mathcal{H}', \quad (4.126)$$

Table 4.3 Characteristic functions

$F_1(q, Q, t)$	$F_2(q, P, t)$	$F_3(p, Q, t)$	$F_4(p, P, t)$
$p_k = \dfrac{\partial F_1}{\partial q_k}$	$p_k = \dfrac{\partial F_2}{\partial q_k}$	$q_k = -\dfrac{\partial F_3}{\partial p_k}$	$q_k = -\dfrac{\partial F_4}{\partial p_k}$
$P_k = -\dfrac{\partial F_1}{\partial Q_k}$	$Q_k = \dfrac{\partial F_2}{\partial P_k}$	$P_k = -\dfrac{\partial F_3}{\partial Q_k}$	$Q_k = \dfrac{\partial F_4}{\partial P_k}$

$$\mathcal{H}'(P, Q, t) - \mathcal{H}(p, q, t) = \frac{\partial F_1}{\partial t} = \frac{\partial F_2}{\partial t} = \frac{\partial F_3}{\partial t} = \frac{\partial F_4}{\partial t}$$

or:

$$
\begin{aligned}
q_k &= -\frac{\partial}{\partial p_k} F_4(p, P, t) \\
Q_k &= \frac{\partial}{\partial P_k} F_4(p, P, t)
\end{aligned}
\qquad (k = 1, \ldots, n), \qquad (4.127a)
$$

$$\mathcal{H}'(P, Q, t) - \mathcal{H}(p, q, t) = \frac{\partial F_4(p, P, t)}{\partial t}. \qquad (4.127b)$$

The different forms of the characteristic function and of the associated condition equations are summarized in Table 4.3.

4.8.3 Conditions for Canonicity Using the Poisson and Lagrange Brackets

Let $F(q, Q, t)$ be the characteristic function of a canonical transformation from the n dynamical variables q_k, p_k, to Q_k, P_k. We have:

$$\frac{d}{dt} Q_k = \sum_{i=1}^{n} \frac{\partial Q_k}{\partial q_i} \dot{q}_i + \sum_{i=1}^{n} \frac{\partial Q_k}{\partial p_i} \dot{p}_i + \frac{\partial Q_k}{\partial t} \qquad (k = 1, \ldots, n), \qquad (4.128)$$

or, through the Hamilton equations (4.45a) and the definition of the Poisson brackets (M.1):

$$\frac{d}{dt} Q_k = \sum_{i=1}^{n} \frac{\partial Q_k}{\partial q_i} \frac{\partial \mathcal{H}}{\partial p_i} - \sum_{i=1}^{n} \frac{\partial Q_k}{\partial p_i} \frac{\partial \mathcal{H}}{\partial q_i} + \frac{\partial Q_k}{\partial t} = (\mathcal{H}, Q_k) + \frac{\partial Q_k}{\partial t}, \qquad (4.129)$$

where $\mathcal{H} = \mathcal{H}(p, q, t)$ is the Hamiltonian function in the old dynamical coordinates. Similarly:

$$\frac{d}{dt}P_k = (\mathcal{H}, P_k) + \frac{\partial P_k}{\partial t} \qquad (k = 1, \ldots, n). \tag{4.130}$$

Given that $\mathcal{H}'(P, Q, t)$ is the Hamiltonian in the new dynamical coordinates, we have (Table 4.3):

$$\mathcal{H}(p, q, t) = \mathcal{H}'(P, Q, t) - \frac{\partial F}{\partial t}, \tag{4.131}$$

and thus, for any value of the index k:

$$\begin{aligned}
(\mathcal{H}, Q_k) &= (\mathcal{H}', Q_k) - (\frac{\partial F}{\partial t}, Q_k) \\
(\mathcal{H}, P_k) &= (\mathcal{H}', P_k) - (\frac{\partial F}{\partial t}, P_k)
\end{aligned} \qquad (k = 1, \ldots, n). \tag{4.132}$$

Through these expressions, the (4.129) and (4.130) become:

$$\begin{aligned}
\frac{dQ_k}{dt} &= (\mathcal{H}', Q_k) + \frac{\partial Q_k}{\partial t} - (\frac{\partial F}{\partial t}, Q_k) \\
\frac{dP_k}{dt} &= (\mathcal{H}', P_k) + \frac{\partial P_k}{\partial t} - (\frac{\partial F}{\partial t}, P_k)
\end{aligned} \qquad (k = 1, \ldots, n). \tag{4.133}$$

Since the considered transformation is canonical, the latter hold for the Hamilton equations (4.108), with which we can rewrite the first members of (4.133):

$$\begin{aligned}
\frac{\partial \mathcal{H}'}{\partial P_k} &= (\mathcal{H}', Q_k) + \frac{\partial Q_k}{\partial t} - (\frac{\partial F}{\partial t}, Q_k) \\
\frac{\partial \mathcal{H}'}{\partial Q_k} &= -(\mathcal{H}', P_k) - \frac{\partial P_k}{\partial t} + (\frac{\partial F}{\partial t}, P_k)
\end{aligned} \qquad (k = 1, \ldots, n). \tag{4.134}$$

Taking advantage of the property given by (M.10), we can develop the Poisson brackets appearing at the right-hand side of (4.134):

$$\begin{aligned}
(\mathcal{H}', Q_k) &= \sum_{i=1}^{n}(Q_i, Q_k)\frac{\partial \mathcal{H}'}{\partial Q_i} + \sum_{i=1}^{n}(P_i, Q_k)\frac{\partial \mathcal{H}'}{\partial P_i}, \\
(\mathcal{H}', P_k) &= \sum_{i=1}^{n}(Q_i, P_k)\frac{\partial \mathcal{H}'}{\partial Q_i} + \sum_{i=1}^{n}(P_i, P_k)\frac{\partial \mathcal{H}'}{\partial P_i},
\end{aligned} \tag{4.135}$$

and, making the characteristic function depend on the new dynamical coordinates:

$$\begin{aligned}
\left(\frac{\partial F}{\partial t}, Q_k\right) &= \sum_{i=1}^{n}(Q_i, Q_k)\frac{\partial^2 F}{\partial Q_i \partial t} + \sum_{i=1}^{n}(P_i, Q_k)\frac{\partial^2 F}{\partial P_i \partial t}, \\
\left(\frac{\partial F}{\partial t}, P_k\right) &= \sum_{i=1}^{n}(Q_i, P_k)\frac{\partial^2 F}{\partial Q_i \partial t} + \sum_{i=1}^{n}(P_i, P_k)\frac{\partial^2 F}{\partial P_i \partial t}.
\end{aligned} \tag{4.136}$$

The latter can be further simplified by recalling that the function F is an exact differential, and therefore:

$$\frac{\partial^2 F}{\partial Q_i \partial t} = \frac{\partial^2 F}{\partial t \partial Q_i} = -\frac{\partial P_i}{\partial t},$$
$$\frac{\partial^2 F}{\partial P_i \partial t} = \frac{\partial^2 F}{\partial t \partial P_i} = \frac{\partial Q_i}{\partial t},$$

(4.137)

where we made use of Table 4.3. In conclusion, the (4.136) become:

$$\left(\frac{\partial F}{\partial t}, Q_k\right) = -\sum_{i=1}^{n}(Q_i, Q_k)\frac{\partial P_i}{\partial t} + \sum_{i=1}^{n}(P_i, Q_k)\frac{\partial Q_i}{\partial t},$$
$$\left(\frac{\partial F}{\partial t}, P_k\right) = -\sum_{i=1}^{n}(Q_i, P_k)\frac{\partial P_i}{\partial t} + \sum_{i=1}^{n}(P_i, P_k)\frac{\partial Q_i}{\partial t},$$

(4.138)

which, replaced in (4.134) together with the (4.135), give:

$$\frac{\partial \mathcal{H}'}{\partial P_k} - \frac{\partial Q_k}{\partial t} = \sum_{i=1}^{n}(Q_i, Q_k)\left[\frac{\partial \mathcal{H}'}{\partial Q_i} + \frac{\partial P_i}{\partial t}\right] +$$
$$+ \sum_{i=1}^{n}(P_i, Q_k)\left[\frac{\partial \mathcal{H}'}{\partial P_i} - \frac{\partial Q_i}{\partial t}\right],$$
$$-\frac{\partial \mathcal{H}'}{\partial Q_k} - \frac{\partial P_k}{\partial t} = \sum_{i=1}^{n}(Q_i, P_k)\left[\frac{\partial \mathcal{H}'}{\partial Q_i} + \frac{\partial P_i}{\partial t}\right] +$$
$$+ \sum_{i=1}^{n}(P_i, P_k)\left[\frac{\partial \mathcal{H}'}{\partial P_i} - \frac{\partial Q_i}{\partial t}\right].$$

(4.139)

The $2n$ equality (4.139) must be collectively satisfied and therefore, given the independence of the dynamical coordinates, it is necessary that, for all the values of the indexes between 1 and n, it is:

$$(Q_i, Q_k) = 0,$$
$$(P_i, P_k) = 0,$$
$$(P_i, Q_k) = \delta_{ik},$$

(4.140)

where δ_{ik} is the Kronecker symbol. We now prove that the condition of canonicity expressed by the (4.140), besides being necessary, is also sufficient. We begin using the (4.140) to simplify the (4.129) and (4.130). Making use of the property (M.10) applied to the Hamiltonian $\mathcal{H}$ (i.e., placing $f = \mathcal{H}$), which is thought to be a function of q_k and P_k, we have immediately:

$$\frac{dQ_k}{dt} = \frac{\partial\mathcal{H}}{\partial P_k} + \frac{\partial Q_k}{\partial t},$$
$$\frac{dP_k}{dt} = -\frac{\partial\mathcal{H}}{\partial Q_k} + \frac{\partial P_k}{\partial t}. \tag{4.141}$$

We must prove that these equations occur in canonical form, namely that:

$$\frac{dQ_k}{dt} = \frac{\partial\mathcal{H}}{\partial P_k} + \frac{\partial Q_k}{\partial t} = \frac{\partial\mathcal{H}'}{\partial P_k},$$
$$\frac{dP_k}{dt} = -\frac{\partial\mathcal{H}}{\partial Q_k} + \frac{\partial P_k}{\partial t} = -\frac{\partial\mathcal{H}'}{\partial Q_k}, \tag{4.142}$$

where $\mathcal{H}'$ is the Hamiltonian function in the new dynamical coordinates. For this purpose it is sufficient to find any characteristic function f transforming the old into the new coordinates; if it exists, it implies that the transformation is canonical. We place:

$$\mathcal{H}' = \mathcal{H} + \frac{\partial f}{\partial t}, \tag{4.143}$$

which, substituted in (4.142), gives:

$$\frac{\partial Q_k}{\partial t} = \frac{\partial}{\partial P_k}(\mathcal{H}' - \mathcal{H}) = \frac{\partial^2 f}{\partial P_k \partial t} = \frac{\partial^2 f}{\partial t \partial P_k},$$
$$\frac{\partial P_k}{\partial t} = -\frac{\partial}{\partial Q_k}(\mathcal{H}' - \mathcal{H}) = -\frac{\partial^2 f}{\partial Q_k \partial t} = -\frac{\partial^2 f}{\partial t \partial Q_k}, \tag{4.144}$$

having admitted that the function f satisfies the Schwarz theorem on the inversion of the differentiation order (see Sect. 4.2). Integrating with respect to time, it results:

$$Q_k = \frac{\partial f}{\partial P_k} + b_k,$$
$$P_k = -\frac{\partial f}{\partial Q_k} + a_k, \tag{4.145}$$

where a_k and b_k are arbitrary integration constants. Finally, placing:

$$F = f + \sum_{k=1}^{n}(-a_k Q_k + b_k P_k), \tag{4.146}$$

from (4.143) and (4.145) we obtain:

$$\frac{\partial F}{\partial t} = \frac{\partial f}{\partial t} = \mathcal{H}' - \mathcal{H}, \tag{4.147a}$$

$$\frac{\partial F}{\partial Q_k} = \frac{\partial f}{\partial Q_k} - a_k = -P_k$$
$$\frac{\partial F}{\partial P_k} = \frac{\partial f}{\partial P_k} + b_k = Q_k \qquad (k = 1, \dots, n), \tag{4.147b}$$

which, compared with the relations in Table 4.3, prove that F is just the characteristic function of a canonical transformation from the dynamical coordinates q_k, p_k to Q_k, P_k.

In summary, a necessary and sufficient condition for a set of dynamical coordinates to be canonical is that the relations (4.140) are satisfied. We could have anticipated this result given the fact that the generalized coordinates q_k and p_k, with which we have obtained the Hamilton equations (Sect. 4.4), were quite generic and nonetheless they naturally satisfy the condition (4.140), as proved by (M.9).

The condition of canonicity via the Poisson brackets highlights a second feature of canonical coordinates (obviously contained in the Hamilton equations). For each value of the index k, one and only one generalized momentum conjugates with the generalized coordinate Q_k, the one with which the Poisson bracket equals unity.

Taking advantage of the relations (M.26) in the Poisson and Lagrange brackets, it is easy to prove that the canonicity conditions imply:

$$[Q_i, Q_k] = 0, \qquad [P_i, P_k] = 0, \qquad [P_i, Q_k] = \delta_{ik}, \tag{4.148}$$

for any value of the indexes between 1 and n. We will limit to prove that the determinant of matrix:

$$\|B\| = \begin{Vmatrix} (P_1, Q_1) & \cdots & (P_1, Q_n) & (P_1, P_1) & \cdots & (P_1, P_n) \\ \cdots & \cdots & \cdots & \cdots & \cdots & \cdots \\ (P_n, Q_1) & \cdots & (P_n, Q_n) & (P_n, P_1) & \cdots & (P_n, P_n) \\ (Q_1, Q_1) & \cdots & (Q_1, Q_n) & (Q_1, P_1) & \cdots & (Q_1, P_n) \\ \cdots & \cdots & \cdots & \cdots & \cdots & \cdots \\ (Q_n, Q_1) & \cdots & (Q_n, Q_n) & (Q_n, P_1) & \cdots & (Q_n, P_n) \end{Vmatrix}, \tag{4.149}$$

is $|B| = (-1)^n$, leaving to the reader the elementary demonstration. In fact, owing to (4.140), all the elements of the matrix (4.149) are null, except those of the main diagonal. Therefore:

$$|B| = \prod_{k=1}^{n} (P_k, Q_k)(Q_k, P_k) = \prod_{k=1}^{n} \left[-(P_k, Q_k)^2 \right] = (-1)^n. \tag{4.150}$$

Finally, we note that, placed $f = \mathcal{H}$ in (M.10), it is:

$$\frac{\partial \mathcal{H}}{\partial p_k} = (\mathcal{H}, q_k)$$
$$\frac{\partial \mathcal{H}}{\partial q_k} = -(\mathcal{H}, p_k)$$
$$(k = 1, \ldots, n), \tag{4.151}$$

or, through (4.45a):

$$\dot{q}_k = (\mathcal{H}, q_k)$$
$$\dot{p}_k = (\mathcal{H}, p_k)$$
$$(k = 1, \ldots, n). \tag{4.152}$$

This form of the Hamilton equations is conserved in the canonical transformations. In fact:

$$(\mathcal{H}', Q_k) = \sum_{i=1}^{n} (Q_i, Q_k) \frac{\partial \mathcal{H}'}{\partial Q_i} + \sum_{i=1}^{n} (P_i, Q_k) \frac{\partial \mathcal{H}'}{\partial P_i},$$

$$(\mathcal{H}', P_k) = \sum_{i=1}^{n} (Q_i, P_k) \frac{\partial \mathcal{H}'}{\partial Q_i} + \sum_{k=1}^{n} (P_i, P_k) \frac{\partial \mathcal{H}'}{\partial P_i}, \tag{4.153a}$$

from where, through the canonicity conditions (4.140):

$$(\mathcal{H}', Q_k) = \frac{\partial \mathcal{H}'}{\partial P_k} = \dot{Q}_k,$$

$$(\mathcal{H}', P_k) = -\frac{\partial \mathcal{H}'}{\partial Q_k} = \dot{P}_k. \tag{4.153b}$$

4.8.4 The Canonical Invariants

To any mechanical system with n degrees of freedom we associate a hyperspace at $2n$ dimensions, the so-called phase space, defined in such a way that each of its coordinated axes represents one of the $2n$ dynamical coordinates q_k, p_k. In classical mechanics, the state of the system is identified, at each moment, by a single point of this space, and the evolution of the system itself is a continuous curve. This representation invites us to demonstrate a theorem due to Poincaré: the n-tuple integral:

$$\Gamma = \int \cdots \int \prod_{k=1}^{n} dq_k dp_k, \tag{4.154}$$

extended to any volume of the phase space is invariant with respect to a canonical transformation, namely:

$$\Gamma = \int \cdots \int \prod_{k=1}^{n} dq_k dp_k = \int \cdots \int \prod_{k=1}^{n} dQ_k dP_k. \qquad (4.155)$$

From mathematical analysis (transformation properties of the integration variables), we know that:

$$\Gamma = \int \cdots \int \prod_{k=1}^{n} dq_k dp_k = \int \cdots \int |J| \prod_{k=1}^{n} dQ_k dP_k, \qquad (4.156)$$

where $|J|$ is the modulus of the Jacobian determinant, i.e., the determinant of the matrix:

$$\|J\| = \begin{Vmatrix} \dfrac{\partial Q_1}{\partial q_1} & \cdots & \dfrac{\partial Q_1}{\partial q_n} & \dfrac{\partial Q_1}{\partial p_1} & \cdots & \dfrac{\partial Q_1}{\partial p_n} \\[2mm] \cdots & \cdots & \cdots & \cdots & \cdots & \cdots \\[2mm] \dfrac{\partial Q_n}{\partial q_1} & \cdots & \dfrac{\partial Q_n}{\partial q_n} & \dfrac{\partial Q_n}{\partial p_1} & \cdots & \dfrac{\partial Q_n}{\partial p_n} \\[2mm] \dfrac{\partial P_1}{\partial q_1} & \cdots & \dfrac{\partial P_1}{\partial q_n} & \dfrac{\partial P_1}{\partial p_1} & \cdots & \dfrac{\partial P_1}{\partial p_n} \\[2mm] \cdots & \cdots & \cdots & \cdots & \cdots & \cdots \\[2mm] \dfrac{\partial P_n}{\partial q_1} & \cdots & \dfrac{\partial P_n}{\partial q_n} & \dfrac{\partial P_n}{\partial p_1} & \cdots & \dfrac{\partial P_n}{\partial p_n} \end{Vmatrix}. \qquad (4.157)$$

The proof of the theorem is therefore reduced to prove that the determinant of the matrix $\|J\|$ has a unitary modulus. To this end, through n exchanges of columns and sign changes in the matrix (4.157), we construct the matrix:

$$\|J'\| = \begin{Vmatrix} \dfrac{\partial Q_1}{\partial p_1} & \cdots & \dfrac{\partial Q_1}{\partial p_n} & -\dfrac{\partial Q_1}{\partial q_1} & \cdots & -\dfrac{\partial Q_1}{\partial q_n} \\[2mm] \cdots & \cdots & \cdots & \cdots & \cdots & \cdots \\[2mm] \dfrac{\partial Q_n}{\partial p_1} & \cdots & \dfrac{\partial Q_n}{\partial p_n} & -\dfrac{\partial Q_n}{\partial q_1} & \cdots & -\dfrac{\partial Q_n}{\partial q_n} \\[2mm] \dfrac{\partial P_1}{\partial p_1} & \cdots & \dfrac{\partial P_1}{\partial p_n} & -\dfrac{\partial P_1}{\partial q_1} & \cdots & -\dfrac{\partial P_1}{\partial q_n} \\[2mm] \cdots & \cdots & \cdots & \cdots & \cdots & \cdots \\[2mm] \dfrac{\partial P_n}{\partial p_1} & \cdots & \dfrac{\partial P_n}{\partial p_n} & -\dfrac{\partial P_n}{\partial q_1} & \cdots & -\dfrac{\partial P_n}{\partial q_n} \end{Vmatrix}. \qquad (4.158)$$

Since the previous manipulations do not alter the value of determinant, it is:

$$\|J\| = \|J'\|. \qquad (4.159)$$

Multiplying the two matrices row by column, we obtain:

$$\|D'\| = \begin{Vmatrix} (Q_1, Q_1) & \cdots & (Q_1, Q_n) & (Q_1, P_1) & \cdots & (Q_1, P_n) \\ \cdots & \cdots & \cdots & \cdots & \cdots & \cdots \\ (Q_n, Q_1) & \cdots & (Q_n, Q_n) & (Q_n, P_1) & \cdots & (Q_n, P_n) \\ (P_1, Q_1) & \cdots & (P_1, Q_n) & (P_1, P_1) & \cdots & (P_1, P_n) \\ \cdots & \cdots & \cdots & \cdots & \cdots & \cdots \\ (P_n, Q_1) & \cdots & (P_n, Q_n) & (P_n, P_1) & \cdots & (P_n, P_n) \end{Vmatrix}, \tag{4.160}$$

which coincides with the matrix $\|D\|$ given by (4.149) but for n exchanges between the lines. Therefore:

$$|D'| = |D|, \tag{4.161}$$

or, through (4.150):

$$|D'| = (-1)^n(-1)^n = 1. \tag{4.162}$$

Finally, through (4.159):

$$|D'| = |J| \times |J'| = J^2 = 1, \tag{4.163}$$

and then $|J| = 1$, which is what we wanted to prove.

A second class of canonical invariants is given by Poisson brackets. Let $f(q, p, t)$ and $g(q, p, t)$ be any two functions of $2n$ dynamical coordinates q_k and p_k ($k = 1, \ldots, n$), and $f'(Q, P, t)$, $g'(Q, P, t)$ the functions obtained from them through a canonical transformation in the new coordinates Q_k and P_k. We show that:

$$(f, g) = \sum_{k=1}^{n} \left(\frac{\partial f}{\partial p_k} \frac{\partial g}{\partial q_k} - \frac{\partial f}{\partial q_k} \frac{\partial g}{\partial p_k} \right) =$$

$$= (f', g') = \sum_{k=1}^{n} \left(\frac{\partial f'}{\partial P_k} \frac{\partial g'}{\partial Q_k} - \frac{\partial f'}{\partial Q_k} \frac{\partial g'}{\partial P_k} \right). \tag{4.164}$$

By applying the relation (M.13) to the functions f and g thought as dependent on Q and P, we have:

$$(f, g) = \sum_{i=1}^{n} \sum_{k=1}^{n} \left[(Q_i, Q_k) \frac{\partial f'}{\partial Q_i} \frac{\partial g'}{\partial Q_k} + (Q_i, P_k) \frac{\partial f'}{\partial Q_i} \frac{\partial g'}{\partial P_k} + \right.$$

$$\left. + (P_i, Q_k) \frac{\partial f'}{\partial P_i} \frac{\partial g'}{\partial Q_k} + (P_i, P_k) \frac{\partial f'}{\partial P_i} \frac{\partial g'}{\partial P_k} \right]. \tag{4.165}$$

Since, by assumption, the transformation is canonical, it must satisfy the conditions (4.140), and therefore the equality (4.165) simplifies in:

$$(f, g) = \sum_{k=1}^{n} \left(\frac{\partial f'}{\partial P_k} \frac{\partial g'}{\partial Q_k} - \frac{\partial f'}{\partial Q_k} \frac{\partial g'}{\partial P_k} \right) = (f', g'), \qquad (4.166)$$

which is what we wanted to prove.

4.8.5 The Infinitesimal Canonical Transformations

Consider a canonical transformation defined by the characteristic function, of the type of (4.120a):

$$F_2(q, P) = \sum_{k=1}^{n} q_k P_k. \qquad (4.167)$$

Using the relations of Table 4.3:

$$p_k = \frac{\partial F_2}{\partial q_k} = P_k$$
$$\qquad\qquad\qquad (k = 1, \ldots, n), \qquad (4.168a)$$
$$Q_k = \frac{\partial F_2}{\partial P_k} = q_k$$
$$\mathcal{H}'(Q, P, t) = \mathcal{H}(q, p, t). \qquad (4.168b)$$

Therefore, the characteristic function (4.167) makes an identity transformation, that is, a transformation leaving both the dynamical coordinates and the Hamiltonian function unchanged.

Having said this, we consider a canonical infinitesimal transformation, able of generating new dynamical coordinates differing from the old ones by infinitesimal quantities:

$$Q_k = q_k + \delta q_k$$
$$\qquad\qquad (k = 1, \ldots, n). \qquad (4.169)$$
$$P_k = p_k + \delta p_k$$

The meaning of this transformation, also called contact transformation, must be understood well. The infinitesimals δq_k and δp_k have no other meaning than a difference between the new and the old canonical dynamic coordinates; for example, they have nothing to do with virtual displacements. So, if we neglect, as we will do later, all the terms higher than the first in the infinitesimals δq_k and δp_k, it is predictable that the characteristic function of the transformation given by (4.169) differs by an infinitesimal amount from the characteristic function of an identity transformation. We therefore put:

$$F_2(q, P) = \sum_{k=1}^{n} q_k P_k + \epsilon \, G(q, P), \qquad (4.170)$$

where ϵ is an infinitesimal parameter. From the relations of Table 4.3 we then have:

$$\frac{\partial F_2}{\partial q_k} = p_k = P_k + \epsilon \, \frac{\partial G}{\partial q_k}, \tag{4.171}$$

or:

$$P_k - p_k = \delta p_k = -\epsilon \, \frac{\partial G}{\partial q_k}. \tag{4.172}$$

Moreover:

$$\frac{\partial F_2}{\partial P_k} = Q_k = q_k + \epsilon \, \frac{\partial G}{\partial P_k}. \tag{4.173}$$

Since the second term of (4.173) is linear in ϵ and since P_k differs from p_k by an infinitesimal $\big($cf. (4.169)$\big)$, the equality:

$$\epsilon \, \frac{\partial G}{\partial P_k} = \epsilon \, \frac{\partial G}{\partial p_k}, \tag{4.174}$$

is correct up to the first order, which is what we wanted to obtain. Therefore:

$$\frac{\partial F_2}{\partial P_k} = Q_k = q_k + \epsilon \, \frac{\partial G}{\partial p_k}, \tag{4.175}$$

or:

$$Q_k - q_k = \delta q_k = \epsilon \, \frac{\partial G}{\partial p_k}. \tag{4.176}$$

Finally, it is evident that $\mathcal{H}'(Q, P, t) = \mathcal{H}(q, p, t)$, since F_2 does not depend on time.

Although, strictly speaking, F_2 $\big($the function (4.170)$\big)$ should retain the role of characteristic function of this transformation, it is however common practice to call by this name also the function G that defines, through (4.172) and (4.176), the infinitesimal variations δq_k and δp_k. This use is justified by the fact that the difference between F_2 and ϵG is the characteristic function of an identity transformation.

An interesting application of the previous results is obtained by placing:

$$\begin{aligned} \epsilon &= dt, \\ G &= \mathcal{H}. \end{aligned} \tag{4.177}$$

Using the relations (4.172) and (4.176) and remembering the Hamilton equations (4.45a), we have:

$$\begin{aligned} \delta q_k &= \dot{q}_k \, dt = dq_k \\ \delta p_k &= \dot{p}_k \, dt = dp_k \end{aligned} \qquad (k = 1, \ldots, n), \tag{4.178}$$

and thus, from (4.169):

$$\begin{aligned} Q_k &= q_k + dq_k \\ P_k &= p_k + dp_k \end{aligned} \qquad (k = 1, \ldots, n), \tag{4.179}$$

or:

$$Q_k = q_k \, (t + dt) = q_k(t) + dq_k$$
$$P_k = p_k \, (t + dt) = p_k(t) + dp_k \qquad (k = 1, \ldots, n). \qquad (4.180)$$

These equations show that the characteristic function:

$$F_2(q, P) = \sum_{k=1}^{n} q_k \, P_k + \mathcal{H}dt, \qquad (4.181)$$

transforms the dynamical coordinates into new coordinates Q_k and P_k representing the values assumed by the old coordinates in consequence of an infinitesimal increment of time dt. In other words, the dynamical evolution of a mechanical system in the infinitesimal interval of time dt can be described by an infinitesimal contact transformation generated by the Hamiltonian function.

Now, since the result of two canonical transformations applied in succession is equivalent to that of a single transformation, the values of the dynamical coordinates $q_k(t)$ and $p_k(t)$ at any time t can be obtained from the initial ones, $q_k(t_o)$, $p_k(t_o)$, through a canonical transformation that is a continuous function of time. Conversely, there must be one canonical transformation able to reduce to the initial values, which are obviously constant, those assumed by the dynamical coordinates at each time t. The knowledge of such a transformation corresponds to the solution of the problem of the motion of the system, as it implies that we have $2n$ functions:

$$q_k = q_k\big(t, q_1(t_o), \ldots, q_n(t_o), p_1(t_o), \ldots, p_n(t_o)\big),$$
$$p_k = p_k\big(t, q_1(t_o), \ldots, q_n(t_o), p_1(t_o), \ldots, p_n(t_o)\big),$$

which are precisely the sought solutions of the Hamilton equations. This last consideration offers us the cue for important developments, which we will meet in Sect. 4.9.

Through the study of the transformation operated by the characteristic function (4.181) on any function $f(q, p)$, we now examine the connections between Poisson brackets and infinitesimal transformations. It is appropriate to clarify here the meaning of the term 'transformation'. In general, a canonical transformation:

$$q_k = q_k(Q_1, \ldots, Q_n, P_1, \ldots, P_n),$$
$$p_k = p_k(Q_1, \ldots, Q_n, P_1, \ldots, P_n), \qquad (4.182)$$

applied to the function f, simply modifies its functional dependence, i.e., its analytical form. On the other hand, since the (4.182) do not explicitly contain the time, the transformation does not alter the numerical value of f. We put this concept in terms of phase space. There the function f realizes a correspondence between a numerical set (co-domain of the function) and the points of the phase space. This correspondence is obviously independent of any one reference system transformation.

The nature of the transformation defined by (4.181) is different. In this case, in fact, the transformation is intended as a simple substitution of q_k with Q_k, and p_k with P_k, anywhere in the analytical expression of f. In other words, the functional dependence remains the same for the function f, whether we use the old or the new dynamical coordinates. In the phase space the transformation operates so as to move the representative point in which the function is evaluated. In particular, if the infinitesimal canonical transformation is generated by the Hamiltonian function, the result of the replacement of the old variables with the new ones is to bring the value the function f takes at the time t to that it will have at the time $t + dt$.

To summarize, by a generic infinitesimal canonical transformation with a Hamiltonian generator, it results:

$$\delta f = f(Q, P) - f(q, p) = f(q + \delta q, p + \delta p) - f(q, p). \tag{4.183}$$

With a Taylor development truncated at the first order in the infinitesimals δq_k and δp_k, we obtain:

$$\delta f = \sum_{k=1}^{n} \left(\frac{\partial f}{\partial q_k} \delta q_k + \frac{\partial f}{\partial p_k} \delta p_k \right), \tag{4.184}$$

or, from (4.172) and (4.176):

$$\delta f = \epsilon \sum_{k=1}^{n} \left(\frac{\partial f}{\partial q_k} \frac{\partial G}{\partial p_k} - \frac{\partial f}{\partial p_k} \frac{\partial G}{\partial q_k} \right), \tag{4.185}$$

and, through the Poisson brackets:

$$\delta f = \epsilon \, (G, f). \tag{4.186}$$

Consequently, the change of the Hamiltonian function due to an infinitesimal canonical transformation is given by:

$$\delta \mathcal{H} = \epsilon \, (G, \mathcal{H}). \tag{4.187}$$

If the generating function is the Hamiltonian itself, then:

$$\delta \mathcal{H} = \epsilon \, (\mathcal{H}, \mathcal{H}) = 0. \tag{4.188}$$

Again, if the generating function is a constant of motion a, which does not explicitly contain the time ($\partial a / \partial t = 0$), in consequence of (4.61) it is:

$$\delta \mathcal{H} = \epsilon \, (a, \mathcal{H}) = 0. \tag{4.189}$$

Thus, the constants of motion that do not explicitly depend on time are functions generating infinitesimal canonical transformations which do not modify the Hamiltonian. Taking advantage of this property it is possible to identify some constants of motion directly when the Hamiltonian function presents special symmetries. This is compliant and generalizes the considerations exposed in Sect. 3.1.1.

We quote in the end the famous theorem of Liouville,[18] fundamental in the study of systems with many bodies. Let $\psi(q, p, t)$ be the spatial distribution function in the phase space; it expresses the number density ν of system particles per unit volume of phase space, i.e.:

$$dv = \psi(q, p, t) \, d\Gamma, \tag{4.190}$$

where $d\Gamma = \prod_{k=1}^{n} dq_k dp_k$, is a volume element. The Liouville theorem states that:

$$\frac{d\psi}{dt} = 0. \tag{4.191}$$

For a complete discussion and for applications to the dynamics of stellar systems see [7].

4.8.6 Canonical Systems of Constants of Motion

We now demonstrate that the constants of motion for a Hamiltonian function $\mathcal{H}(q, p, t)$ of a mechanical system with n degrees of freedom can always be chosen so as to satisfy the conditions of canonicity (4.140). For this purpose, we first illustrate the procedure by which, given any function $\alpha_1(q, p)$ depending on $2n$ dynamical coordinates q_k and p_k, it can be associated to $n - 1$ functions $\alpha_i(q, p)$ and n functions $\beta_j(q, p)$ in such a way that the following conditions are satisfied:

$$(\alpha_i, \alpha_j) = (\beta_i, \beta_j) = 0, \qquad (\beta_i, \alpha_j) = \delta_{ij}. \tag{4.192}$$

Let $\beta_1(q, p)$ be any one of the $2n$ independent solutions of the linear partial differential equation:

$$(\beta_1, \alpha_1) = 1, \tag{4.193}$$

in the independent variables q_k and p_k. Consider the linear homogeneous partial differential equation:

$$(b, \alpha_1) = 0. \tag{4.194}$$

It has $2n - 1$ independent solutions; α_1 is one of them. We indicate the remaining $2n - 2$ by $b_i(q, p)$, with $i = 1, 2,...,2n$. In conclusion, any function of α_1 and of the $2n - 2$ functions b_i, is also a solution of the (4.194). Moreover, from the property (M.3) we obtain:

---

[18] Joseph Liouville (1809–1882): French mathematician and politician, founder of the *Journal de Mathématiques Pures et Appliquées*, also known as being the first to promote the unpublished work of Évarist Galois (1811–1832).

$$\left(b, (\alpha_1, \beta_1)\right) + \left(\alpha_1, (\beta_1, b)\right) + \left(\beta_1, (b, \alpha_1)\right) =$$
$$= \left(\alpha_1, (\beta_1, b)\right) + \left(\beta_1, (b, \alpha_1)\right) = 0, \tag{4.195}$$

where we have used the (4.193). If the function b is a solution of the (4.194), the previous relation reduces to:

$$\left(\alpha_1, (\beta_1, b)\right) = 0, \tag{4.196}$$

showing that the function (β_1, b) is also a solution of (4.194). So, even (β_1, b), as any other integral of (4.194), depends on $\alpha_1, b_1, \ldots, b_{2n-2}$ only, that is:

$$(\beta_1, b) = f_i(\alpha_1, b_1, \ldots, b_{2n-2}). \tag{4.197}$$

This allows us to prove that the $2n - 2$ non-trivial solutions of the linear homogeneous partial differential equation:

$$(\beta_1, c) = 0, \tag{4.198}$$

are functions of $\alpha_1, b_1, \ldots, b_{2n-2}$ only. In other words, they do not depend on β_1. Indeed, by expanding the Poisson bracket at first member of (4.198) in the hypothesis that c also depends [19] on β_1, we have:

$$(\beta_1, c) = (\beta_1, \alpha_1)\frac{\partial c}{\partial \alpha_1} + (\beta_1, \beta_1)\frac{\partial c}{\partial \beta_1} + \sum_{j=1}^{2n-2}(\beta_1, b_j)\frac{\partial c}{\partial b_j} = 0, \tag{4.199}$$

or, through (4.193):

$$(\beta_1, c) = \frac{\partial c}{\partial \alpha_1} + \sum_{j=1}^{2n-2}(\beta_1, b_j)\frac{\partial c}{\partial b_j} = 0. \tag{4.200}$$

Recalling that the functional dependence given by (4.197) holds for the coefficients of $\partial c/\partial b_i$, the second term of (4.200) can be thought of as independent of β_1. That said, we denote by c_i ($i = 1, 2, \ldots, 2n$) the independent solutions of (4.200). Since they are functions of $\alpha_1, b_1, \ldots, b_{2n-2}$, it results:

$$(c_i, \alpha_1) = (\alpha_1, \alpha_1)\frac{\partial c_i}{\partial \alpha_1} + \sum_{j=1}^{2n-2}(b_j, \alpha_1)\frac{\partial c_i}{\partial b_j}, \tag{4.201}$$

or, through (4.192) and (4.194):

---

[19] Note that $\alpha_1, \beta_1, b_1, \ldots, b_{2n-2}$ are $2n$ independent functions of the dynamical coordinates q_k, p_k. Therefore, they can always be thought of as independent variables, replacing the dynamical coordinates.

$$(c_i, \alpha_1) = 0 \qquad\qquad (i = 1, \ldots, 2n - 2). \qquad (4.202)$$

We now prove that the $2n - 2$ functions c_i share the property that, for any combination of the indexes, the bracket (c_i, c_j) depends on the functions c only. Since α_1, β_1, $c_1, \ldots, c_{2n-2}$, are $2n$ independent functions, they can always serve to represent, as independent variables, any function of the dynamical coordinates, including therefore (c_i, c_j). In order to demonstrate the property enunciated above, it is enough to prove that (c_i, c_j) does not depend on either α_1 or β_1. To this end we develop the Poisson bracket $\big(\beta_1, (c_i, c_j)\big)$. It is:

$$\big(\beta_1, (c_i, c_j)\big) = (\beta_1, \alpha_1)\frac{\partial(c_i, c_j)}{\partial\alpha_1} + (\beta_1, \beta_1)\frac{\partial(c_i, c_j)}{\partial\beta_1} +$$

$$+ \sum_{k=1}^{2n-2}(\beta_1, c_k)\frac{\partial(c_i, c_j)}{\partial c_k} = 0, \qquad (4.203)$$

from which, through (4.192), (4.193) and (4.198), we obtain the expression:

$$\big(\beta_1, (c_i, c_j)\big) = \frac{\partial(c_i, c_j)}{\partial\alpha_1} = 0, \qquad (4.204)$$

proving that (c_i, c_j) is independent of α_1. In a similar way we demonstrate the independence from β_1 starting with the relation:

$$\big(\alpha_1, (c_i, c_j)\big) = 0. \qquad (4.205)$$

In conclusion, (c_i, c_j) is a function of $c_1, c_2, \ldots, c_{2n-2}$ only. For this property we say that the functions c form a group.

We now illustrate the strategy for obtaining, starting from α_1, the $2n - 1$ functions of the dynamical coordinates that satisfy the conditions (4.140). We choose, quite arbitrarily, any of the c functions, for example c_1, and call it α_2. Based on (4.198) and (4.202), we have:

$$(\alpha_1, \alpha_2) = 0, \qquad (4.206a)$$

$$(\beta_1, \alpha_2) = 0; \qquad (4.206b)$$

in other words, the function α_2 satisfies the canonicity conditions (4.140) relatively to α_1 and β_1. We must now associate it with a function β_2 that is 'orthogonal' to α_1 and β_1. Let $\beta_2(\alpha_2, c_2, \ldots, c_{2n-2})$ be any one of the $2n - 2$ solutions of the linear partial differential equation:

$$(\beta_2, \alpha_2) = \sum_{i=1}^{2n-2}(c_i, \alpha_2)\frac{\partial\beta_2}{\partial c_i} = 1, \qquad (4.207)$$

in the $2n - 2$ independent variables $c_1, c_2, \ldots, c_{2n-2}$, with $c_1 = \alpha_2$. Since β_2 depends on the functions c_i, (4.200) and (4.202) are satisfied. In other terms, it is:

$$(\beta_1, \beta_2) = (\beta_2, \alpha_1) = 0, \tag{4.208}$$

in agreement with the conditions (4.192). We indicate with $b'_1, b'_2, \ldots, b'_{2n-4}$, the $2n - 4$ functions that, besides α_2, are solutions of the linear homogeneous partial differential equation:

$$(b', \alpha_2) = \sum_{i=1}^{2n-2} (c_i, \alpha_2) \frac{\partial b'}{\partial c_i} = 0, \tag{4.209}$$

in the independent variables $\alpha_2, c_2, \ldots, c_{2n-2}$. By the same arguments we used to show that (β_1, b_i) does not depend on β_1 (see pag. 134), it is easy to prove that the functions (β_2, b_i) do not depend on β_2. Thus, in analogy with (4.197), we set:

$$(\beta_2, b'_i) = f'_i(\alpha_2, b'_1, \ldots, b'_{2n-4}). \tag{4.210}$$

Looking back at the path outlined above, let us consider the linear homogeneous partial differential equation:

$$(\beta_2, c') = \frac{\partial c'}{\partial \alpha_2} + \sum_{i=1}^{2n-4} (\beta_2, b'_i) \frac{\partial c'}{\partial b'_i} = 0. \tag{4.211}$$

Since, as the (4.211) shows, the coefficients of $\partial c'/\partial b'_i$ do not depend on β_2, the $2n - 4$ solutions $c'_1, c'_2, \ldots, c'_{2n-4}$, are only functions of $\alpha_2, b'_1, \ldots, b'_{2n-4}$, that is:

$$c'_i = c'_i(\alpha_2, b'_1, \ldots, b'_{2n-4}). \tag{4.212}$$

As before, it is possible to show that the $2n - 4$ functions c'_i form a group, in the sense that the functions (c'_i, c'_j) do not depend on $\alpha_1, \alpha_2, \beta_1$ and β_2, but only on the c functions themselves. We leave the reader this verification and, quite arbitrarily, place $c'_1 = \alpha_3$. By means of the (4.206a), it results:

$$(\alpha_2, \alpha_3) = (\alpha_2, c'_1) = \sum_{i=1}^{2n-4} (\alpha_2, b'_i) \frac{\partial c'_1}{\partial b'_i} = 0. \tag{4.213}$$

From (4.211) we have:

$$(\beta_2, \alpha_3) = 0. \tag{4.214}$$

Finally, since the functions b'_i depend only on the $2n - 2$ functions c_i, from the (4.200) and (4.202) we obtain:

$$(\beta_1, \alpha_3) = (\alpha_3, \alpha_1) = 0. \tag{4.215}$$

Fig. 4.2 Scheme of the iterative cycles to find a set of $2n$ canonically conjugated functions, α_i and β_i, starting from a function α_1 of the $2n$ dynamical variables q_k, p_k

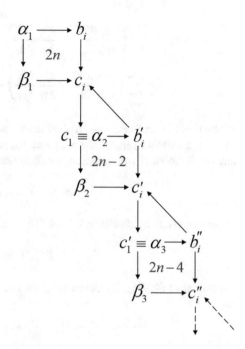

We interrupt here the iterative process that we have followed for two successive cycles. It is clear that n cycles will be required to complete it, because the order of the differential equation defining β_i (originally of order $2n$) is lowered by 2 at each cycle. The overall strategy is now clear. At each subsequent cycle, we look for two functions, β_i and α_{i+1} that are orthogonal to all the previous ones, except for α_i (see Fig. 4.1).

We now return to the problem formulated at the beginning of this section. Given the Hamiltonian function $\mathcal{H}(q, p, t)$ of $2n$ dynamical coordinates, we define a new function (Fig. 4.2):

$$\mathcal{H}'(q, p) = \mathcal{H}(q, p, t) + p_{n+1}, \qquad (4.216)$$

where p_{n+1} is an independent variable that we arbitrarily conjugate to time by placing: $q_{n+1} = t$. The function $\mathcal{H}'$ will thus depend on $2n + 2$ coordinates $q_1, q_2, \ldots, q_n, q_{n+1} = t, p_1, p_2, \ldots, p_n, p_{n+1}$ only. Starting from this function, we apply the procedure described in the first part of this section to find the $2n + 1$ functions $\alpha_2, \ldots, \alpha_{n+1}, \beta_1, \beta_2, \ldots, \beta_{n+1}$, that are canonically associated to $\alpha_1 = \mathcal{H}'$, so to satisfy the conditions (4.192). The (4.193) becomes:

$$(\beta_1, \mathcal{H}')_\star = 1, \qquad (4.217)$$

or, through (4.216):

$$(\beta_1, \mathcal{H}')_\star = \sum_{i=1}^{n+1} \left(\frac{\partial \beta_1}{\partial p_i} \frac{\partial \mathcal{H}'}{\partial q_i} - \frac{\partial \beta_1}{\partial q_i} \frac{\partial \mathcal{H}'}{\partial p_i} \right) =$$

$$= \sum_{i=1}^{n} \left(\frac{\partial \beta_1}{\partial p_i} \frac{\partial \mathcal{H}}{\partial q_i} - \frac{\partial \beta_1}{\partial q_i} \frac{\partial \mathcal{H}}{\partial p_i} \right) - \frac{\partial \beta_1}{\partial t} + \frac{\partial \beta_1}{\partial p_{n+1}} \frac{\partial \mathcal{H}}{\partial t} = 1, \qquad (4.218)$$

where the subscript $\star$ attached to the Poisson bracket remarks that it must be referred to $2n + 2$ variables, including q_{n+1} and p_{n+1}. Remember that we are interested in any one solution of this equation, which is readily found:

$$\beta_1 = -t = -q_{n+1}, \qquad (4.219)$$

as we can verify by replacing it in (4.218). This done, consider the equation:

$$(b, \mathcal{H}')_\star = 0, \qquad (4.220)$$

equivalent to (4.194). By developing the Poisson bracket through the (4.216), we obtain:

$$(b, \mathcal{H}')_\star = (b, \mathcal{H}) + (b, p_{n+1})_\star = (b, \mathcal{H}) - \frac{\partial b}{\partial t}. \qquad (4.221)$$

Finally, if $\partial b / \partial p_{n+1} = 0$, the previous equation gives:

$$(b, \mathcal{H}')_\star = (b, \mathcal{H}) - \frac{\partial b}{\partial t} = 0. \qquad (4.222)$$

This linear homogeneous partial differential equation in the variables $q_1, \ldots, q_n$, $p_1, \ldots, p_n$, and t, admits $2n$ independent solutions. The comparison with (4.61) shows that they coincide with the $2n$ constants of motion, $a_i(q, p, t) = $ const, relative to the Hamiltonian $\mathcal{H}$. Therefore, each solution b of the (4.220) must be a function of the constants of motion a_i, and therefore it must itself be a constant of motion. But, for the function $\mathcal{H}' = \alpha_1$, the canonicity conditions (4.192) are reduced to:

$$\begin{aligned} (\alpha_i, \mathcal{H}')_\star = 0 \qquad & (i = 1, \ldots, n+1), \\ (\beta_j, \mathcal{H}')_\star = 0 \qquad & (j = 2, \ldots, n+1), \end{aligned} \qquad (4.223)$$

that is, to the conditions (4.222). In particular, the $2n$ canonically associated functions $\alpha_2, \ldots, \alpha_{n+1}, \beta_2, \ldots, \beta_{n+1}$, depend on the $2n$ constants of motion a_i relative to $\mathcal{H}$, and therefore they are themselves constants of motion. This proves that it is always possible to construct constants of motion canonically conjugated.

Before moving on to a practical application of this property, we observe that, if $\mathcal{H}'$ is still a Hamiltonian function in canonical form, from the Hamilton equations (4.45a) and from (4.216) we have:

$$\frac{dq_{n+1}}{dt} = \frac{\partial \mathcal{H}'}{\partial p_{n+1}} = \frac{\partial (\mathcal{H} + p_{n+1})}{\partial p_{n+1}} = 1,$$

$$\frac{dp_{n+1}}{dt} = -\frac{\partial \mathcal{H}'}{\partial q_{n+1}} = -\frac{\partial \mathcal{H}}{\partial q_{n+1}}. \tag{4.224}$$

The first gives: $q_{n+1} = t$, which is consistent with our assumptions. Consequently, the second equation becomes:

$$\frac{dp_{n+1}}{dt} = -\frac{\partial \mathcal{H}}{\partial t} = -\frac{d\mathcal{H}}{dt}, \tag{4.225}$$

where we have used the relation (4.46). By integrating, we have:

$$p_{n+1} = -\mathcal{H}. \tag{4.226}$$

This shows that the Hamiltonian $\mathcal{H}$ is canonically conjugated to time t. Moreover, substituting in (4.216) the expression of p_{n+1} given by (4.226), we have:

$$\mathcal{H}' = \mathcal{H} + p_{n+1} = \mathcal{H} - \mathcal{H} = 0. \tag{4.227}$$

We will resume this result at Sect. 4.9, where we deal with the Jacobi equation.

4.8.7 A System of Canonical Elements for the Elliptical Orbit

We will now use the results of previous section to obtain a system of 6 canonical elements replacing the classic elements of the elliptical orbit, $a, e, \delta_o = -nt_o, \Omega, \omega$, and i. The relations (4.79a) show that the semi-major axis length, a, forms a null Poisson bracket with any other element except δ_o (cf. Table 4.2):

$$(a, \delta_o) = -\frac{2}{n\,a}. \tag{4.228}$$

Following the scheme traced in Sect. 4.8.6, it is convenient to identify α_1 with any function of a. With all the arbitrariness that is allowed, we place[20]

$$\alpha_1 = \mathcal{E} = -G\,\frac{m_1 + m_2}{2\,a}, \tag{4.229}$$

observing that, according to (2.68), it coincides with the total specific energy. Thus:

$$(\delta_o, \alpha_1) = (\delta_o, a)\,\frac{\partial \alpha_1}{\partial a} = \sqrt{\frac{G\,(m_1 + m_2)}{a^3}} = n, \tag{4.230}$$

---

[20] Note that here the total energy, $\mathcal{E}$, is given per unit mass of μ, at variance with formula (2.67) in Chap. 2.

where we have used the (M.10) and the expression of $(\delta_\circ, a)$ from Table 4.2, with the the mean motion given by (2.83). The equality (4.230) suggests us to choose:

$$\beta_1 = \delta_\circ \sqrt{\frac{a^3}{G\,(m_1 + m_2)}} = \frac{\delta_\circ}{n} = -t_\circ, \tag{4.231}$$

since in this way:

$$(\beta_1, \alpha_1) = \left(-t_\circ, -G\,\frac{m_1 + m_2}{2a} \right) = 1. \tag{4.232}$$

Let us now consider the equation equivalent to (4.194):

$$(b, \alpha_1) = (b, a)\,\frac{\partial \alpha_1}{\partial a} = 0. \tag{4.233}$$

It admits $2n - 2 = 4$ independent solutions[21] in addition to the semi-major axis length itself. Recalling that (Sect. 4.7):

$$(e, a) = (\Omega, a) = (\omega, a) = (i, a) = 0, \tag{4.234}$$

we can set: $b_1 = e, b_2 = \Omega, b_3 = \omega$, and $b_4 = i$. Thus, the solutions of the equation:

$$(\beta_1, c) = (-t_\circ, c) = (\frac{\delta_\circ}{n}, c) = 0, \tag{4.235}$$

equivalent to (4.202), are functions of $\alpha_1 = -\dfrac{G\,(m_1 + m_2)}{2a}$, and of $b_1 = e, b_2 = \Omega$, $b_3 = \omega$, and $b_4 = i$ only, that is:

$$(\beta_1, c) = (\beta_1, \alpha_1)\,\frac{\partial c}{\partial \alpha_1} + (\beta_1, e)\,\frac{\partial c}{\partial e} + (\beta_1, \Omega)\,\frac{\partial c}{\partial \Omega} +$$
$$+ (\beta_1, \omega)\,\frac{\partial c}{\partial \omega} + (\beta_1, i)\,\frac{\partial c}{\partial i} = 0. \tag{4.236}$$

Using the (4.231) and the relations of Table 4.2, the coefficients of the previous equation become:

---

[21] Recall that the two-body problem has only 3 degrees of freedom as it reduces to the motion of a single point of mass μ subjected to a central force with pole at the second body.

$$(\beta_1, \alpha_1) = 1,$$

$$(\beta_1, e) = \frac{1}{n}(\delta_\circ, e) = \frac{1 - e^2}{n^2 a^2 e},$$

$$(\beta_1, \Omega) = 0, \qquad\qquad (4.237)$$

$$(\beta_1, \omega) = 0,$$

$$(\beta_1, i) = 0,$$

with which the (4.236) becomes:

$$(\beta_1, c) = \frac{\partial c}{\partial \alpha_1} + \frac{1 - e^2}{n^2 a^2 e} \frac{\partial c}{\partial e} = 0, \qquad\qquad (4.238)$$

in the independent variables $\alpha_1, e, \Omega, \omega$, and i. As it is easy to prove by direct substitution, $2n - 2 = 4$ possible independent solutions are:

$$c_1 = \sqrt{G(m_1 + m_2)\, a\, (1 - e^2)},$$

$$c_2 = \sqrt{G\, (m_1 + m_2)\, a}\ \cos i,$$

$$c_3 = \Omega, \qquad\qquad (4.239)$$

$$c_4 = \omega.$$

With all the allowed arbitrariness, we define:

$$\alpha_2 = c_1 = \sqrt{G\, (m_1 + m_2)\, a\, (1 - e^2)}. \qquad\qquad (4.240)$$

Recalling the (2.65), the previous function can be written as:

$$\alpha_2 = \sqrt{G\, (m_1 + m_2)\, p} = h, \qquad\qquad (4.241)$$

which thus coincides with the total angular momentum. We now find one solution of the equation:

$$(\beta_2, \alpha_2) = 1. \qquad\qquad (4.242)$$

Since $\beta_2 = \beta_2(\alpha_2, c_2, c_3, c_4)$, then:

$$(\beta_2, \alpha_2) = (\alpha_2, \alpha_2)\frac{\partial \beta_2}{\partial \alpha_2} + \sum_{i=2}^{4}(c_i, \alpha_2)\frac{\partial \beta_2}{\partial c_i} = 1. \qquad\qquad (4.243)$$

The coefficients can be made explicit using the relations (4.239) and those in Table 4.2. It results:

$$(\alpha_2, \alpha_2) = 0,$$

$$(c_2, \alpha_2) = (i, \alpha_2)\frac{\partial c_2}{\partial i} = (i, a)\frac{\partial c_2}{\partial i}\frac{\partial \alpha_2}{\partial a} + (i, e)\frac{\partial c_2}{\partial i}\frac{\partial \alpha_2}{\partial e} = 0,$$

$$(c_3, \alpha_2) = (\Omega, \alpha_2) = (\Omega, a)\frac{\partial \alpha_2}{\partial a} + (\Omega, e)\frac{\partial \alpha_2}{\partial e} = 0, \qquad (4.244)$$

$$(c_4, \alpha_2) = (\omega, \alpha_2) = (\omega, a)\frac{\partial \alpha_2}{\partial a} + (\omega, e)\frac{\partial \alpha_2}{\partial e} = (\omega, e)\frac{\partial \alpha_2}{\partial e} =$$

$$= -\frac{\sqrt{1 - e^2}}{n\,a^2 e}\frac{\partial}{\partial e}\sqrt{G\,(m_1 + m_2)\,a\,(1 - e^2)} = 1.$$

Thus, the (4.242) becomes simply:

$$(\beta_2, \alpha_2) = \frac{\partial \beta_2}{\partial c_4} = \frac{\partial \beta_2}{\partial \omega} = 1, \qquad (4.245)$$

of which:

$$\beta_2 = \omega = \pi - \Omega, \qquad (4.246)$$

is an obvious solution.

We now look for the $2n - 4 = 2$ independent functions that, in addition to α_2, are solutions of the equation:

$$(b', \alpha_2) = 0. \qquad (4.247)$$

The relations (4.244) show that they are:

$$b'_1 = c_2 = \sqrt{G\,(m_1 + m_2)\,a}\,\cos i,$$
$$b'_2 = c_3 = \Omega. \qquad (4.248)$$

Therefore, the two solutions of the equation:

$$(\beta_2, c') = 0, \qquad (4.249)$$

will be functions of α_2, b'_1, and b'_2 only. Consequently:

$$(\beta_2, c') = (\beta_2, \alpha_2)\frac{\partial c'}{\partial \alpha_2} + (\beta_2, b'_1)\frac{\partial c'}{\partial b'_1} + (\beta_2, b'_2)\frac{\partial c'}{\partial b'_2} = 0. \qquad (4.250)$$

But:

$$(\beta_2, \alpha_2) = 1,$$

$$(\beta_2, b'_1) = (\omega, b'_1) = (\omega, a)\frac{\partial b'_1}{\partial a} + (\omega, i)\frac{\partial b'_1}{\partial i} = \frac{\cos i}{\sqrt{1 - e^2}} = -\frac{b'_1}{\alpha_2}, \qquad (4.251)$$

$$(\beta_2, b'_2) = (\omega, \Omega) = 0,$$

and thus (4.249) reduces to:

$$(\beta_2, c') = \frac{\partial c'}{\partial \alpha_2} + \frac{b_1'}{\alpha_2} \frac{\partial c'}{\partial b_1'} = 0, \tag{4.252}$$

which, of course, admits b_1' and b_2' as independent solutions. Therefore:

$$\begin{aligned} c_1' &= b_2' = \Omega, \\ c_2' &= b_1' = \sqrt{G\,(m_1 + m_2)\,a}\,\cos i. \end{aligned} \tag{4.253}$$

Having placed arbitrarily:

$$\alpha_3 = c_1' = \Omega, \tag{4.254}$$

as a last step we look for the solution of the equation:

$$(\beta_3, \alpha_3) = (\beta_3, \Omega) = 1. \tag{4.255}$$

It will be a function of α_3 and c_2' only, so:

$$(\beta_3, \alpha_3) = (\alpha_3, \alpha_3)\frac{\partial \beta_3}{\partial \alpha_3} + (c_2', \alpha_3)\frac{\partial \beta_3}{\partial c_2'} = 1, \tag{4.256}$$

or:

$$(\beta_3, \alpha_3) = (c_2', \alpha_3)\frac{\partial \beta_3}{\partial c_2'} = (a, \Omega)\frac{\partial c_2'}{\partial a}\frac{\partial \beta_3}{\partial c_2'} + (i, \Omega)\frac{\partial c_2'}{\partial i}\frac{\partial \beta_3}{\partial c_2'} =$$
$$= (i, \Omega)\frac{\partial c_2'}{\partial i}\frac{\partial \beta_3}{\partial c_2'} = \frac{1}{\sqrt{1 - e^2}}\frac{\partial \beta_3}{\partial c_2'} = 1. \tag{4.257}$$

The sought solution is therefore:

$$\beta_3 = c_2'\sqrt{1 - e^2} = \sqrt{G\,(m_1 + m_2)\,a\,(1 - e^2)}\,\cos i, \tag{4.258}$$

or, with the (2.65):

$$\beta_3 = h\cos i, \tag{4.259}$$

which coincides with the component of the angular momentum in the direction of the polar axis.

In conclusion, we have found a canonical system whose relation to the original orbital elements is summarized by the matrix:

$$\left\| \begin{array}{ll} \alpha_1 = -G\,\dfrac{m_1 + m_2}{2\,a} & \beta_1 = -t_\circ \\[2mm] \alpha_2 = h & \beta_2 = \omega = \pi - \Omega \\[2mm] \alpha_3 = \Omega & \beta_3 = h\cos i \end{array} \right\| \qquad (4.260)$$

The introduction of these elements allows us to exploit the canonicity properties (4.192). It is clear that, due to the arbitrariness of some of the choices made in the the iterative procedure just followed, the system of canonical constants (4.260) is not the only one possible for the elliptical orbit, but just one of the simplest.

4.9   The Equation of Jacobi

The idea behind the Jacobi equation is contained in the considerations made at the end of Sect. 4.8.6. Let $\mathcal{H}(q, p, t)$ be the Hamiltonian function of a mechanical system with n degrees of freedom. If the functions:

$$\begin{aligned} Q_i &= Q_i(q, p, t) \\ P_i &= P_i(q, p, t) \end{aligned} \qquad (i = 1, \ldots, n), \qquad (4.261)$$

are independent, they constitute a system of equations for a possible transformation of dynamic coordinates. We impose that this transformation is canonical and, in addition, we require that it makes the Hamiltonian $\mathcal{H}'(Q, P, t)$ identically null in the new dynamic coordinates. Before verifying whether such a transformation exists and, if so, what conditions the generating function must satisfy, we want to examine the consequences of having:

$$\mathcal{H}'(Q, P, t) \equiv 0. \qquad (4.262)$$

If the coordinates Q and P are canonical, the Hamilton equations assume the form:

$$\begin{aligned} \dot{Q}_i &= \frac{\partial}{\partial P_i} \mathcal{H}'(Q, P, t) = 0 \\[2mm] \dot{P}_i &= -\frac{\partial}{\partial Q_i} \mathcal{H}'(Q, P, t) = 0 \end{aligned} \qquad (i = 1, \ldots, n), \qquad (4.263)$$

or:

$$\begin{aligned} Q_i &= \alpha_i = \text{const} \\ P_i &= \beta_i = \text{const} \end{aligned} \qquad (i = 1, \ldots, n). \qquad (4.264)$$

This result identifies the $2n$ functions of the system (4.261), which are independent by hypothesis, with an equal number of constant functions of the dynamical coordinates q and p, and possibly of time t. In other words, the functions:

$$\alpha_i = \alpha_i(q, p, t)$$
$$\beta_i = \beta_i(q, p, t) \qquad (i = 1, \ldots, n), \qquad (4.265)$$

constitute a system of $2n$ independent first integrals of the Hamiltonian equations. According to the arguments of Sect. 4.5, this is sufficient to state that the system (4.265) solves completely the mechanical problem proposed by $\mathcal{H}$. It should be emphasized once again that the existence of a transformation capable of nullifying identically the Hamilton $\mathcal{H}'$ is not sufficient for our purpose. We require the transformation to be canonical so that the form of the Hamilton equations is conserved. Only in this case, in fact, the condition: $\mathcal{H}' \equiv 0$, necessarily implies the constancy of Q and P.

So, let $F(q, Q, t)$ be the generating function of a canonical transformation between the old coordinates, q_k, p_k $(k = 1, \ldots, n)$, and the new ones, Q_k, P_k. The results of Sect. 4.8.5, summarized in Table 4.3, show that the function F establishes the following relations:

$$p_k = \frac{\partial}{\partial q_k} F(q, Q, t)$$
$$\qquad\qquad\qquad (k = 1, \ldots, n), \qquad (4.266a)$$
$$P_k = -\frac{\partial}{\partial Q_k} F(q, Q, t)$$

$$\mathcal{H}'(Q, P, t) = \mathcal{H}(q, p, t) + \frac{\partial}{\partial t} F(q, Q, t). \qquad (4.266b)$$

The system of n independent equations (4.266a) can be solved with respect to the unknowns Q_i $(i = 1, \ldots, n)$. We obtain functions of the type (4.261) which define the new coordinates Q_i using the old dynamical coordinates. Similarly, the set of n functions (4.266a) defines the new moments P_i $(i = 1, \ldots, n)$ using the coordinates q and Q. Since the latter are functions of q and p, the n functions (4.266a) are equivalent to the functions P_i of (4.261). With that, the canonical transformation is perfectly identified. We are left with the relation (4.266b), which allows us to find the expression of the Hamiltonian function in the new coordinates. Suppose now that the characteristic function is such that $\mathcal{H}' \equiv 0$. We will call it the principal Hamilton function and, following the literature, indicate it with the symbol S. Let then be:

$$F(q, Q, t) = S(q, \alpha, t), \qquad (4.267)$$

where we used the condition $Q_i = \alpha_i$ which follows from being $\mathcal{H}' \equiv 0$. Since the new moments are constant, the relations (4.266a) become:

$$p_k = \frac{\partial}{\partial q_k} S(q, \alpha, t)$$
$$\qquad\qquad\qquad (k = 1, \ldots, n). \qquad (4.268)$$
$$\beta_k = -\frac{\partial}{\partial \alpha_k} S(q, \alpha, t)$$

Furthermore, the principal Hamilton function must be solution of the homogeneous first order partial differential equation in the $n + 1$ variables $q_1, q_2, \ldots, q_n$, and t:

$$\frac{\partial}{\partial t} S(q, \alpha, t) + \mathcal{H}(q, \frac{\partial S}{\partial q_k}, t) = 0, \tag{4.269}$$

which is obtained from (4.266b) by setting $\mathcal{H}' \equiv 0$, and replacing p_k with the expression given by (4.268). In conclusion, (4.269), named after Jacobi, defines the canonical transformation which nullifies the Hamiltonian identically and, consequently, transforms the old dynamic coordinates into an equal number of canonically conjugated constants of motion. From the mathematical point of view we note that, in the case where the homogeneous equation (4.269) provides a solution, this must contain n independent constants $\alpha_1, \alpha_2, \ldots, \alpha_n$. We place then:

$$Q_k = \alpha_k \qquad (k = 1, \ldots, n), \tag{4.270}$$

with which we obtain n independent constant coordinates. The corresponding conjugate moments result from the application of the relation (4.268).

Since the Jacobi equation is a linear differential equation in $n + 1$ variables, at a first look it might appear that the procedure now described halves the number of integrations needed to solve the mechanical problem. In reality, together with the equation (4.269) that possibly allows us to differentiate the main function of Hamilton, and so the constant coordinates α_i, we have to consider the n first order differential equations (4.268) which define the constants of motion β_i conjugated to α_i, with which the numbers check out.

In order to show what the meaning of the Hamilton function is, we calculate the total derivative of $S(q, \alpha, t)$ with respect to time:

$$\frac{dS}{dt} = \sum_{k=1}^{n} \frac{\partial S}{\partial q_k} \dot{q}_k + \frac{\partial S}{\partial t}, \tag{4.271}$$

or, through (4.268), (4.269) and (4.43):

$$\frac{dS}{dt} = \sum_{k=1}^{n} p_k \dot{q}_k - \mathcal{H} = \mathcal{L}(q, \dot{q}, t), \tag{4.272}$$

which, once integrated, provides:

$$S(q, \alpha, t) = \int_{t_o}^{t} \mathcal{L}(q, \dot{q}, t') dt'. \tag{4.273}$$

Unfortunately, this expression does not help us calculate S since, in order to integrate the Lagrangian function, we should know the functions $q_k(t)$; in other words, we should have already solved the problem of motion.

We have seen that, as it was to be expected, the formalism of Jacobi does not decrease the number of integrations necessary to solve the mechanical problem. Indeed we will see in the example at Sect. 4.12.1, that this formalism does not even simplify the calculations, and often complicates them. In fact, the power of the Hamilton-Jacobi formalism is not in a trivial, although useful reduction of the complexity of the calculations.

4.10 Special Cases of the Jacobi Equation

We now consider two particular cases of the Jacobi equation (4.269). The first one is refers to those problems in which the total mechanical energy is conserved. The second provides an effective method of calculus applicable to Hamiltonians of particular form.

4.10.1 When the Hamiltonian does not Depend on Time

If the Hamiltonian function $\mathcal{H}$ is independent of time, from the equality (4.46) it is:

$$\frac{\partial \mathcal{H}}{\partial t} = \frac{d\mathcal{H}}{dt} = 0, \tag{4.274}$$

or:

$$\mathcal{H}(q, p) = -\alpha_1 = \text{const}, \tag{4.275}$$

where the choice of the constant is arbitrary. In this case the Jacobi equation becomes:

$$\frac{\partial}{\partial t} S(q, \alpha, t) - \alpha_1 = 0, \tag{4.276}$$

and therefore time can be separated by writing the solution of (4.276) as:

$$S(q, \alpha, t) = W(q, \alpha) + \alpha_1 t, \tag{4.277}$$

where, as usual, the functions α_i are integration constants. The auxiliary function $W(q, \alpha)$, called characteristic Hamilton function, does not explicitly contain time. It does not appear in the Jacobi equation (which is identically satisfied), but only in equations (4.268) and (4.269), which become:

$$p_k = \frac{\partial S}{\partial q_k} = \frac{\partial W}{\partial q_k},$$

$$\beta_k = -\frac{\partial S}{\partial \alpha_k} = \begin{cases} -\dfrac{\partial W}{\partial \alpha_1} - t & \text{for } k = 1, \\[2ex] -\dfrac{\partial W}{\partial \alpha_k} & \text{for } k \neq 1. \end{cases} \tag{4.278}$$

These relations show that the function W is itself characteristic of a canonical transformation from the old coordinates, q_k and p_k, to new ones, Q_k and P_k, given by:

$$Q_k = \begin{cases} \alpha_1 = -\mathcal{H}(q, p) & \text{for } k = 1, \\[1ex] \alpha_k & \text{for } k \neq 1, \end{cases} \tag{4.279}$$

and, from (4.118):

$$P_k = \begin{cases} -\dfrac{\partial W}{\partial Q_1} = -\dfrac{\partial W}{\partial \alpha_1} = t + \beta_1 & \text{for } k = 1, \\[2ex] \beta_k & \text{for } k \neq 1. \end{cases} \tag{4.280}$$

Finally, since W does not depend on time, the new Hamiltonian $\mathcal{H}'(Q, P)$ coincides with the old one:

$$\mathcal{H}'(Q, P) = \mathcal{H}(q, p) = -\alpha_1, \tag{4.281}$$

or, through the (4.279) and (4.280):

$$\mathcal{H}'(\alpha_1, \ldots, \alpha_n, \beta_1 + t, \beta_2, \ldots, \beta_n) = -\alpha_1. \tag{4.282}$$

In summary, if the Hamiltonian is independent of time, it is possible to operate a canonical transformation generated by the characteristic Hamiltonian function in dynamical coordinates given by (4.279) and (4.280).

4.10.2 Separation of Variables

As any other partial differential equation, the Jacobi equation is usually rather difficult to solve (if the solution exists). It becomes a practical tool when there are the conditions to apply the variable separation technique. In this case, in fact, the search for the solution can always be traced back to a quadrature.

Suppose that one of the variables, say q_1, appears in the Jacobi equation (4.269) only in a certain combination $h_1(q_1, \partial S/\partial q_1)$, not containing any other coordinate, including time t. In other words, let us assume that the Jacobi equation:

$$\frac{\partial S}{\partial t} + \mathcal{H}\left(q_i, \frac{\partial S}{\partial q_i}, t\right) = 0, \tag{4.283}$$

is in the form in which:

$$\frac{\partial S}{\partial t} = s(q_2, \ldots, q_n, t), \tag{4.284}$$

and, furthermore, that:

$$\mathcal{H} = \mathcal{H}\left(q_2, \ldots, q_n, \frac{\partial S}{\partial q_2}, \ldots, \frac{\partial S}{\partial q_n}, t, h_1\left(q_1, \frac{\partial S}{\partial q_1}\right)\right). \tag{4.285}$$

We search for a solution in the form:

$$S(q, t) = S'(q_2, \ldots, q_n, t) + S_1(q_1). \tag{4.286}$$

Substituting in (4.283), we obtain the equation:

$$\frac{\partial}{\partial t} S'(q_2, \ldots, q_n, t) +$$

$$+ \, \mathcal{H}\left(q_2, \ldots, q_n, \frac{\partial S'}{\partial q_2}, \ldots, \frac{\partial S'}{\partial q_n}, t, h_1\left(q_1, \frac{d S_1(q_1)}{d q_1}\right)\right) = 0; \tag{4.287}$$

it shows that necessarily:

$$h_1\left(q_1, \frac{d S_1(q_1)}{d q_1}\right) = \alpha_1 = \text{const.} \tag{4.288}$$

In fact, since the n variables q_i are independent, it is possible to vary q_1 keeping all the others constant; but, if this operation were able to change the value of h_1, the equation (4.287) would no longer be satisfied.

By replacing the function h_1 in (4.287) with the constant solution α_1, we have:

$$\frac{\partial}{\partial t} S'(q_2, \ldots, q_n, t) + \mathcal{H}\left(q_2, \ldots, q_n, \frac{\partial S'}{\partial q_2}, \ldots, \frac{\partial S'}{\partial q_n}, \alpha_1, t\right) = 0. \tag{4.289}$$

With this procedure, the first order partial differential equation (4.283) in the n variables q_i and the time t, is reduced to the two equations (4.288) and (4.289). The first of them is an ordinary first order differential equation; it provides the function $S_1(q_1)$ by a quadrature. The second is a partial differential equation of the same type as (4.283), but containing one less of the independent variables.

The procedure can be iterated if a second variable in the equation (4.289), say q_2, appears only in a certain combination $h_2(q_2, \partial S'/\partial q_2)$, containing neither coordinates nor time. If we then place:

$$S(q, t) = S''(q_3, \ldots, q_n, t) + S_1(q_1) + S_2(q_2), \tag{4.290}$$

we obtain the new equations:

$$h_2\left(q_2, \frac{dS_2(q_2)}{dq_2}\right) = \alpha_2 = \text{const}, \tag{4.291a}$$

$$\frac{\partial}{\partial t}S''(q_3, \ldots, q_n, t) + \mathcal{H}\left(q_3, \ldots, q_n, \frac{\partial S''}{\partial q_3}, \ldots, \frac{\partial S''}{\partial q_n}, \alpha_1, \alpha_2, t\right) = 0. \tag{4.291b}$$

If all the n coordinates q_i and the time t can be separated, then the Jacobi equation is reduced entirely to quadratures, and its solution is of the type:

$$S(q_1, \ldots, q_n, t) = \sum_{k=1}^{n} S_k(q_k) + S_{n+1}(t), \tag{4.292}$$

where the functions S_k are solutions of the ordinary differential equations:

$$h_k\left(q_k, \frac{dS_k(q_k)}{dq_k}\right) = \alpha_k = \text{const}, \tag{4.293}$$

$$\frac{d}{dt}S_{n+1}(t) + \mathcal{H}(\alpha_1, \ldots, \alpha_n, t) = 0. \tag{4.294}$$

In particular, if the Hamiltonian is independent of time, condition (4.275) holds, and therefore the equation (4.294) reduces to:

$$\frac{d}{dt}S_{n+1}(t) - \alpha_1 = 0, \tag{4.295}$$

which is readily integrated in:

$$S_{n+1}(t) = \alpha_1 t + \gamma_1. \tag{4.296}$$

In conclusion, we explicitly note that the conditions required for the application of the variable separation method are closely connected with the chosen coordinate system. This fact will be apparent in the example presented in the next section.

4.11   The Method of Hamilton-Jacobi Applied to the Two-Body Problem

As an instructive exercise, we take up here the two-body problem, already formulated and solved in Chap. 2, addressing it with the Hamilton-Jacobi technique. We adopt a Cartesian orthogonal reference system with axes of any orientation but the origin in one of the two massive points, say P_1. Let then be:

$$q_1' = x,$$
$$q_2' = y, \qquad \qquad (4.297)$$
$$q_3' = z,$$

the coordinates of the other massive body P_2. The corresponding conjugated moments are those that we already calculated in Sect. 4.7:

$$p_1' = \dot{x} = \dot{q}_1',$$
$$p_2' = \dot{y} = \dot{q}_2', \qquad \qquad (4.298)$$
$$p_3' = \dot{z} = \dot{q}_3'.$$

Since the system is conservative, we obtain right away the Hamilton function from the expression of total specific energy (2.8):

$$\mathcal{H}(q', p') = \frac{1}{2}(p_1'^2 + p_2'^2 + p_3'^2) - G\,\frac{m_1 + m_2}{\sqrt{q_1'^2 + q_2'^2 + q_3'^2}} = -\alpha_1. \qquad (4.299)$$

The corresponding Jacobi equation uses of the Hamilton characteristic function (4.277), as $\mathcal{H}$ does not explicitly contains time. Replacing the moments p_k' appearing in (4.299) with the expressions:

$$p_k' = \frac{\partial}{\partial q_k'} W(q_1', q_2', q_3', \alpha_1, \alpha_2, \alpha_3) \qquad (k = 1, \dots, n), \qquad (4.300)$$

equivalent to (4.117), we obtain the equation:

$$\mathcal{H}(p', q') = \frac{1}{2}\left[\left(\frac{\partial W}{\partial q_1'}\right)^2 + \left(\frac{\partial W}{\partial q_2'}\right)^2 + \left(\frac{\partial W}{\partial q_3'}\right)^2\right] +$$
$$- G\,\frac{m_1 + m_2}{\sqrt{q_1'^2 + q_2'^2 + q_3'^2}} = -\alpha_1, \qquad (4.301)$$

where the variable separation technique cannot be applied due to the presence of the term: $r = \sqrt{q_1'^2 + q_2'^2 + q_3'^2}$. Because of that, it is worth reconsidering the choice of the coordinate system, looking for one in which r appears as an independent coordinate. The most obvious choice is the spherical system.
We will then place:

$$q_1 = r,$$
$$q_2 = \theta, \qquad \qquad (4.302)$$
$$q_3 = \phi.$$

Fig. 4.3 Link between
Cartesian and spherical
coordinates

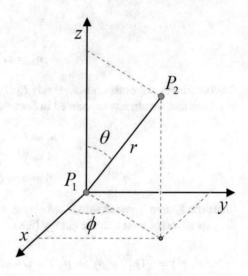

Since (cf. Fig. 4.3):

$$x = r \, \cos\phi \, \sin\theta,$$
$$y = r \, \sin\phi \, \sin\theta, \qquad\qquad\qquad (4.303)$$
$$z = r \, \cos\theta,$$

by differentiating and summing up the squares, after obvious simplifications, we obtain:

$$\dot{x}^2 + \dot{y}^2 + \dot{z}^2 = \dot{r}^2 + r^2 \dot{\theta}^2 + r^2 \sin^2\theta \, \dot{\phi}^2, \qquad\qquad (4.304)$$

or, through (4.302):

$$\dot{x}^2 + \dot{y}^2 + \dot{z}^2 = \dot{q}_1^2 + q_1^2 \dot{q}_2^2 + q_1^2 \sin^2 q_2 \, \dot{q}_3^2. \qquad\qquad (4.305)$$

The Lagrangian function of the system becomes then:

$$\mathcal{L}(q, \dot{q}) = \frac{1}{2} (\dot{x}^2 + \dot{y}^2 + \dot{z}^2) + G \frac{m_1 + m_2}{(x^2 + y^2 + z^2)^{1/2}} =$$

$$= \frac{1}{2} (\dot{q}_1^2 + q_1^2 \dot{q}_2^2 + q_1^2 \sin^2 q_2 \, \dot{q}_3^2) + G \frac{m_1 + m_2}{q_1}. \qquad (4.306)$$

Through the (4.37), the latter allows to obtain the conjugated moments:

$$p_1 = \frac{\partial \mathcal{L}}{\partial \dot{q}_1} = \dot{q}_1,$$

$$p_2 = \frac{\partial \mathcal{L}}{\partial \dot{q}_2} = q_1^2\,\dot{q}_2, \qquad (4.307)$$

$$p_3 = \frac{\partial \mathcal{L}}{\partial \dot{q}_3} = q_1^2\,\sin^2 q_2\,\dot{q}_3,$$

with which we built the Hamiltonian function:

$$\mathcal{H}(p,q) = \sum_{k=1}^{3} p_k \dot{q}_k - \mathcal{L}(q,\dot{q}) =$$

$$= \frac{1}{2}\left(p_1^2 + \frac{p_2^2}{q_1^2} + \frac{p_3^2}{q_1^2\,\sin^2 q_2}\right) - G\,\frac{m_1 + m_2}{q_1} = -\alpha_1. \qquad (4.308)$$

Using (4.300), we built the corresponding Jacobian function:

$$\frac{1}{2}\left[\left(\frac{\partial W}{\partial q_1}\right)^2 + \frac{1}{q_1^2}\left(\frac{\partial W}{\partial q_2}\right)^2 + \frac{1}{q_1^2\,\sin^2 q_2}\left(\frac{\partial W}{\partial q_3}\right)^2\right] +$$

$$- G\,\frac{m_1 + m_2}{q_1} = -\alpha_1. \qquad (4.309)$$

Since the coordinate q_3 appears only in the term:

$$h_3 = \left(\frac{\partial W}{\partial q_3}\right)^2, \qquad (4.310)$$

in searching for a solution of equation (4.309) we can apply the variable separation technique in the form:

$$W(q_1, q_2, q_3) = W'(q_1, q_2) + W_3(q_3). \qquad (4.311)$$

From (4.288) and (4.289) we then have:

$$h_3 = \left(\frac{d}{dq_3} W_3(q_3)\right)^2 = \alpha_3^2, \qquad (4.312)$$

where α_3^2 is an arbitrary (necessarily not negative) constant, and:

$$\frac{1}{2}\left[\left(\frac{\partial W}{\partial q_1}\right)^2 + \frac{1}{q_1^2}\left\{\left(\frac{\partial W}{\partial q_2}\right)^2 + \frac{\alpha_3^2}{\sin^2 q_2}\right\}\right] - G\,\frac{m_1 + m_2}{q_1} = -\alpha_1. \qquad (4.313)$$

In this last equation, the variable q_2 appears only in the combination:

$$h_2 = \left(\frac{\partial W'}{\partial q_2}\right)^2 + \frac{\alpha_3^2}{\sin^2 q_2}. \tag{4.314}$$

We then set:

$$W'(q_2, q_3) = W_1(q_1) + W_2(q_2), \tag{4.315}$$

or:

$$W(q_1, q_2, q_3) = W_1(q_1) + W_2(q_2) + W_3(q_3), \tag{4.316}$$

obtaining:

$$h_2 = \left(\frac{d}{dq_2} W_2(q_2)\right)^2 + \frac{\alpha_3^2}{\sin^2 q_2} = \alpha_2^2, \tag{4.317}$$

where α_2^2 is an arbitrary constant, and:

$$\frac{1}{2}\left[\left(\frac{dW_1}{dq_1}\right)^2 + \frac{\alpha_2^2}{q_1^2}\right] - G\frac{m_1 + m_2}{q_1} = -\alpha_1. \tag{4.318}$$

With this the Jacobi equation reduces to the three ordinary differential equations (4.312), (4.317), and (4.318). Indeed, as shown by the relation (4.316), the solution of (4.309) is simply the sum of the solutions of these three independent equations.

By interpreting the Jacobi equation as a canonical transformation in the new constant coordinates Q_k and P_k ($k = 1, \ldots, 3$), according to (4.279) we place:

$$\begin{aligned} Q_1 &= \alpha_1, \\ Q_2 &= \alpha_2, \\ Q_3 &= \alpha_3. \end{aligned} \tag{4.319}$$

We can readily see the meaning of the new coordinates. Remembering the relation (4.300), from (4.312) it is:

$$\left(\frac{d}{dq_3} W_3(q_3)\right)^2 = p_3^2 = \alpha_3^2, \tag{4.320}$$

or, through the third of equations (4.307):

$$\left(q_1^2 \sin^2 q_2 \, \dot{q}_3\right)^2 = \alpha_3^2. \tag{4.321}$$

Finally, from the equivalence (4.302) between spherical and generalized coordinates we obtain the expression:

$$\alpha_3 = \pm\left(r^2 \sin^2\theta \, \dot{\phi}\right), \tag{4.322}$$

which shows that the constant α_3 corresponds, except for the sign, to the z component of the angular momentum of the P_2 relative to P_1. Since this axis is arbitrarily oriented, the condition (4.322) involves the invariance of the instantaneous plane of motion; the double sign is instead the consequence of the two possible directions P_2 can assume in its orbital motion. From (4.317), by the definition of the momentum p_2 given in (4.300) and (4.307), we have:

$$q_1^4 \, \dot{q}_2^2 + \frac{\alpha_3^2}{\sin^2 q_2} = \alpha_2^2, \tag{4.323}$$

or, through (4.302) and (4.322):

$$\alpha_2 = \pm r^2 \sqrt{\dot{\theta}^2 + \sin^2\theta \; \dot{\phi}^2}. \tag{4.324}$$

This relation proves the constancy of the total angular momentum module. Finally, we recall that, according to (4.309), α_1 represents the total specific energy of the system.

We now express the constants α_k, i.e., the new constant canonical coordinates Q_k, through the usual conic elements of the orbits, arbitrarily choosing positive values in the relations (4.322) and (4.324):

$$
\begin{aligned}
Q_1 &= \alpha_1 = G \, \frac{m_1 + m_2}{2a}, \\
Q_2 &= \alpha_2 = h = \sqrt{G \, (m_1 + m_2) \, a \, (1 - e^2)}, \\
Q_3 &= \alpha_3 = h \, \cos i = \sqrt{G \, (m_1 + m_2) \, a \, (1 - e^2)} \, \cos i.
\end{aligned}
\tag{4.325}
$$

To derive the corresponding conjugated moments, we use the relations (4.280):

$$
\begin{aligned}
P_1 &= t + \beta_1 = -\frac{\partial W}{\partial Q_1} = -\frac{\partial W}{\partial \alpha_1}, \\
P_2 &= \beta_2 = -\frac{\partial W}{\partial Q_2} = -\frac{\partial W}{\partial \alpha_2}, \\
P_3 &= \beta_3 = -\frac{\partial W}{\partial Q_3} = -\frac{\partial W}{\partial \alpha_3}.
\end{aligned}
\tag{4.326}
$$

The characteristic function W, given by (4.316), can be written in the form:

$$W(q_1, q_2, q_3, \alpha_1, \alpha_2, \alpha_3) = \sum_{k=1}^{3} \int^{q_k} \frac{dW_k}{dq_k} dq_k, \tag{4.327}$$

or, through the relations (4.312), (4.317), and (4.318):

$$W(q_1, q_2, q_3, \alpha_1, \alpha_2, \alpha_3) = \int^{q_1} dq_1 \sqrt{-2\alpha_1 + 2G \frac{m_1 + m_2}{q_1} - \frac{\alpha_2^2}{q_1^2} +}$$

$$+ \int^{q_2} dq_2 \sqrt{\alpha_2^2 - \frac{\alpha_3^2}{\sin^2 q_2}} + \int^{q_3} \alpha_3 \, dq_3. \tag{4.328}$$

Replacing this expression in (4.326) and making use of the equivalences (4.302), we have:

$$P_1 = t + \beta_1 = \int^r \frac{r \, dr}{\sqrt{-2\alpha_1 r^2 + 2G(m_1 + m_2) r - \alpha_2^2}},$$

$$P_2 = \beta_2 = \int^r \frac{\alpha_2/r \, dr}{\sqrt{-2\alpha_1 r^2 + 2G(m_1 + m_2) r - \alpha_2^2}} +$$

$$- \int^\theta \frac{\alpha_2 \sin \theta \, d\theta}{\sqrt{\alpha_2^2 \sin^2\theta - \alpha_3^2}}, \tag{4.329}$$

$$P_3 = \beta_3 = \int^\theta \frac{\alpha_3 \, d\theta}{\sin \theta \sqrt{\alpha_2^2 \sin^2\theta - \alpha_3^2}} - \int^\phi d\phi.$$

These relations show that the moments P_k can be obtained by quadrature. We can save ourselves the tedious calculation of the integrals that appear in (4.329). Indeed, the moments P_k conjugated to the coordinates Q_k given by (4.325):

$$P_1 = t - t_\circ,$$
$$P_2 = \omega = \pi - \Omega, \tag{4.330}$$
$$P_3 = h \cos i,$$

have been already obtained by another way in Sect. 4.8.7. As we will see in the following, the dynamic coordinates given by (4.325) and (4.330) have a significant role in the theory of special perturbations.

4.12 The Variation of Elements

A powerful method to find an approximate solution of N-body problem was developed by Lagrange specifically to investigate the motion of the System Solar planets (cf. Sect. 3.6). In this section we prove the famous Lagrange equations for the variations of the orbital elements and then make some applications.

Let be $\mathcal{H}_\circ(p, q, t)$ the Hamiltonian of a mechanical system with n degrees of freedom. Suppose further that the same mechanical system is subject to new conditions (for example, by adding other forces) and indicate the corresponding new

Hamiltonian with $\mathcal{H}(q, p, t)$. We assume also that the two Hamiltonians are such that:

$$\mathcal{H}(p, q, t) = \mathcal{H}_o(p, q, t) + \mathcal{R}(p, q, t). \tag{4.331}$$

The function $\mathcal{R}$ is called the perturbing Hamiltonian. We will show below that the solution to the problem related to $\mathcal{H}$ is given by:

$$\frac{da_i}{dt} = \sum_{k=1}^{2n} (a_k, a_i) \frac{\partial \mathcal{R}(a, t)}{\partial a_k} \qquad (i = 1, \ldots, 2n), \tag{4.332}$$

where the $2n$ variables a_i represent, at any moment, a possible set of motion constants for the Hamiltonian $\mathcal{H}_o(p, q, t)$. In particular, if the constants are canonical, distinguishing by α_k and β_k the coordinates from the conjugated moments, for the conditions (4.192) the equations (4.332) simplify to:

$$\begin{aligned} \frac{d\alpha_k}{dt} &= \frac{\partial \mathcal{R}(\beta, \alpha, t)}{\partial \beta_k} \\ \frac{d\beta_k}{dt} &= -\frac{\partial \mathcal{R}(\beta, \alpha, t)}{\partial \alpha_k} \end{aligned} \qquad (i = 1, \ldots, n). \tag{4.333}$$

The system of Hamilton equations for $\mathcal{H}$:

$$\begin{aligned} \dot{q}_k &= \frac{\partial \mathcal{H}}{\partial p_k} = \frac{\partial \mathcal{H}_o}{\partial p_k} + \frac{\partial \mathcal{R}}{\partial p_k} \\ \dot{p}_k &= -\frac{\partial \mathcal{H}}{\partial q_k} = -\frac{\partial \mathcal{H}_o}{\partial q_k} - \frac{\partial \mathcal{R}}{\partial q_k} \end{aligned} \qquad (k = 1, \ldots, n), \tag{4.334}$$

is equivalent to the system of differential equations (H.14) treated in Appendix H, if we identify the partial derivatives of $\mathcal{H}_o$ with the functions f_k and the partial derivatives of $\mathcal{R}$ with the functions r_k. Therefore, the solution of the system (4.334) will be given by functions such as:

$$\begin{aligned} q_k &= q_k(t, a_1, a_2, \ldots, a_{2n}) \\ p_k &= p_k(t, a_1, a_2, \ldots, a_{2n}) \end{aligned} \qquad (k = 1, \ldots, n). \tag{4.335}$$

The constants a_k relative to $\mathcal{H}_o$ are now thought of as functions of time, which must satisfy the conditions (H.17):

$$\sum_{k=1}^{2n} \frac{\partial q_i}{\partial a_k} \frac{da_k}{dt} = \frac{\partial \mathcal{R}}{\partial p_i}$$

$$\sum_{k=1}^{2n} \frac{\partial p_i}{\partial a_k} \frac{da_k}{dt} = -\frac{\partial \mathcal{R}}{\partial q_i}$$

$$(i = 1, \ldots, n). \qquad (4.336)$$

We multiply the first ones of (4.336) by $\partial p_i/\partial a_j$ and the second ones by $-\partial q_i/\partial a_j$, and then we sum up according to the index i. It results:

$$\sum_{k=1}^{2n} \frac{da_k}{dt} \sum_{i=1}^{n} \left(\frac{\partial p_i}{\partial a_j} \frac{\partial q_i}{\partial a_k} - \frac{\partial q_i}{\partial a_j} \frac{\partial p_i}{\partial a_k} \right) = \sum_{i=1}^{n} \left(\frac{\partial \mathcal{R}}{\partial p_i} \frac{\partial p_i}{\partial a_j} + \frac{\partial \mathcal{R}}{\partial q_i} \frac{\partial q_i}{\partial a_j} \right), \qquad (4.337)$$

or, noting that the second summation on the left side of the equation is a Lagrange bracket and that on the right side is the derivative of $\mathcal{R}$ with respect to a_j:

$$\sum_{k=1}^{2n} [a_j, a_k] \frac{da_k}{dt} = \frac{\partial \mathcal{R}}{\partial a_j} \qquad (j = 1, \ldots, 2n). \qquad (4.338)$$

Recalling the relations between Lagrange and Poisson brackets, it is immediate to invert the system (4.338) with respect to the time derivatives of the functions a_j. What we obtain is just the (4.332). Finally, from this we easily obtain the (4.333) if the $2n$ functions a_j form a canonical set for the Hamiltonian $\mathcal{H}_o$, since then the canonical conditions (4.334) hold.

One may ask how it is possible that variable functions can be at the same time constant solutions of the unperturbed problem. The explanation lies in the fact that functions a_k (or equivalent canonical functions) can be thought of as constants of motion of the unperturbed problem relative to initial conditions variable with time. This is precisely the starting point of Lagrange's thought, which is expressed in the notion of osculating orbit. The notion is so important that it deserves an illustrative example. First, however, let us explicit the (4.332) for the constants of the elliptical orbit, a, e, δ_o, Ω, ω, and i. Making use of expressions of the Poisson brackets calculated in Sect. 4.7, we have:

$$\frac{da}{dt} = (\delta_o, a)\,\frac{\partial \mathcal{R}}{\partial \delta_o} = \frac{2}{na}\frac{\partial \mathcal{R}}{\partial \delta_o}, \tag{4.339a}$$

$$\frac{de}{dt} = (\delta_o, e)\,\frac{\partial \mathcal{R}}{\partial \delta_o} + (\omega, e)\,\frac{\partial \mathcal{R}}{\partial \omega} = \frac{\sqrt{1-e^2}}{na^2 e}\left(\sqrt{1-e^2}\,\frac{\partial \mathcal{R}}{\partial \delta_o} - \frac{\partial \mathcal{R}}{\partial \omega}\right), \tag{4.339b}$$

$$\frac{d\delta_o}{dt} = (e, \delta_o)\,\frac{\partial \mathcal{R}}{\partial e} + (a, \delta_o)\,\frac{\partial \mathcal{R}}{\partial a} = -\frac{1}{na^2}\left(\frac{1-e^2}{e}\,\frac{\partial \mathcal{R}}{\partial e} + 2a\,\frac{\partial \mathcal{R}}{\partial a}\right), \tag{4.339c}$$

$$\frac{d\Omega}{dt} = (i, \Omega)\,\frac{\partial \mathcal{R}}{\partial i} = \frac{1}{na^2\sqrt{1-e^2}\,\sin i}\,\frac{\partial \mathcal{R}}{\partial i}, \tag{4.339d}$$

$$\frac{d\omega}{dt} = (e, \omega)\,\frac{\partial \mathcal{R}}{\partial e} + (i, \omega)\,\frac{\partial \mathcal{R}}{\partial i} = \frac{\sqrt{1-e^2}}{na^2}\left(\frac{1}{e}\frac{\partial \mathcal{R}}{\partial e} - \frac{\cot i}{1-e^2}\frac{\partial \mathcal{R}}{\partial i}\right), \tag{4.339e}$$

$$\frac{di}{dt} = (\Omega, i)\,\frac{\partial \mathcal{R}}{\partial \Omega} + (\omega, i)\,\frac{\partial \mathcal{R}}{\partial \omega} =$$

$$= \frac{1}{na^2\sqrt{1-e^2}}\left(-\frac{1}{\sin i}\frac{\partial \mathcal{R}}{\partial \Omega} + \cot i\,\frac{\partial \mathcal{R}}{\partial \omega}\right). \tag{4.339f}$$

4.12.1 Example to Illustrate the Method of the Variation of Constants

The mathematical and conceptual basis of the perturbation theory is often obscured by the large number of variables at play and by the considerable complexity of the formalism. To grasp the essential characteristics of the method, it is convenient to resort to a simple example. Following Forest Ray Moulton ([2] at pp. 367–72), we look for the solution of the equation:

$$\ddot{x} + \omega^2 x = -\mu\dot{x}^3 + v\cos(ft), \tag{4.340}$$

where ω^2, μ, v, and f are positive constants. If μ and v are null, equation (4.340) reduces to that of a harmonic motion, as that of a simple unperturbed pendulum. Therefore, the terms at the right-hand side of (4.340) represent perturbations on a purely harmonic motion. The first of them is velocity-dependent, as it may happen to a pendulum under the action of a resistant medium. The third power of the velocity is not chosen for physical reasons but only to satisfy the needs of our example; an even power would have made the strength of the perturbation independent of the direction of motion, while the first power would have allowed immediate integration. The second term at the right-hand side in equation (4.340) represents a periodic perturbation depending only on time.

Following the methodology of the theory of special perturbations, we transform the differential equation of the second order (4.340) in a system of two first order differential equations, placing:

$$y = x, \qquad z = \dot{x}, \tag{4.341}$$

with which we obtain:

$$\begin{cases} \dot{y} - z = 0, \\ \dot{z} + \omega^2 y = -\mu z^3 + \nu \cos(ft). \end{cases} \tag{4.342}$$

The corresponding equations of the unperturbed motion are:

$$\begin{cases} \dot{y} - z = 0, \\ \dot{z} + \omega^2 y = 0, \end{cases} \tag{4.343}$$

whose general integral is:

$$\begin{aligned} y &= \alpha \, \cos(\omega t) + \beta \, \sin(\omega t), \\ z &= -\omega \, \alpha \, \sin(\omega t) + \omega \beta \, \cos(\omega t), \end{aligned} \tag{4.344}$$

where α and β are arbitrary integration constants. In the language of celestial mechanics, they are the motion constants for the pendulum of our example.

Now consider the problem posed by the system (4.342). We look for solutions with the same form of (4.344), relative to the unperturbed motion, with the condition that α and β are now functions of time. From the mathematical point of view, the (4.344) are simply relations between two sets of independent variables, the original ones, y and z, and the new variables, α and β. They make possible the transformation of equations (4.342) from one system to another. It is evident that this statement remains true whatever the meaning of the relations (4.344) be, but since they are solutions of the unperturbed system (for constant α and β), we expect to obtain some advantage by operating this particular transformation of variables.

Based on the above considerations, in (4.342) we replace the variables y and z of (4.344) with α and β, thought of as functions of time. We obtain:

$$\dot{\alpha} \cos(\omega t) + \dot{\beta} \sin(\omega t) = 0,$$
$$\tag{4.345}$$
$$- \dot{\alpha} \sin(\omega t) + \dot{\beta} \cos(\omega t) = \mu \, \omega^2 \Big[\alpha \, \sin(\omega t) - \beta \, \cos(\omega t) \Big]^3 + \frac{\nu}{\omega} \cos(ft).$$

This linear system in $\dot{\alpha}$ and $\dot{\beta}$ can be solved with respect to these variables since the functional determinant is different from zero. It is:

$$\dot{\alpha} = -\mu \, \omega^2 \Big[\alpha \sin(\omega t) - \beta \cos(\omega t) \Big]^3 \sin(\omega t) - \frac{\nu}{\omega} \cos(ft) \sin(\omega t),$$
$$\tag{4.346}$$
$$\dot{\beta} = \mu \, \omega^2 \Big[\alpha \sin(\omega t) - \beta \cos(\omega t) \Big]^3 \cos(\omega t) + \frac{\nu}{\omega} \cos(ft) \cos(\omega t).$$

In searching for the solution, the system of equations (4.346) presents the same difficulties as the (4.342), where it comes from. Indeed, the functions α and β show the same complex algebraic structure with which y and z appear in (4.342). One wonders, then, where the benefits of the transformation is.

To find them, consider the case where the parameters μ and ν have very small values (relative to something we will specify later). This means that, correspondingly, $\dot{\alpha}$ and $\dot{\beta}$ have small values, so α and β vary slowly. Consequently, these two functions can be calculated with sufficient accuracy over a considerably wide interval of time if the system equations (4.346) are integrated treating α and β as constants at the second members. Note that this approximation does not affect the solution of the unperturbed problem, which remains as accurate as it was.

To better understand this point, which is the core of the method, we consider the simple equation:

$$\dot{\alpha} = \mu \, \alpha \Big[1 + \omega \, \cos(\omega t) \Big], \qquad (4.347)$$

whose solution is:

$$\alpha = \alpha_\circ \exp \Big\{ \mu \Big[t + \sin(\omega t) \Big] \Big\}, \qquad (4.348)$$

where $\alpha_\circ$ is an integration constant. By developing the second member of (4.348) in a power series of the parameter μ, the expression of α becomes:

$$\alpha = \alpha_\circ \Big\{ 1 + \mu \Big[t + \sin(\omega t) \Big] + \frac{\mu^2}{2} \Big[t + \sin(\omega t) \Big]^2 + \cdots \Big\}. \qquad (4.349)$$

Now, if μ is very small and t not too large[22] (the limit on time would not be required if this appeared only as an argument to the trigonometric functions), the expansion (4.349) can be truncated at the first order, keeping a good approximation. The same result could have been achieved by direct integration of (4.347) with the condition $\alpha = \alpha_\circ$ for the second term.

We then return to the (4.346) to integrate them in the approximation which treats as constants the functions α and β at the second members (we remark this fact by adding the suffix $\star$). The laborious quadrature returns:

---

[22] Note that in this case the variable t plays the function of a time interval, counted from the instant $t_\circ = 0$.

$$\alpha = \alpha_o - \mu\omega^2 \left\{ \frac{3\alpha_\star}{8} \left[\alpha_\star^2 + \beta_\star^2 \right] t + \frac{\beta_\star}{8\omega} \left[3\alpha_\star^2 + \beta_\star^2 \right] \left[\cos(2\omega t) - 1 \right] + \right.$$

$$- \frac{\beta_\star}{32\omega} \left[3\alpha_\star^2 - \beta_\star^2 \right] \left[\cos(4\omega t) - 1 \right] +$$

$$- \frac{\alpha_\star^3}{4\omega} \sin(2\omega t) + \frac{\alpha_\star}{32\omega} \left[\alpha_\star^2 - 3\beta_\star^2 \right] \sin(4\omega t) \right\} +$$

$$+ \frac{v}{2\omega(f+\omega)} \left\{ \cos\left[(f+\omega)\, t \right] - 1 \right\} +$$

$$- \frac{v}{2\omega(f-\omega)} \left\{ \cos\left[(f-\omega)\, t \right] - 1 \right\},$$

$$\text{(4.350)}$$

$$\beta = \beta_o + \mu\omega^2 \left\{ -\frac{3\beta_\star}{8} \left[\alpha_\star^2 + \beta_\star^2 \right] t - \frac{\alpha_\star}{8\omega} \left[\alpha_\star^2 + 3\beta_\star^2 \right] \left[\cos(2\omega t) - 1 \right] + \right.$$

$$+ \frac{\alpha_\star}{32\omega} \left[\alpha_\star^2 - 3\beta_\star^2 \right] \left[\cos(4\omega t) - 1 \right] +$$

$$- \frac{\beta_\star^3}{4\omega} \sin(2\omega t) + \frac{\beta_\star}{32\omega} \left[3\alpha_\star^2 - \beta_\star^2 \right] \sin(4\omega t) \right\} +$$

$$+ \frac{v}{2\omega(f+\omega)} \sin\left[(f+\omega)\, t \right] +$$

$$+ \frac{v}{2\omega(f-\omega)} \sin\left[(f-\omega)\, t \right],$$

where α_o and β_o are the values of α and β at the initial time $t = 0$. Finally, replacing the (4.350) in the (4.344), we obtain the approximate expressions of x and y, valid for values of t not too distant from the initial instant of time.

By examining the second terms of each of the (4.350), we note that time t appears as a factor in one of the terms, while in all the others it is an argument of periodic functions. The presence of a term directly proportional to t may suggests that α and β increase (or decrease) indefinitely with time. This is a hasty conclusion, though, which does not take into account the fact that the (4.350) are approximate solutions, acceptable within a limited time interval. It could be that the exact solution might contain powers of t greater than the first and that the sum of all these terms would still converge for any value of t. Think, for example, of an expansion of the sine function: $\sin t = \sum_{i=0}^{\infty} (-1)^i \frac{t^{2i+1}}{(2i+1)!}$. In the first order approximation, it results $\sin t = t$, an expression that, whether taken at plain face, would lead to the absurd that $\sin t > 1$ for $t > 1$ radiants. On the other hand it is not possible to exclude, in the absence of further investigations, that the terms proportional to t indicate really an indefinite increase of α and β over time. If this fact is verified, these terms are named secular.

The second terms of (4.350) also contain periodic terms of periods $\frac{2\pi}{\omega}$, $\frac{\pi}{2\omega}$, $\frac{2\pi}{(f+\omega)}$, $\frac{2\pi}{(f-\omega)}$. If $f \simeq \omega$, the terms with the cosine of $(f-\omega)t$ are named long period terms.

4.13 Apsidal Precession

Consider a classic problem of two bodies P_1 and P_2 with $m_1 \gg m_2$ and $\mathcal{E} < 0$ and assume that their otherwise elliptical orbit is perturbed by a force (per unit of reduced mass):

$$\mathbf{f}_p = -\alpha\, m_1 \frac{k}{r^{k+1}} \frac{\mathbf{r}}{r} \qquad (k \geq 2), \qquad (4.351)$$

where α is a constant (we place the gravitational constant $G = 1$) and $\mathbf{r} = \overrightarrow{P_1 P_2}$. Being purely radial, the perturbing force $\mathbf{f}_p$ does not alter the centrality of the motion of P_2, which therefore remains on a plane; in other words, the inclination does not change. We chose this plane as fundamental for the reference system, so the constants needed to describe the motion reduce to four; they can be, for instance, the independent orbital elements a, e, t_o, and $\pi = \Omega + \omega$ (the longitude at the ascending node and the argument of the pericenter cannot be independently defined because, with our choice of the reference system, the line of the nodes remains undetermined). Let us study the effect of the perturbation produced by (4.351) on line of apsides of the Keplerian orbit (that is, on the orientation of the major axis of the elliptical orbit), calculating the variation $\delta\pi$ of the longitude of the pericenter, π, in the time interval of one period.

Adding (4.339d) to (4.339e), and recalling that, with our assumption on $\mathbf{f}_p$, the inclination i is invariant, we have (for the unit mass $\mu = 1$):

$$\frac{d\pi}{dt} = (e, \omega) \frac{\partial \mathcal{R}}{\partial e} = \frac{\sqrt{1 - e^2}}{n\, a^2\, e} \frac{\partial \mathcal{R}}{\partial e}. \qquad (4.352)$$

The perturbation term $\mathcal{R}$ of Hamiltonian $\mathcal{H} = \mathcal{H}_0 + \mathcal{R}$ coincides with the potential of the force $\mathbf{f}_p = \nabla \mathcal{R} = \nabla\left(\dfrac{\alpha m_1}{r^k}\right)$, and therefore:

$$\frac{d\pi}{dt} = \frac{\sqrt{1 - e^2}}{n\, a^2\, e} \left(-\frac{k\,\alpha\, m_1}{r^{k+1}}\right) \frac{\partial r}{\partial e}. \qquad (4.353)$$

From (2.52), $r = a(1 - e \cos E)$, we have:

$$\frac{\partial r}{\partial e} = a\left(-\cos E + e \sin E \frac{\partial E}{\partial e}\right), \qquad (4.354)$$

and, from the Kepler equation (2.54), $E - e \sin E = M = n(t - t_0)$:

$$\left(1 - e \cos E\right) \frac{dE}{de} - \sin E = 0, \qquad (4.355)$$

with which:

$$\frac{\partial r}{\partial e} = a \left(-\cos E + \frac{e \sin^2 E}{1 - e \cos E} \right) = a \frac{e - \cos E}{1 - e \cos E}. \tag{4.356}$$

In conclusion:

$$\frac{d\pi}{dt} = \frac{k \alpha m_1 \sqrt{1 - e^2}}{n e a^{k+2}} \frac{\cos E - e}{(1 - e \cos E)^{k+2}}. \tag{4.357}$$

In order to integrate the equation (4.357), we use the approximation illustrated in the example of Sect. 4.12.1, justifiable on the basis of the Poincaré theorem. The idea is the following: the motion of P_2 is perturbed, but in a short interval of time the deviations from the Keplerian orbit are modest if the perturbation is small. In this case the orbit elements do not change significantly during one revolution, and can be considered constant in the second term of (4.357); see also the example at page 197. Only the eccentric anomaly E and the radius r (in turn dependent on E) remain functions of time. Under these working assumptions, we integrate (4.357) over a complete revolution, and denote the corresponding change in the longitude of the pericenter by $\delta \pi$.

It is convenient to use the eccentric anomaly as the integration variable; it appears both explicitly and implicitly as an argument of r. Differentiating the Kepler equation with respect to time in the hypothesis that a, n, and $t_\circ$ stay constant (enough) within a period, we have:

$$\left(1 - e \cos E \right) \frac{dE}{dt} = n, \tag{4.358}$$

and therefore, making also use of (2.83):

$$\frac{d\pi}{dE} = \frac{d\pi}{dt} \frac{dt}{dE} = \frac{k \alpha m_1 \sqrt{1 - e^2}}{n^2 e a^{k+2}} \frac{\cos E - e}{(1 - e \cos E)^{k+1}}. \tag{4.359}$$

Then, using Kepler's third law: $n^2 a^3 = m_1$ (remember that $G = 1$)

$$\begin{aligned}
\frac{d\pi}{dE} &= k \alpha \frac{\sqrt{1 - e^2}}{a^{k-1} e} \frac{\cos E - e}{(1 - e \cos E)^{k+1}} = \\
&= k \alpha \frac{\sqrt{1 - e^2}}{a^{k-1} e^2} \frac{e \cos E - e^2}{(1 - e \cos E)^{k+1}} = \\
&= k \alpha \frac{\sqrt{1 - e^2}}{a^{k-1} e^2} \frac{e \cos E - 1 + 1 - e^2}{(1 - e \cos E)^{k+1}} = \\
&= k \alpha \frac{\sqrt{1 - e^2}}{a^{k-1} e^2} \left[-\frac{1 - e \cos E}{(1 - e \cos E)^{k+1}} + \frac{1 - e^2}{(1 - e \cos E)^{k+1}} \right] = \\
&= k \alpha \frac{\sqrt{1 - e^2}}{a^{k-1} e^2} \left[(1 - e^2) A_{k+1} - A_k \right], \tag{4.360}
\end{aligned}$$

where:

$$A_k = \frac{1}{(1 - e \cos E)^k}. \tag{4.361}$$

The limits for a revolution still remain to be defined. We take $E = 0$ and $E = 2\pi$. In the perturbed motion, the variable E defined in the Kepler equation in which a, e, and t_o, are explicit functions of time, is called osculating eccentric anomaly. It is called anomalistic a revolution confined within the values $E = 0$ and $E = 2\pi$, corresponding to two successive transits through the pericenter. In the quadrature of the right-hand side of (4.360), we find integrals of the type:

$$B_k = \int_0^{2\pi} A_k \, dE = \int_0^{2\pi} \frac{dE}{(1 - e \cos E)^k}. \tag{4.362}$$

Through the relations (2.51b), we have:

$$B_k = \frac{1}{(1 - e^2)^{(2k-1)/2}} \int_0^{2\pi} (1 + e \cos \theta)^{k-1} d\theta, \tag{4.363}$$

which, in the special cases of $k = 2, 3, 4$, easily integrates into:

$$B_2 = \frac{2\pi}{(1 - e^2)^{3/2}}, \tag{4.364a}$$

$$B_3 = \frac{\pi (2 + e^2)}{(1 - e^2)^{5/2}}, \tag{4.364b}$$

$$B_4 = \frac{\pi (2 + 3e^2)}{(1 - e^2)^{7/2}}. \tag{4.364c}$$

(Do not confuse here the symbol π at the numerator of B_k with the longitude; it is Archimedes's constant $3.14159\ldots$.)

The change of the longitude of pericenter, π, for a complete revolution is then:

$$\delta\pi = k \, \alpha \frac{\sqrt{1 - e^2}}{a^{k-1} e^2} \left[-B_k + (1 - e^2) B_{k+1} \right], \tag{4.365}$$

and in particular, for $k = 2$:

$$\delta\pi = \frac{2\pi \, \alpha}{(1 - e^2) \, a}, \tag{4.366}$$

and for $k = 3$:

$$\delta\pi = \frac{6\pi \, \alpha}{(1 - e^2)^2 \, a^2}. \tag{4.367}$$

Summarizing, if on the plane of the orbits of two bodies we introduce a small perturbing radial force that varies as an inverse power low of the distance, it produces an advancement of the periastron. For the same force, the effect of the perturbation within a period is larger for a smaller semi-axis and/or a larger flattening.

4.14 The Two-Body Problem in General Relativity

Following up on the topic discussed in the previous section, here we evaluate the relativistic effects on the perihelion advance of a planetary orbit.[23] Applied to Mercury, this effect has provided one the main tests of general relativity and chronologically the first to be carried out, of the three originally proposed in 1915 by Einstein.[24] During the 19-th century, in fact, it came to be determined that Mercury's perihelion precession is equal to 5600 seconds of arc (≈ 1.6 degrees) per century,[25] all accounted for by the methods of classical celestial mechanics, excluding 43 arc seconds that required further assumptions to be made. This tension between theory and fact, often the bearer of important advances in science, required a drastic change of paradigm. The tiny little drift could be explained in 1919 in the framework of general relativity and not, as proposed in 1859 by the discoverer of Neptune, Urbain Le Verrier, assuming the perturbing action of an unseen planet placed between Mercury and the Sun (that Le Verrier, without waiting for confirmation, had even already baptized with the name of Vulcan, the god of fire).

We start recalling that, in classical mechanics, the Lagrangian of a free particle is given by the kinetic energy:

$$\mathcal{L} = \frac{m}{2}\, v^2. \tag{4.368}$$

In the relativistic case, the Lagrangian of the same particle of mass m in a field with metric g_{ik} has the following expression:

$$\mathcal{L} = \frac{m}{2}\, g_{ik}\, \frac{\partial x^i}{\partial \tau}\, \frac{\partial x^k}{\partial \tau} \qquad\qquad i = 0, \ldots, 3. \tag{4.369}$$

Here g_{ik} is a metric tensor and τ the proper time (i.e., the time as measured by a clock following a timelike world line). The Lagrangian equation (4.24) applied to this function gives the geodesic equation.

---

[23] This section requires some knowledge of general relativity but the topic has been treated in a way that does not presume the knowledge of tensor calculus.

[24] The other two were the bending of light in gravitational fields (with an angular value which is twice that of the Newtonian case), observed in 1919 during a Solar eclipse by Arthur Eddington (1882–1944), and the gravitational redshift (relative frequency shift of an electromagnetic wave due to the gravitational field of a compact object), clearly measured only in 1954.

[25] Note that only 574 out of the 5600 seconds of arc per century are caused by physical effects; the bulk is due to the motion of the reference system with respect to an inertial one (astronomers adopt the International Celestial Reference System (ICRS) that is almost coincident with the equatorial system at the year 2000.0).

We want to consider the motion of the test particle m around a body of mass $M \gg m$. The spherically symmetric solution of the Einstein equation in vacuum[26] is described by the Schwarzschild metric:

$$ds^2 = \left(1 - \frac{r_g}{r}\right) c^2 dt^2 - \frac{dr^2}{\left(1 - \frac{r_g}{r}\right)} - r^2 \left(d\theta^2 + \sin^2\theta d\phi^2\right), \qquad (4.370)$$

where x^0 is time coordinate and (r, θ, ϕ) are spherically coordinates:

$$x^0 = ct, \qquad x^1 = r, \qquad x^2 = \theta, \qquad x^3 = \phi. \qquad (4.371)$$

The components of the metric tensor are:

$$g_{00} = 1 - \frac{r_g}{r},$$

$$g_{11} = -\left(1 - \frac{r_g}{r}\right)^{-1},$$

$$g_{22} = -r^2, \qquad (4.372)$$

$$g_{33} = -r^2 \sin^2\theta,$$

and the gravitational radius for the mass M is:

$$r_g = \frac{2GM}{c^2}. \qquad (4.373)$$

In the following, we put for simplicity the speed of light $c = 1$ and the mass of the test particle $m = 1$.

Taking into account the components of the metric tensor (4.372), the Lagrangian (4.369) is:

$$\mathcal{L} = \frac{1}{2}\left[\left(1 - \frac{r_g}{r}\right)\left(\frac{dx^0}{d\tau}\right)^2 - \frac{1}{1 - \frac{r_g}{r}}\left(\frac{dr}{d\tau}\right)^2 + \right.$$

$$\left. - r^2\left(\frac{d\theta}{d\tau}\right)^2 - r^2 \sin^2\theta \left(\frac{d\phi}{d\tau}\right)^2\right]. \qquad (4.374)$$

Using (4.37) we can write the 4 momenta of the system which correspond to the 4 coordinates:

$$p_0 = \frac{\partial \mathcal{L}}{\partial \dot{x}^0} = \left(1 - \frac{r}{r_g}\right)\frac{dx^0}{d\tau}, \qquad (4.375a)$$

---

[26] Einstein equation in vacuum is $R_{ik} = 0$, where R_{ik} is the Ricci tensor. Gregorio Ricci Curbastro (1853–1925) was an Italian mathematician, famous for the invention of the tensor calculus.

$$p_r = \frac{\partial L}{\partial \dot{r}} = -\left(1 - \frac{r}{r_g}\right)^{-1} \frac{dr}{d\tau}, \tag{4.375b}$$

$$p_\theta = \frac{\partial L}{\partial \dot{\theta}} = -r^2 \frac{d\theta}{d\tau}, \tag{4.375c}$$

$$p_\phi = \frac{\partial L}{\partial \dot{\phi}} = -r^2 \sin^2 \theta \frac{d\phi}{d\tau}, \tag{4.375d}$$

where the upper dot denotes the derivation by the proper time τ. Through (4.38), consider the derivative of momentum:

$$\dot{p}_\theta = \frac{\partial L}{\partial \theta} = r^2 \sin \theta \cos \theta \dot{\phi}^2. \tag{4.376}$$

Through (4.375c), we can also see that:

$$\dot{p}_\theta = \frac{d}{d\tau}\left(-r^2 \frac{d\theta}{d\tau}\right) = -2r\dot{r}\dot{\theta} - r^2\ddot{\theta}. \tag{4.377}$$

As initial conditions we consider the equatorial plane ($\theta = \pi/2$) and $\dot{\theta} = 0$, which, from (4.376), give $\dot{p}_\theta = 0$. Since the left part of (4.377) is null, the condition $\dot{\theta} = 0$ implies $\ddot{\theta} = 0$. In this case the orbit lies on a plane with $\theta = \pi/2$. This result is similar to Newton's.

Since the Lagrangian does not depend on x^0 and ϕ, the corresponding momenta are constant. Integrating (4.375a) and (4.375d) with $\theta = \pi/2$, we have:

$$\left(1 - \frac{r_g}{r}\right)\frac{dx^0}{d\tau} = \text{const}_1 = \mathcal{E}, \tag{4.378}$$

$$r^2 \frac{d\phi}{d\tau} = \text{const}_2 = J. \tag{4.379}$$

Here the constants are the energy of the particle and the angular momentum respectively. Massive test particles move in timelike geodesics,[27] so $g_{il}\frac{dx^i}{d\tau}\frac{dx^l}{d\tau} = 1$, and the Lagrangian is $2\mathcal{L} = 1$. Substituting (4.378) and (4.379) in (4.374) and taking into account the condition for timelike geodesics $2\mathcal{L} = 1$ together with $\theta = \pi/2$, we get:

$$\left(\frac{dr}{d\tau}\right)^2 + \left(1 - \frac{r_g}{r}\right)\left(1 + \frac{J^2}{r^2}\right) = \mathcal{E}^2. \tag{4.380}$$

---

[27] A timelike interval must satisfy the condition $ds^2 \geq 0$, i.e., the interval between two event is real. If there is a reference system in which two events occur in the same place, the interval is always a timelike one because always $dl < cdt$.

To obtain the equation of the trajectory, we need to express r making ϕ explicit:

$$\frac{dr}{d\tau} = \frac{dr}{d\phi}\frac{d\phi}{d\tau} = \frac{dr}{d\phi}\frac{J}{r^2},$$ (4.381)

and:

$$\left(\frac{dr}{d\phi}\right)^2 = \frac{r^4}{J^2}\left[\mathcal{E}^2 - \left(1 - \frac{r_g}{r}\right)\left(1 + \frac{J^2}{r^2}\right)\right].$$ (4.382)

The equation of the trajectory is:

$$r = \int d\phi \frac{r^2}{J}\left[\mathcal{E}^2 - \left(1 - \frac{r_g}{r}\right)\left(1 + \frac{J^2}{r^2}\right)\right]^{1/2}.$$ (4.383)

This integral cannot be solved by elementary functions and requires elliptical integrals (see Sect. N.4). Placing $u = 1/r$, from which $dr = -du/u^2$, (4.382) takes the form:

$$\left(\frac{du}{d\phi}\right)^2 = \frac{1}{J^2}\left[\mathcal{E}^2 - (1 - r_g u)\left(1 + u^2 J^2\right)\right],$$ (4.384)

which, after simplifications, becomes:

$$\left(\frac{du}{d\phi}\right)^2 = r_g u^3 - u^2 + \frac{r_g}{J^2}u - \frac{1}{J^2}\left(1 - \mathcal{E}^2\right).$$ (4.385)

In the non-relativistic case, it can be proven that the corresponding equation has the form:

$$\left(\frac{du}{d\phi}\right)^2 = -u^2 + \frac{2GM}{J^2}u + \frac{\mathcal{E}}{J^2}.$$ (4.386)

There is a clear difference between the two cases, mainly the presence of an additional term. So, we can see (cf. Bertrand's theorem at Sect. 2.2) that in the relativistic case the orbit will not be closed, at variance with the classical two-body problem.[28]
We further change the variables placing:

$$u = \frac{1}{p}\left(1 + e\cos\theta'\right),$$ (4.387)

which is the orbit equation (2.64) in the classical two-body problem, where $p = a(1 - e^2)$ is the focal parameter, e the eccentricity, a the major semiaxis length, and θ' the true anomaly (we added the superscript to remark the difference with θ, which here is just the polar angle of the spherical reference frame). This form is convenient as we can compare the revolution by θ' with the corresponding angle ϕ.

---

[28] Note that the precession obtained in the previous section corresponds to a perturbed two-body problem.

Consequently, the left part of (4.385) is:

$$\left(\frac{du}{d\phi}\right)^2 = \left(\frac{du}{d\theta'}\frac{d\theta'}{d\phi}\right)^2 = \frac{e^2}{p^2}\sin^2\theta'\left(\frac{d\theta'}{d\phi}\right)^2, \tag{4.388}$$

so that the equation (4.385) has the following form:

$$\frac{e^2}{p^2}\sin^2\theta'\left(\frac{d\theta'}{d\phi}\right)^2 = r_g f(u), \tag{4.389}$$

where:

$$f(u) = u^3 - \frac{1}{r_g}u^2 + \frac{u}{J^2} - \frac{1}{r_g J^2}(1 - \mathcal{E}^2). \tag{4.390}$$

Taking into account Vieta's[29] theorem, the expression (4.390) writes as:

$$f(u) = (u - u_1)(u - u_2)(u - u_3) =$$
$$= u^3 - u^2(u_1 + u_2 + u_3) + u(u_1 u_2 + u_1 u_3 + u_2 u_3) - u_1 u_2 u_3, \tag{4.391}$$

where u_1, u_2, u_3 are the roots of equation: $f(u) = 0$. Comparing (4.390) with (4.391), we obtain:

$$u_1 + u_2 + u_3 = \frac{1}{r_g}, \tag{4.392a}$$

$$u_1 u_2 + u_1 u_3 + u_2 u_3 = \frac{1}{J^2}, \tag{4.392b}$$

$$u_1 u_2 u_3 = \frac{1}{r_g J^2}(1 - \mathcal{E}^2). \tag{4.392c}$$

Since we consider the equation $(du/d\phi)^2 = r_g f(u)$, then $f(u) \geq 0$ and $f(u) = 0$ corresponds to extremes of u. Therefore, we can parametrise the two roots as:

$u_1 = u_{min}(\theta = \pi) = (1 - e)/p$ (apoapsis),
$u_2 = u_{max}(\theta = 0) = (1 + e)/p$ (periapsis).

The third root is obtained from (4.392a) using the expressions above:

$u_3 = 1/r_g - 2/p,$

and:

$$f(u) = (u - u_1)(u - u_2)(u - u_3) = -\frac{e^2}{p^3}\sin^2\theta'\left(3 + e\cos\theta' - \frac{p}{r_g}\right), \tag{4.393}$$

---

[29] Vieta's formulas relate the coefficients of a polynomial to sums and products of its roots. Named after François Viète (1540–1603), more commonly referred to by the Latinate form of his name, "Franciscus Vieta", the formulas are used specifically in algebra.

where we applied the (4.387). Substituting it in (4.389), we obtain:

$$\left(\frac{d\theta'}{d\phi}\right)^2 = 1 - \frac{r_g}{p}\left(3 + e\cos\theta'\right). \tag{4.394}$$

The solution of this differential equation gives us the orbit for this relativistic case. The function $\theta'(\phi)$ can be expressed through elliptical integrals. To see it, we put $\cos\theta' = 2\cos^2(\theta'/2) - 1$ in (4.394), obtaining:

$$\left(\frac{d\theta'}{d\phi}\right)^2 = (1 - 3b + be)\left(1 - k^2\cos^2(\theta'/2)\right), \tag{4.395}$$

where:

$$k^2 = \frac{2be}{1 - 3b + be}, \tag{4.396}$$

with $b = r_g/p$. Note that $1 - 3b + be > 0$ and $k^2 < 1$. We rewrite equation (4.395) in the form:

$$\frac{d\theta'}{\sqrt{1 - k^2\cos^2(\theta'/2)}} = \pm\sqrt{1 - 3b + be}\, d\phi. \tag{4.397}$$

The integral of the left part of (4.397) can be expressed through the elliptical integral using the following substitution; $\beta = \pi/2 - \theta'/2$. So, we have:

$$\phi = \pm\frac{2\,K\left(\dfrac{\pi}{2} - \dfrac{\theta'}{2}, k\right)}{\sqrt{1 - 3b + be}}, \tag{4.398}$$

where:

$$K(\beta, k) = \int_0^\beta \frac{d\beta'}{\sqrt{1 - k^2\sin^2\beta'}}, \tag{4.399}$$

is the incomplete elliptic integral of the first kind (see Sect. N.4) and k its modulus determined by (4.396).

To obtain the value of the perihelion shift, we need to invert equation (4.398) with respect to θ':

$$\theta' = \pi - 2\operatorname{am}\left(\frac{\phi}{2\sqrt{1 - 3b + be}}, k\right), \tag{4.400}$$

where $\operatorname{am}(\beta, k)$ is the Jakobi amplitude which is the inverse function relative to the incomplete elliptic integral of the first kind: $K(\operatorname{am}(\beta, k)) = \beta$. By substituting it in (4.387), we finally obtain the equation of the orbit:

$$r = \frac{p}{1 + e \cos \left[\pi - 2 \operatorname{am} \left(\dfrac{\phi}{2\sqrt{1 - 3b + be}}, k \right) \right]}.$$ (4.401)

We now use (4.394) in the form:

$$\frac{d\theta'}{\sqrt{1 - \dfrac{r_g}{p}(3 + e \cos \theta')}} = d\phi.$$ (4.402)

We consider the case where r_g/p is small (which obviously applies to the planets of the Solar System) and expand (4.402) in series retaining only the first term:

$$d\theta' \left[1 + \frac{r_g}{2p} (3 + e \cos \theta') \right] \approx d\phi.$$ (4.403)

After integration, we obtain:

$$\phi - \phi_0 = \left(1 + \frac{3r_g}{2p} \right) \theta' + \frac{r_g}{2p} e \sin \theta'.$$ (4.404)

One revolution corresponds to $\theta' = [0, 2\pi]$. For $\theta' = 0$, the angle is $\phi = \phi_0$; for $\theta' = 2\pi$, we can see from (4.404) that:

Fig. 4.4 Orbit in the relativistic two-body problem, showing the advance of perihelion

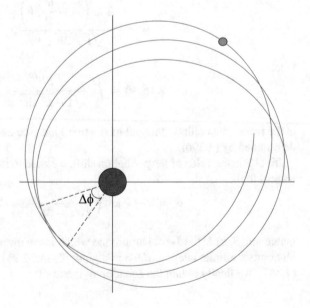

$$\phi - \phi_0 = 2\pi \left(1 + \frac{3r_g}{2p} \right) = 2\pi + \Delta\phi, \tag{4.405}$$

where:

$$\Delta\phi = \frac{3\pi r_g}{p} = \frac{3\pi r_g}{a(1 - e^2)}. \tag{4.406}$$

The angle (4.406) is the relativistic shift of the perihelion of the planet. It is the consequence of the fact that the orbit, at variance with the classical case where it is just Keplerian, here is open with a rosette pattern (Fig. 4.4) because of the curvature of the space induced by the central mass.

The relativistic shift $\Delta\phi$ for one revolution of Mercury due the Sun (using $r_g = 2.95$ km for the Sun and for Mercury $a_{\mercury} = 57.91 \times 10^6$ km, $e_{\mercury} = 0.206$, and orbital period $P_{\mercury} = 87.6$ days) is $\Delta\phi \approx 5.01 \times 10^{-7}$ rad; over 100 years, it returns exactly $\Delta\phi = 43''$.

A recent spectacular application of the formula (4.406) concerns the star named S2 orbiting around the supermassive black hole (SMBH) that sits at the center of the Milky Way, in Sagittarius (Fig. 4.5). The orbital characteristics of the star (semi-

Fig. 4.5 Apparent orbit of the star S2 around the supermassive black hole at the center of the Milky Way, from observations with ESO telescopes and instruments (listed in the legend) over a period of more than 25 years. Note the position of the SMBH which, because of projection, is well off the focus of the apparent orbit (cf. Appendix J). *Credit* ESO/MPE/GRAVITY Collaboration

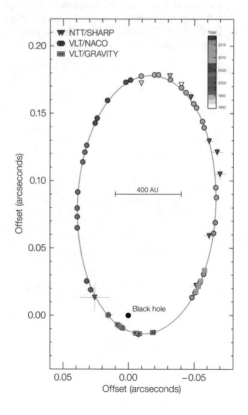

major axis $a_{S2} = 971$ AU and eccentricity $e_{S2} = 0.88$) derived from very difficult astrometric and spectroscopic measurements (the system is at a distance of 7.93 kpc from the Earth, where a_{S2} appears of only 0.123 arc seconds), together with the huge mass of the galactic SMBH, $M_{BH} = 4.1 \times 10^6\, M_\odot$, implying a gravitational radius $r_g = 1.21 \times 10^7$ km, according to (4.406) determine a remarkable apsidal advance of the stellar orbit, equal to $\Delta\phi \simeq 12'$ over a period of $P_{S2} = 16.05$ years. Direct measures have confirmed the predictions based on general relativity, which in this way has received a further robust confirmation (cf. [8]).

References

1. E. Finlay-Freundlich, *Celestial Mechanics* (Pergamon Press, New York, 1958)
2. F.R. Moulton, *An Introduction to Celestial Mechanics* (The MacMillan Company, New York, 1960)
3. J.E. Marsden, T.S. Ratiu, *Introduction to Mechanics and Symmetry, in Applied Mathematics*, vol. 17 (Springer, Berlin, 1994)
4. H. Goldstein, *Classical Mechanics* (Addison-Wesley, Boston, 1980)
5. L.D. Landau, E.M. Lifshitz, *Mechanics* (Elsevier, Amsterdam, 1982)
6. R.A. Broucke, On the matrizant of the two-body problem. Astron. Astrophys. **6**, 173–182 (1970)
7. J. Binney, S. Tremaine, *Galactic Dynamics* (Princeton University Press, Princeton, 1987)
8. R. Abuter et al., Detection of the Schwarzschild precession in the orbit of the star S2 near the Galactic Center massive black hole. Astron. Astrophys. **636**, L5 (2020)

Chapter 5
Gravitational Potential

Lewis Carroll *Alice's Adventures in the Wonderland (1865)*
*"I could have done it in a much more complicated way", said
the Red Queen, immensely proud.*

In the previous chapters we have limited ourselves to consider only massive points, i.e., fictitious bodies that attract each other according to a law of force with a seemingly simple mathematical expression: Newton's law of gravitation.[1] This approximation is successfully applied to many real situations. For instance, it is appropriate to deal with the motions of the planets of the Solar System, treated as point-like bodies given the enormous distances among them and the Sun (relative to their sizes). However, the use of massive particles instead of real bodies (finite and even asymmetric) prevents us from modeling phenomena such as spin-orbit interactions.[2] Usually, the point-like approximation fails more and more as the distances among the various bodies, compared to their sizes, reduce. Obviously, Newton's interaction law continues to hold true even when we consider forms of bodies more complex than massive points, but only between pairs of elements of infinitesimal volume in

---

[1] We remember that Newton's gravitational force acts instantaneously; it propagates at infinite velocity and does not require any mediator. It was precisely this action at distance that Newton himself did not trust. Einstein was also unhappy with the infinite propagation velocity which was against special relativity. It was the need to eliminate this discrepancy that led him to reconsider gravitation and formulate in 1915–16 the so-called theory of general relativity.

[2] The best know example of the spin-orbit interaction is the identity of the rotation and revolution periods of the Moon which is the reason why the satellite shows always the same side to the Earth (but for the libration, i.e., the wavering of the Moon perceived by Earth-bound observers and caused by changes in perspective, by the ellipticity of the orbit, and by an intrinsic oscillation). Other famous spin-orbit resonances happen with Sun-Mercury, Earth-Venus, and the Medicean planets. See also Chap. 1 and https://history.nasa.gov/SP-345/ch8.htm.

© The Author(s), under exclusive license to Springer Nature Switzerland AG 2022   213
E. Bannikova and M. Capaccioli, *Foundations of Celestial Mechanics*, Graduate Texts
in Physics, https://doi.org/10.1007/978-3-031-04576-9_5

which the same bodies can be ideally subdivided. The evaluation of the global effect (which is the sum of these interactions) goes through an integration process that often reaches a high level of mathematical difficulty. This led to the development of tools that we need to know in order to be able to manipulate problems with complex potentials: pluripotent instruments, which had and still have useful applications in several branches of mathematical physics. This chapter is devoted to the presentation of some general methods to describe the properties of the gravitational field generated by a body of any finite shape, with some relevant examples.

5.1 The Theorem of Gauss or of the Flux

In an orthogonal Cartesian reference system $O[x, y, z]$ we consider the function $\rho(x, y, z)$ which maps the spatial matter density within the volume Σ bounded by the closed surface S. In other words, ρ may describe the mass distribution in a finite body[3] or in that part of a finite body falling inside S. We indicate with:

$$m(\Sigma) = \iiint_{\Sigma} \rho \, d\Sigma, \qquad (5.1)$$

the total mass enclosed by S, and with:

$$\mathbf{g}(P) = G \iiint_{\Sigma} \rho \frac{\mathbf{r}}{r^3} \, d\Sigma, \qquad (5.2)$$

the gravitational acceleration vector (gravitational force per unit mass) generated at any point $P(x, y, z)$ of S from the mass m enclosed by S. The flux $\mathcal{F}$ of $\mathbf{g}$ though S is the integral of the component g_n of $\mathbf{g}$ orthogonal to S, extended to the entire surface:

$$\mathcal{F}(S) = \iint_{S} g_n \, dS = \iint_{S} \mathbf{g} \cdot d\mathbf{S}, \qquad (5.3)$$

with the usual rule that $d\mathbf{S}$ is an infinitesimal vector perpendicular to the corresponding surface element and oriented outwards; its modulus $|d\mathbf{S}|$ is the measure of the surface of the infinitesimal area element. The Gauss or continuity theorem (also known as Gauss flux theorem) states that the flux $\mathcal{F}$ of the field $\mathbf{g}$ through the surface S depends on the total mass m only and is:

$$\mathcal{F}(S) = \iint_{S} \mathbf{g} \cdot d\mathbf{S} = 4\pi \, G \, m. \qquad (5.4)$$

---

[3] What follows remains true for a discrete distribution of massive points too; obviously the integrals change into summations.

Replacing into (5.3) the expression (5.2) of the acceleration of gravity, we obtain a quintuple integral extended to the entire volume Σ and to the boundary surface S:

$$\mathcal{F}(S) = G \iint_S \left[\iiint_\Sigma \rho \frac{\mathbf{r}}{r^3} d\Sigma \right] \cdot d\mathbf{S}, \tag{5.5}$$

where the modulus r of the vector:

$$\mathbf{r} = (x' - x)\,\mathbf{i} + (y' - y)\,\mathbf{j} + (z' - z)\,\mathbf{k}, \tag{5.6}$$

represents the distance of P from the volume element $d\Sigma(x', y', z')$ belonging to Σ, and $\mathbf{i}$, $\mathbf{j}$ and $\mathbf{k}$ are the unit vectors of the reference system axes. Since the two integration variables are independent one from the other, it is also:

$$\mathcal{F}(S) = G \iiint_\Sigma d\Sigma \, \rho \iint_S \frac{1}{r^3} \mathbf{r} \cdot d\mathbf{S}. \tag{5.7}$$

The scalar $\mathbf{r} \cdot d\mathbf{S}/r^3$ is the elementary solid angle $d\omega$ (Fig. 5.1) under which the volume element $d\Sigma$ sees the element dS of the closed surface. Thus:

$$\iint_S \frac{\mathbf{r} \cdot d\mathbf{S}}{r^3} = \iint_{4\pi} d\omega = 4\pi. \tag{5.8}$$

The expression (5.7) becomes then:

$$\mathcal{F}(S) = 4\pi \, G \iiint_\Sigma \rho \, d\Sigma, \tag{5.9}$$

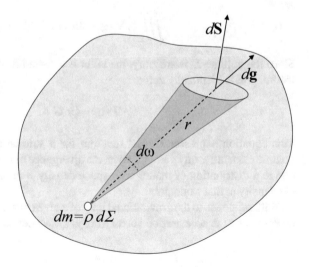

Fig. 5.1 Gauss theorem: integration of the gravitational acceleration vector

which, together with (5.1), demonstrates the Gauss theorem. Note how the result is strictly dependent on the proportionality of the acceleration to the inverse of the square of the distance (in an Euclidean space).

5.2   The Equations of Poisson and Laplace

The Gauss theorem allows us to express the flow of the gravitational acceleration through a surface enclosing a body (or a cluster of bodies) of space density ρ as:

$$\iint_S \mathbf{g} \cdot d\mathbf{S} = 4\pi\, G \iiint_\Sigma \rho\, d\Sigma. \tag{5.10}$$

Obviously ρ can be non-zero even outside the boundaries of Σ, whether S is cutting the material body, but only the inner part counts in (5.10). By applying the divergence theorem (cf. (C.11) in Appendix C), the surface integral in (5.10) can be transformed into a volume integral of the function:

$$\nabla \cdot \mathbf{g}(x, y, z) = \frac{\partial g_x}{\partial x} + \frac{\partial g_y}{\partial y} + \frac{\partial g_z}{\partial z}, \tag{5.11}$$

where g_x, g_y, and g_z, are the components of $\mathbf{g}$:

$$\iint_S \mathbf{g} \cdot d\mathbf{S} = \iiint_\Sigma \nabla \cdot \mathbf{g}\, d\Sigma, \tag{5.12}$$

so that the (5.10) transforms into:

$$\iiint_\Sigma \left[\nabla \cdot \mathbf{g} - 4\pi\, G\, \rho \right] d\Sigma = 0. \tag{5.13}$$

Since the volume Σ is arbitrary, the latter is satisfied if and only if the argument of the integral is identically zero:

$$\nabla \cdot \mathbf{g} = 4\pi\, G\, \rho. \tag{5.14}$$

This equation expresses Gauss' theorem for a surface enclosing an infinitesimal volume element: at any point in space, the divergence of the gravitational acceleration due to a distribution of matter with space density ρ is proportional to the value of the density at that same point.

With reference to the gravitational field of a given distribution of matter $\rho(x, y, z)$, described by the acceleration vector $\mathbf{g}(x, y, z)$, we denote by $\mathcal{U}(x, y, z)$ (as we have already done at Sect. 1.2.1) the function named potential[4] defined, up to an additive constant, by:

$$\mathbf{g} = -\nabla \mathcal{U}, \qquad (5.15)$$

where:

$$\nabla \mathcal{U} = \frac{\partial \mathcal{U}}{\partial x}\mathbf{i} + \frac{\partial \mathcal{U}}{\partial y}\mathbf{j} + \frac{\partial \mathcal{U}}{\partial z}\mathbf{k}. \qquad (5.16)$$

It can be shown that the potential exists for any distribution of matter. In fact, whatever it is the matter density ρ within the volume Σ, the function:

$$\mathcal{U} = G \iiint_{\Sigma} \frac{\rho \, d\Sigma}{r}, \qquad (5.17)$$

satisfies (5.15). It is:

$$\nabla \mathcal{U} = \nabla \left[G \iiint_{\Sigma} \frac{\rho \, d\Sigma}{r} \right], \qquad (5.18)$$

and, since the gradient does not depend on the integration variable:

$$\nabla \mathcal{U} = G \iiint_{\Sigma} \rho \, d\Sigma \, \nabla \left(\frac{1}{r} \right) = -G \iiint_{\Sigma} \frac{\rho \, d\Sigma}{r^3} \mathbf{r}. \qquad (5.19)$$

It is apparent from (5.19) why the potential function is defined up to an additive constant. As a rule, we agree to remove this indeterminacy imposing that $\mathcal{U}$ vanishes at an infinite distance from the body that generates it; a condition verified by (5.19). Also note that the gravitational potential does not explicitly depend on time (as it is instead for the electromagnetic potential, for instance).

Replacing in (5.14) the expression of $\mathbf{g}$ given by (5.15), we obtain the famous Poisson equation:

$$\nabla^2 \mathcal{U}(x, y, z) = \frac{\partial^2 \mathcal{U}}{\partial x^2} + \frac{\partial^2 \mathcal{U}}{\partial y^2} + \frac{\partial^2 \mathcal{U}}{\partial z^2} = -4\pi \, G \, \rho(x, y, z), \qquad (5.20)$$

where the operator ∇^2 is called Laplacian or squared nabla. Its properties are reviewed in Appendix C together with other vector operators.

---

[4] This function was introduced by Laplace in a memory appeared in 1785; the name of potential with which it now universally known was subsequently adopted by the British George Green. The expression 'potential energy' was instead coined by the Scottish engineer and physicist William Rankine (1820–1872) who, together with the German Rudolf Clausius (1822–1888) and the English William Thomson, Lord Kelvin (1824–1907), gave fundamental contributions to thermodynamics. The concept of 'potentiality' is far older and goes back to the speculations of Aristotle. The Greek philosopher identified matter with 'power' and form with 'act'. Potency is the possibility of changing and becoming something else; the act identifies both the activity to bring about change and reality produced by that activity. It precedes potency with respect to substance and value.

If the density ρ is zero, as is the case in space outside material bodies, from (5.20) we obtain the Laplace equation:

$$\nabla^2 \mathcal{U} = 0. \tag{5.21}$$

Most of this chapter will be devoted to the study of this equation. In fact, the solution of this homogeneous differential equation provides the most general expression for the potential in the outer space of any body and therefore, through (5.15), also the expression of the relative force field.

To introduce us to the problem, let us first consider an elementary case, the study of the potential due to a massive point Q placed outside the origin of the reference system. This simple shift of the coordinate system from that centered at Q is sufficient to complicate the expression of the distance that appears at the denominator of Newton's law, and therefore the expression of the potential itself. Clearly, in the case of a single massive point, the obstacle can be overcome by shifting back the center of the coordinate system at the point Q. But this possibility vanishes when, for example, there are two or more massive points involved in the computation of the potential and they are distinct. Thus, the case we propose in the next section is anything but an academic exercise.

5.3   The Potential of a Massive Point

Given a the spherical reference system of coordinates, the gravitational potential generated in $P(r, \theta, \phi)$ by the point-like mass m placed in $Q(r_o, \theta_o, \phi_o)$, is:

$$\mathcal{U} = G \frac{m}{\delta}, \tag{5.22}$$

where (Fig. 5.2):

$$\delta = |\overrightarrow{QP}| = (r^2 - 2rr_o \cos \gamma + r_o^2)^{1/2}, \tag{5.23}$$

and, using (4.303):

$$\cos \gamma = \mathbf{r} \cdot \mathbf{r}_o = \frac{x\,x_o}{r\,r_o} + \frac{y\,y_o}{r\,r_o} + \frac{z\,z_o}{r\,r_o} =$$
$$= \cos \theta \, \cos \theta_o + \sin \theta \, \sin \theta_o \, \cos(\phi - \phi_o). \tag{5.24}$$

It can be shown directly that (5.22) satisfies the Laplace equation for any value of $\delta \neq 0$. Unfortunately, the function δ^{-1} is hardly tractable; so it is necessary to search for a different representation of the reciprocal of the distance between P and Q. For this purpose, let it be:

$$f(z, \mu) = (1 - 2\mu z + z^2)^{-1/2} \qquad |z| < 1, \; |\mu| \leq 1. \tag{5.25}$$

Fig. 5.2 Geometry to
compute the potential
generated by a massive point
placed off the center of the
coordinate system

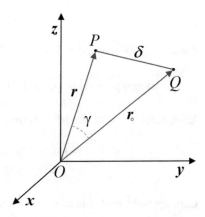

Placing $\mu = \cos \gamma$, and:

$$z = \begin{cases} r/r_\circ & \text{for } r \leq r_\circ, \\ r_\circ/r & \text{for } r \geq r_\circ, \end{cases} \tag{5.26}$$

the expression of the potential in P (5.22) can be re-written as:

$$\mathcal{U} = \begin{cases} G \dfrac{m}{r_\circ} f(z, \mu) & \text{for } r \leq r_\circ, \\[2mm] G \dfrac{m}{r} f(z, \mu) & \text{for } r \geq r_\circ. \end{cases} \tag{5.27}$$

The first of (5.27) is valid for all points inside the sphere of radius $r_\circ$ centered in the
origin O of the coordinate system, while the second one applies to external points.

We will now try to develop the function $f(z, \mu)$ in a power series of z. To this
end, note that:

$$f(z, \mu) = (1 - 2\mu z + z^2)^{-1/2} =$$

$$= \left| 1 - z \left(\mu - \sqrt{\mu^2 - 1} \right) \right|^{-1/2} \times \left| 1 - z \left(\mu + \sqrt{\mu^2 - 1} \right) \right|^{-1/2}, \tag{5.28}$$

as it is proven by running the product directly. By the new variables:

$$\begin{aligned} x &= -z(\mu - \sqrt{\mu^2 - 1}), \\ x' &= -z(\mu + \sqrt{\mu^2 - 1}), \end{aligned} \tag{5.29}$$

the (5.28) clearly shows to be the product of two binomials:

$$f(z, \mu) = (1 + x)^{-1/2}(1 + x')^{-1/2}. \tag{5.30}$$

We firstly demonstrate that:

$$|x| = |x'| < 1. \tag{5.31}$$

Since by hypothesis $|\mu| \le 1$, the quantity $(\mu^2 - 1)^{1/2}$ is purely imaginary; therefore:

$$\left| \mu \pm \sqrt{\mu^2 - 1} \right| = \left[(\mu \pm i\sqrt{1 - \mu^2})(\mu \mp i\sqrt{1 - \mu^2}) \right]^{1/2} = 1. \tag{5.32}$$

Then, since we assumed $|z| < 1$, from the (5.29) it results:

$$\begin{aligned} |x\,| &= |z|\,|(\mu - \sqrt{\mu^2 - 1})| = |z| < 1, \\ |x'| &= |z|\,|(\mu + \sqrt{\mu^2 - 1})| = |z| < 1. \end{aligned} \tag{5.33}$$

Consequently, each one of the two functions $(1 + x)^{-1/2}$ and $(1 + x')^{-1/2}$ of (5.30) can be developed in a binomial series. We remind that the expansion in a Taylor[5] series of the function $f(x) = (1 + x)^m$, where m is any complex number:

$$(1 + x)^m = \binom{m}{0} + \binom{m}{1} x + \cdots + \binom{m}{n} x^n + \cdots , \tag{5.35}$$

with the generalized coefficients being:

$$\binom{m}{n} = \frac{m!}{n!(m - n)!} = \frac{m(m-1)(m-2)\cdots(m-n+1)}{n!}, \tag{5.36}$$

and $\binom{m}{0} = 1$, converges absolutely[6] in the open interval $|x| < 1$. If m is a positive integer, the series reduces to the well-known Newton binomial formula. In conclusion, the development of the binomials of (5.30) converges absolutely to these same functions throughout their common domain. It is:

---

[5] Brook Taylor (1685–1731): British mathematician who, among other achievements, sat in the commission charged to solve the controversy between Newton and Leibniz about the priority in the introduction of the differential methodology in mathematical analysis. He authored the famous equation known as Taylor's theorem:

$$f(x) = \sum_{n=0}^{\infty} \left(\frac{d^n f}{dx^n} \right)_{x_0} \frac{(x - x_0)^n}{n!}, \tag{5.34}$$

where the function $f(x)$ is defined in an open interval $(x_0 - \Delta x, x_0 + \Delta x)$ with real or complex values and it is infinitely differentiable (if $x_0 = 0$, the expansion is named after the Scottish mathematician Colin Maclaurin). The importance of this power series was not recognized until 1772, when Lagrange realized its value, calling it "the main foundation of differential calculus".

[6] A real or complex series is said to converge absolutely if the sum of the absolute values of the summands is a finite (real) number. A key property of absolute series is that rearranging the terms, which is not allowed for conditionally convergent series, does not change the value of the sum.

$$(1+x)^{-1/2} = \sum_{m=0}^{\infty} \binom{-1/2}{m} x^m = 1 - \frac{1}{2}x + \frac{3}{8}x^2 - \frac{5}{16}x^3 + \cdots,$$

$$(1+x')^{-1/2} = \sum_{n=0}^{\infty} \binom{-1/2}{n} x'^m = 1 - \frac{1}{2}x' + \frac{3}{8}x'^2 - \frac{5}{16}x'^3 + \cdots.$$

(5.37)

Let us multiply the two series by the Cauchy product,[7] which ensures that the sum of the product of two absolutely convergent series coincides with the algebraic product of the sums of the series themselves. We then have:

$$f(z, \mu) = (1+x)^{-1/2}(1+x')^{-1/2} = \sum_{k=0}^{\infty} P_k(\mu) z^k.$$

(5.39)

The coefficients $P_k(\mu)$ of the powers of z have the form:

$$P_0(\mu) = 1,$$

(5.40a)

$$P_1(\mu) z = -\frac{1}{2}x - \frac{1}{2}x' = \mu z,$$

(5.40b)

$$P_2(\mu) z^2 = \frac{3}{8}x^2 + \frac{3}{8}x'^2 + \frac{1}{4}xx' = \left(\frac{3}{2}\mu^2 - \frac{1}{2}\right) z^2,$$

(5.40c)

$$\cdots\cdots\cdots$$

They are called Legendre coefficients or Legendre[8] polynomials of degree k. The function (5.39) has the name of generating function of the Legendre polynomials.

Through the development (5.39), the expression (5.27) of the potential of a massive point becomes:

$$\mathcal{U}(r) = G\frac{m}{r_o} \sum_{k=0}^{\infty} P_k(\cos\gamma)\left(\frac{r}{r_o}\right)^k \qquad \text{for } r \leq r_o,$$

(5.41a)

$$\mathcal{U}(r) = G\frac{m}{r} \sum_{k=0}^{\infty} P_k(\cos\gamma)\left(\frac{r_o}{r}\right)^k \qquad \text{for } r \geq r_o.$$

(5.41b)

---

[7] The Cauchy product is a discrete convolution of two infinite series:

$$\sum_{i=0}^{\infty} f_i(x) \times \sum_{j=0}^{\infty} g_j(x) = \sum_{k=0}^{\infty}\sum_{l=0}^{k} f_l(x) g_{k-l}(x).$$

(5.38)

[8] Adrien-Marie Legendre (1752–1833) was a French mathematician. A student of Euler and Lagrange, he contributed to the fields of geometry, differential calculus, function theory, mechanics, and number theory. Concepts such as Legendre polynomials and Legendre transformation bear his name.

You can see Appendix O for an account on the main properties of the Legendre polynomials and a useful digression on the family of orthogonal polynomials.

5.4 The Potential of Spherical Bodies

We consider here a massive infinitely thin spherical layer Σ centered in O, with radius $r_\circ$ and infinitesimal thickness $dr_\circ$, and assume that the volume density $\rho_\circ$ and the surface density $\sigma_\circ = \rho_\circ \, dr_\circ$ are both constant. We apply the (5.17) to compute the potential at a generic point P. Owing to the perfect symmetry of the density, we can always place P on the z-axis of the reference system centered in O (Fig. 5.3). Whatever the value of $r = OP$ is, all the surface elements of the infinitesimal spherical corona of axis OP and mass $dm = 2\pi \, \sigma_\circ r_\circ^2 \, \sin\theta \, d\theta$, have the same distance from P:

$$QP = (r^2 - 2rr_\circ \cos\theta + r_\circ^2)^{1/2}. \tag{5.42}$$

Thus:

$$d\mathcal{U}(r) = G \frac{dm}{QP} = G \frac{2\pi \, \sigma_\circ r_\circ^2 \, \sin\theta \, d\theta}{(r^2 - 2rr_\circ \cos\theta + r_\circ^2)^{1/2}}, \tag{5.43}$$

and, by integrating over the sphere:

$$\mathcal{U}(r) = G \int_0^\pi \frac{dm}{QP} = G \int_0^\pi \frac{2\pi \, \sigma_\circ r_\circ^2 \, \sin\theta \, d\theta}{(r^2 - 2rr_\circ \cos\theta + r_\circ^2)^{1/2}}. \tag{5.44}$$

Fig. 5.3 Potential of a thin spherical layer

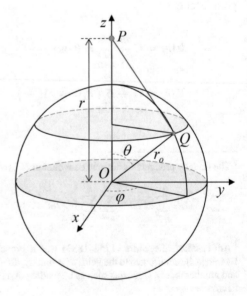

The integral in (5.44) reduces to the known form:

$$\mathcal{U}(r) = -2\pi\, G\,\sigma_o \int_0^\pi \frac{d(\cos\theta)}{(r^2 + r_o^2 - 2rr_o\cos\theta)^{1/2}} =$$
$$= \frac{2\pi\, G\,\sigma_o}{r_o\, r}\left(r_o + r - |r_o - r|\right). \qquad (5.45)$$

So, opening the module, we obtain the expression for the potential of the homogeneous thin spherical layer:

$$\mathcal{U}(r) = \begin{cases} \dfrac{Gm}{r} & \text{for } r \ge r_o, \\[2mm] \dfrac{Gm}{r_o} & \text{for } r \le r_o. \end{cases} \qquad (5.46)$$

This result can be obtained by another way using the expansion in Legendre polynomials. To solve the integral (5.44) we distinguish the two cases: P external to Σ (that is, $r > r_o$) and P internal to Σ ($r < r_o$). In the first case, by placing $z = r_o/r < 1$, we have:

$$\mathcal{U}_e(r) = 2\pi\, G\sigma_o \frac{r_o^2}{r}\int_0^\pi \frac{\sin\theta\, d\theta}{(1 - 2z\cos\theta + z^2)^{1/2}}. \qquad (5.47)$$

From (5.39), remembering that the binomial series is absolutely convergent, it results:

$$\mathcal{U}_e(r) = 2\pi\, G\,\sigma_o \frac{r_o^2}{r}\int_0^\pi -\left[\sum_{k=0}^\infty z^k P_k(\cos\theta)\right] d(\cos\theta) =$$
$$= -2\pi\, G\,\sigma_o \frac{r_o^2}{r}\sum_{k=0}^\infty z^k \int_0^\pi P_k(\cos\theta)\, d(\cos\theta) =$$
$$= 2\pi\, G\,\sigma_o \frac{r_o^2}{r}\sum_{k=0}^\infty z^k \int_{-1}^{+1} P_0(\mu) P_k(\mu)\, d(\mu), \qquad (5.48)$$

where we have placed $\mu = \cos\theta$ and used the identity $P_0(\mu) \equiv 1$ (cf. the property (a) at Appendix O.3) to insert the zero-th order Legendre polynomial in the integral. By this way we can exploit the orthogonality of Legendre polynomials:

$$\int_{-1}^1 P_0(\mu) P_k(\mu)\, d\mu = \begin{cases} 2 & \text{for } k = 0, \\ 0 & \text{for } k \ne 0. \end{cases} \qquad (5.49)$$

In conclusion, by assuming that $r > r_o$, we obtain the potential at any point outside the spherical shell, which is a function of just the radial distance r of P from the shell center:

$$\mathcal{U}_e(r) = 4\pi\, G\sigma_o \frac{r_o^2}{r} = G\frac{m}{r}, \qquad (5.50)$$

as $4\pi\, r_{\mathrm{o}}^2\sigma_{\mathrm{o}}$ coincides with the total mass m within the volume Σ (actually over the thin surface S). The expression just found proves that the potential in a point outside a thin and homogeneous spherical layer coincides with that of a point of equal mass placed at the center of the shell.

We now consider the second case, when P is inside Σ. Placing $z = r/r_{\mathrm{o}} < 1$ and repeating the procedure adopted for the first case, we have:

$$\mathcal{U}_i(r) = 4\pi\, G\,\sigma_{\mathrm{o}}\frac{r_{\mathrm{o}}^2}{r_{\mathrm{o}}} = G\,\frac{m}{r_{\mathrm{o}}}. \tag{5.51}$$

Thus, the potential at any point internal to a thin and homogeneous spherical layer is constant. These two results have been put together in Fig. 5.6. Note the discontinuity of $\nabla\mathcal{U}$ at the surface of the layer, at the radius $r = r_{\mathrm{o}}$.

The constant value of the inner potential of the spherical layer tells us that the cumulative force acting on an internal point is zero, whatever the position of the point. This can be explained in terms of simple physical considerations, similar to those that made Newton himself to discover this fundamental property. Consider a point P located within the homogeneous spherical layer and draw two opposite cones, both emerging from P, with the same small solid angles (Fig. 5.4):

$$d\Omega = \frac{dS_1}{r_1^2} = \frac{dS_2}{r_2^2}. \tag{5.52}$$

These cones cut two opposite spherical caps with masses dm_1 and dm_2 proportional to the areas dS_1 and dS_2. Then the ratio of the moduli of the gravitational forces acting in opposite directions on P due to the attraction generated by their mass elements is:

$$\frac{F_1}{F_2} = \frac{dm_1}{R_1^2}\frac{R_2^2}{dm_2} = \frac{dS_1}{R_1^2}\frac{R_2^2}{dS_2} = 1, \tag{5.53}$$

where we used the (5.52). The preceding considerations can be repeated for the case of an ellipsoidal layer and lead us to the same result, known as Newton's theorem: The attraction produced by a homogeneous ellipsoidal layer on a point located in its inner cavity is zero.

We can now generalize the previous result to a solid sphere with mass m, radius r_{o}, and a constant volume density ρ_{o}. The strategy is to decompose the sphere into a set of homogeneous shells, each characterized by a radius r', and make use of the results just obtained for the inner and outer potentials of homogeneous layers.

We start considering the inner potential of the solid sphere, i.e. the potential at a point $P(r)$ which is inside of its volume at a distance $r \le r_{\mathrm{o}}$ from the center. The spherical surface containing P divides the volume in two regions: $[0, r]$ and $[r, r_{\mathrm{o}}]$ (Fig. 5.5). The point P is external to the shells with radii $r' \le r$; hence for them we resort to the expression (5.50) which in infinitesimal form writes:

Fig. 5.4 Attraction by opposite elements of a homogeneous thin spherical layer at an inner point P. The conic elements, drawn deliberately large for graphical convenience, should actually be infinitesimal

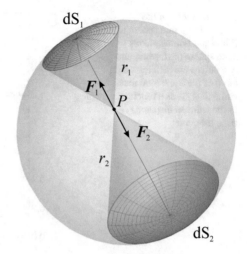

Fig. 5.5 Scheme of a homogeneous solid sphere divided in the set of infinitely thin spherical layers

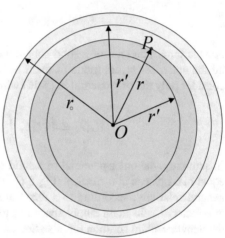

$$d\mathcal{U} = 4\pi G \rho_\circ \frac{r'^2}{r} dr'. \tag{5.54}$$

But P is also internal to the spheres with radii $r \le r' \le r_\circ$. In this case we use the (5.51):

$$d\mathcal{U} = 4\pi G \rho_\circ \frac{r'^2}{r'} dr' = 4\pi G \rho_\circ r' dr'. \tag{5.55}$$

Summing up the (5.54) and the (5.55) duly integrated within the corresponding radial ranges, we find that potential inside the solid sphere:

$$\mathcal{U}_i(r) = 4\pi G \rho_\circ \int_0^r \frac{r'^2}{r} dr' + 4\pi G \rho_\circ \int_r^{r_\circ} r' dr' = \frac{2}{3}\pi G \rho_\circ \left(3r_\circ^2 - r^2\right). \tag{5.56}$$

Fig. 5.6 Trend of the
potential of a thin spherical
(*blue*) and homogeneous
solid sphere (*red*, where the
two curves are different, then
blue) of the same total mass
and radius r_o, as a function
of the reduced distance r/r_o
from the center

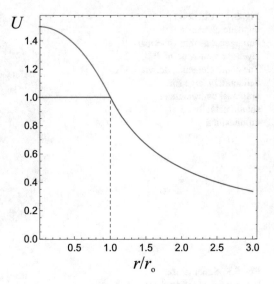

For the outer potential of a solid sphere we take into account the contributions of
all the spherical layers (none of them being 'external'); in other words, we must
consider only the first integral in (5.56) with upper limit r_o:

$$\mathcal{U}_e(r) = 4\pi G \rho_o \int_0^{r_o} \frac{r'^2}{r} dr' = \frac{Gm}{r}. \tag{5.57}$$

In summary, the outer potential of a solid sphere is equivalent to that of a point of
equal mass placed at the center (Fig. 5.6). There is some analogy of this result with
general relativity: according to a theorem due to Birkhoff,[9] a spherically symmetric
gravitational field in an empty space has to be static with metric corresponding to
the Schwarzschild solution (cf. also Sect. 4.14).

We now want to address this same problem in the framework of the so-called
Dirichlet[10] problem, which is to find a function solving a specified partial differential
equation inside a given region under prescribed boundary values (cf. Appendix P).
This problems applies to several cases, but it was originally posed for the Laplace
equation. Let the density filling a definite volume Σ be such that the corresponding
potential satisfies Poisson's equation. If we find a function $\mathcal{U}$ which is regular at
infinity, continuous in the whole area together with its first derivative, and such that
it satisfies the equation:

---

[9] George David Birkhoff (1884–1944): leading American mathematician, best known for what is
now called the ergodic theorem.
[10] Lejeune Dirichlet (1805–1859), German mathematician, best known for the modern definition
of a function.

$$\nabla^2 \mathcal{U} = \begin{cases} -4\pi G \rho_\circ & \text{inside the region } \Sigma, \\ 0 & \text{outside the region } \Sigma, \end{cases} \tag{5.58}$$

then this function coincides with the potential. Actually, the theorem tells us that the potential is a single solution of equation (5.58) satisfying all the above conditions.

Now consider again a solid sphere of radius $r_\circ$ and mass m, with constant density $\rho_\circ$. Due to spherical symmetry, the potential depends on the radial coordinate only: $\mathcal{U} = \mathcal{U}(r)$. So, the Laplace equation assumes the following form[11]:

$$\frac{1}{r^2} \frac{\partial}{\partial r} \left(r^2 \frac{\partial \mathcal{U}_e}{\partial r} \right) = 0, \tag{5.59}$$

where $\mathcal{U}_e$ is the potential in the outer region ($r > r_\circ$). By direct integration we obtain:

$$\mathcal{U}_e(r) = -\frac{c_1}{r} + c_2. \tag{5.60}$$

The constants in (5.60) are set by physical considerations: $c_1 = -Gm$ and $c_2 = 0$, as we want the potential to vanish at infinity.

The gravitational potential in the inner region should satisfy Poisson equation:

$$\frac{1}{r^2} \frac{\partial}{\partial r} \left(r^2 \frac{\partial \mathcal{U}_i}{\partial r} \right) = -4\pi G \rho_\circ, \tag{5.61}$$

which solves in:

$$\mathcal{U}_i(r) = -\frac{2\pi}{3} G \rho_\circ r^2 - \frac{c_1}{r} + c_2. \tag{5.62}$$

Taking into account the boundary conditions given by the request of continuity for both potential and force:

$$\begin{aligned} \mathcal{U}_e \Big|_{r=r_\circ} &= \mathcal{U}_i \Big|_{r=r_\circ}, \\ \frac{d\mathcal{U}_e}{dr} \Big|_{r=r_\circ} &= \frac{d\mathcal{U}_i}{dr} \Big|_{r=r_\circ}, \end{aligned} \tag{5.63}$$

we obtain $c_1 = 0$ from the second of (5.63), and from the first:

$$c_2 = \frac{3}{2} \frac{Gm}{r_\circ}. \tag{5.64}$$

In summary, by this approach the gravitational potential of a homogeneous solid sphere has the following form:

---

[11] Cf. Appendix C for the expression of the Laplacian in spherical coordinates.

$$\mathcal{U}_i(r) = \frac{Gm}{2r_\circ^3} \left(3r_\circ^2 - r^2\right) \qquad \text{for} \quad r \leq r_\circ,$$

$$\mathcal{U}_e(r) = \frac{Gm}{r} \qquad\qquad \text{for} \quad r \geq r_\circ,$$

(5.65)

in agreement with (5.56) and (5.57).

5.5   The Equation of Legendre

From the Laplace equation, which is the starting point of our analysis, we come to the study of the so-called Legendre equation, which has particular solutions called spherical harmonics. Their properties will be described both in the following subsection and in Appendix O.3.

5.5.1   Spherical Harmonics

We now return to the Laplace equation (5.21) with the purpose of studying the properties of an important class of solutions. Following J.C. Maxwell, we call a solid spherical harmonic of degree n any solution $\mathcal{U}_n(x, y, z)$ of the Laplace equation that is positively homogeneous of degree n in x, y, and z. Recall that a function of k variables, $f(x_1, x_2, \ldots, x_k)$, is said to be positively homogeneous of degree α, with α real, if at every point of the domain and for each positive value of the parameter t it is:

$$f(tx_1, tx_2, \ldots, tx_k) = t^\alpha f(x_1, x_2, \ldots, x_k).$$

(5.66)

For example, a simple homogeneous function of degree 1 in three variables takes the form: $f_1 = c_1 x_1 + c_2 x_2 + c_3 x_3$; one of degree 2 can be: $f_2 = c_1 x_1^2 + c_2 x_2^2 + c_3 x_3^2 + c_4 x_1 x_2 + c_5 x_2 x_3 + c_6 x_1 x_3$. In a similar way, more complex homogeneous functions of higher degrees can be constructed. For this whole class, the following theorem due to Euler applies: a necessary and sufficient condition for the function $f(x_1, x_2, \ldots, x_k)$ to be positively homogeneous of degree α is that the relation:

$$x_1 \frac{\partial f}{\partial x_1} + x_2 \frac{\partial f}{\partial x_2} + \cdots + x_k \frac{\partial f}{\partial x_k} = \alpha f(x_1, x_2, \ldots, x_k),$$

(5.67)

is identically verified. By transforming the coordinates from the Cartesian system $O[x, y, z]$ into the spherical system $O[r, \theta, \phi]$ (see Fig. 4.3), and by applying Euler's theorem, we verify that a solid spherical harmonic of degree n in the variables x, y, and z, can always be written as:

$$\mathcal{U}_n(x, y, z) = r^n U_n(\theta, \phi).$$

(5.68)

The function $U_n(\theta, \phi)$ is called surface harmonic. It represents the trend, as a function of the angular variables, of the corresponding solid spherical harmonic at the surface of the sphere of unit radius centered in the origin of the coordinate system. At any other spherical surface of radius r and concentric to this one, the trend of the volume harmonic remains the same, but the amplitude is controlled by the factor r^n.

A theorem named after Lord Kelvin holds for solid spherical harmonics: if $\mathcal{U}_n$ is a solid spherical harmonic of degree n, then the function $r^{-2n-1}\mathcal{U}_n$ is also a solid spherical harmonic, but of degree $(-n - 1)$. First we want to prove that, if $m = -2n - 1$, the function $\mathcal{U} = r^m \mathcal{U}_n(x, y, z)$ satisfies the equation of Laplace. It is:

$$\frac{\partial \mathcal{U}}{\partial x} = \frac{\partial}{\partial x}\left(r^m \mathcal{U}_n\right) = r^m \frac{\partial \mathcal{U}_n}{\partial x} + m\, x\, r^{m-2}\, \mathcal{U}_n, \tag{5.69}$$

where we have used the relation:

$$\frac{\partial r}{\partial x} = \frac{\partial}{\partial x}(x^2 + y^2 + z^2)^{1/2} = \frac{x}{r}. \tag{5.70}$$

Moreover:

$$\frac{\partial^2 \mathcal{U}}{\partial x^2} = r^m \frac{\partial^2 \mathcal{U}_n}{\partial x^2} + 2x\, m\, r^{m-2} \frac{\partial \mathcal{U}_n}{\partial x} + m\, r^{m-2}\, \mathcal{U}_n + m(m-2)\, x^2\, r^{m-4} \mathcal{U}_n. \tag{5.71}$$

Adding similar expressions for y and z, and using Euler's theorem, we obtain:

$$\frac{\partial^2 \mathcal{U}}{\partial x^2} + \frac{\partial^2 \mathcal{U}}{\partial y^2} + \frac{\partial^2 \mathcal{U}}{\partial z^2} = \nabla^2 \mathcal{U} = r^m \nabla^2 \mathcal{U}_n + m(m + 2n+1)\, r^{m-2} \mathcal{U}_n =$$

$$= m(m + 2n+1)\, r^{m-2} \mathcal{U}_n, \tag{5.72}$$

as, by assumption, $\nabla^2 \mathcal{U}_n = 0$. The (5.72) shows that the condition $\nabla^2 \mathcal{U} = 0$ is satisfied if $m = -2n - 1$.

It remains to prove that the function $\mathcal{U} = r^{-2n-1}\mathcal{U}_n$ is positively homogeneous of degree $(-n-1)$. To this purpose, we multiply the (5.69) by x and add the analogous expressions for y and z. After the obvious reductions, we obtain:

$$x\frac{\partial \mathcal{U}}{\partial x} + y\frac{\partial \mathcal{U}}{\partial y} + z\frac{\partial \mathcal{U}}{\partial z} = (-n-1)\,\mathcal{U}, \tag{5.73}$$

which, based on Euler's theorem, is sufficient to prove the hypothesis.

We now seek the solution of the Laplace equation via an expansion in a series of spherical harmonics. This requires finding the relationship between these functions and Legendre polynomials.

5.5.2 *Legendre Equation and Spherical Harmonics*

The explicit form of the Laplace equation in spherical coordinates is (see Appendix C):

$$\nabla^2 \mathcal{U} = \frac{\partial^2 \mathcal{U}}{\partial r^2} + \frac{2}{r}\frac{\partial \mathcal{U}}{\partial r} + \frac{1}{r^2}\frac{\partial^2 \mathcal{U}}{\partial \theta^2} + \frac{\cot\theta}{r^2}\frac{\partial \mathcal{U}}{\partial \theta} + \frac{1}{r^2 \sin^2\theta}\frac{\partial^2 \mathcal{U}}{\partial \phi^2} = 0, \qquad (5.74)$$

or, multiplying by r^2:

$$r^2 \nabla^2 \mathcal{U} = r\frac{\partial^2 (r\mathcal{U})}{\partial r^2} + \frac{1}{\sin\theta}\frac{\partial}{\partial \theta}\left(\sin\theta\frac{\partial \mathcal{U}}{\partial \theta}\right) + \frac{1}{\sin^2\theta}\frac{\partial^2 \mathcal{U}}{\partial \phi^2} = 0. \qquad (5.75)$$

Substituting into (5.75) the expression of a solid spherical harmonic $\mathcal{U}_n$ given by (5.68), we obtain the form of Laplace's equation for surface harmonics:

$$n(n+1)U_n + \frac{1}{\sin\theta}\frac{\partial}{\partial \theta}\left(\sin\theta\frac{\partial U_n}{\partial \theta}\right) + \frac{1}{\sin^2\theta}\frac{\partial^2 U_n}{\partial \phi^2} = 0, \qquad (5.76)$$

or, if we place $\mu = \cos\theta$, thus implicitly assuming that $-1 \le \mu \le +1$:

$$n(n+1)U_n + \frac{\partial}{\partial \mu}\left[(1-\mu^2)\frac{\partial U_n}{\partial \mu}\right] + \frac{1}{1-\mu^2}\frac{\partial^2 U_n}{\partial \phi^2} = 0. \qquad (5.77)$$

Note the difference between $\mu = \cos\theta$ and $\mu = \cos\gamma$, the latter being a function of θ and ϕ (cf. 5.24). In order to avoid any ambiguity, in the following we will always refer explicitly to the meaning of the variable μ.

Suppose now that the surface spherical harmonic of degree n allows for the separation of variables:

$$U_n(\theta, \phi) = \Theta(\theta)\,\Phi(\phi) = \Theta(\mu)\,\Phi(\phi). \qquad (5.78)$$

Substituting in (5.77), we have:

$$n(n+1)(1-\mu^2) + \frac{1-\mu^2}{\Theta}\frac{d}{d\mu}\left[(1-\mu^2)\frac{d\Theta}{d\mu}\right] + \frac{1}{\Phi}\frac{d^2\Phi}{d\phi^2} = 0. \qquad (5.79)$$

The first two terms of (5.79) do not depend on ϕ, while the last one does not depend on $\mu = \cos\theta$. Since θ and ϕ are independent variables, the identity (5.79) is satisfied[12] only if:

---

[12] If x and y are independent variables, the identity $f(x) + g(y) = 0$ implies that the functions f and g are separately equal to constants with the same modulus and opposite sign.

$$n(n+1)(1 - \mu^2) + \frac{1 - \mu^2}{\Theta} \frac{d}{d\mu}\left[(1 - \mu^2)\frac{d\Theta}{d\mu}\right] = m^2, \qquad (5.80a)$$

$$\frac{1}{\Phi}\frac{d^2\Phi}{d\phi^2} = -m^2. \qquad (5.80b)$$

The sign of the non-negative constant m^2 has been chosen in so that the function $\Phi(\phi)$ is periodic with period 2π. The integration of (5.80b) gives:

$$\Phi(\phi) = A\cos(m\phi) + B\sin(m\phi), \qquad (5.81)$$

where A and B are arbitrary constants. Multiplying (5.80a) by $\dfrac{\Theta}{(1 - \mu^2)}$ and recalling that $\Theta(\theta) = y(\mu)$, it follows that the function $\Theta(\theta)$ must satisfy the equation:

$$\frac{d}{d\mu}\left[(1 - \mu^2)\frac{dy(\mu)}{d\mu}\right] + \left[n(n+1) - \frac{m^2}{1 - \mu^2}\right]y(\mu) = 0, \qquad (5.82)$$

for all the values of μ in the interval $-1 \le \mu \le +1$ (external solutions have no interest here). Equation (5.82) is known by the name of associated Legendre equation. Its solutions are indicated by the symbol y_n^m in order to emphasize that they contain the two parameters n and m (note the order of these parameters with respect to their position in the Legendre equation).

In conclusion, we can establish that, if $\Theta_n^m(\theta)$ is a solution of the associated Legendre equation (5.82) in the interval $0 \le \theta \le \pi$ ($|\mu| \le 1$), the function:

$$U_n(\theta, \phi) = \left[A\cos(m\phi) + B\sin(m\phi)\right]\Theta_n^m(\theta), \qquad (5.83)$$

is a surface spherical harmonic of degree n, since it satisfies (5.77). Therefore:

$$\mathcal{U}_n(r, \theta, \phi) = r^n\left[A\cos(m\phi) + B\sin(m\phi)\right]\Theta_n^m(\theta), \qquad (5.84)$$

and, owing to the Kelvin theorem (Sect. 5.5.1):

$$\mathcal{U}_{-n-1}(r, \theta, \phi) = r^{-n-1}\left[A\cos(m\phi) + B\sin(m\phi)\right]\Theta_n^m(\theta), \qquad (5.85)$$

are both solid spherical harmonics of degree n and $(-n-1)$ respectively.

5.5.3   Particular Solution of the Legendre Equation: the Associated Functions

To search for some particular solutions of the associated Legendre equation, it is convenient to start with the function:

$$D(x, y, z) = \left[(x - x_o)^2 + (y - y_o)^2 + (z - z_o)^2 \right]^{-1/2}, \tag{5.86}$$

which actually satisfies the Laplace equation. One might prove this property directly, but it would be a futile effort. It is enough to observe that the function D, being the inverse of a distance, has the form of a Newtonian potential for a massive point. Indicated with γ the angle[13] between the vector radii $\mathbf{r}$ and $\mathbf{r}_o$ of the points $P(x, y, z)$ and $Q(x_o, y_o, z_o)$ respectively, the (5.86) writes as:

$$D(r, r_o, \gamma) = (r^2 - 2rr_o \cos \gamma + r_o^2)^{-1/2}. \tag{5.87}$$

Assuming that $r < r_o$ and remembering (5.41a), we have:

$$\nabla^2 D = \nabla^2 \left[\sum_{n=0}^{\infty} P_n(\cos \gamma) \frac{r^n}{r_o^{n+1}} \right] = 0, \tag{5.88}$$

where $P_n(\cos \gamma)$ is a Legendre polynomial of degree n. Expliciting the Laplace operator, the (5.88) becomes:

$$\sum_{n=0}^{\infty} \frac{r^n}{r_o^{n+1}} \left[n(n+1) P_n(\cos \gamma) + \frac{1}{\sin \theta} \frac{\partial}{\partial \theta} \left(\sin \theta \frac{\partial}{\partial \theta} P_n(\cos \gamma) \right) + \right.$$
$$\left. + \frac{1}{\sin^2 \theta} \frac{\partial^2}{\partial \phi^2} P_n(\cos \gamma) \right] = 0. \tag{5.89}$$

Since this power series must be identically null for all $r < r_o$, all of its coefficients must also vanish:

$$n(n+1) P_n(\cos \gamma) + \frac{1}{\sin \theta} \frac{\partial}{\partial \theta} \left(\sin \theta \frac{\partial}{\partial \theta} P_n(\cos \gamma) \right) +$$
$$+ \frac{1}{\sin^2 \theta} \frac{\partial^2}{\partial \phi^2} P_n(\cos \gamma) = 0. \tag{5.90}$$

This equation is identical to (5.76), which proves that $P_n(\cos \gamma)$ is a surface spherical harmonic of integer degree.

---

[13] Note that γ does not coincide with θ unless one of the two points defining the angle (together with the origin of the reference system) does contain the z-axis.

Let us now consider the special case $\gamma = \theta$. In this instance, the third term in (5.90) cancels, and the latter becomes identical to the associated Legendre equation with $m = 0$. In conclusion, we have proved that the Legendre polynomials of argument $\mu = \cos\theta$ form a family of special solutions of the homonym equation:

$$\frac{d}{d\mu}\left[(1-\mu^2)\frac{d}{d\mu}y(\mu)\right] + n(n+1)\, y(\mu) = 0, \tag{5.91}$$

in the interval $|\mu| \le 1$ and for non-negative values of the parameter n.

That said, we return to the associated Legendre equation (5.82). We will consider only integer and non-negative values of the parameters n and m, with the further condition that $m \le n$. Via the auxiliary function $u(\mu)$ defined by the position[14]:

$$y(\mu) = (\mu^2 - 1)^{m/2} u(\mu), \tag{5.92}$$

with some long but straightforward reductions, (5.82) becomes:

$$(1-\mu^2)\frac{d^2 u}{d\mu^2} - 2(m+1)\,\mu\,\frac{du}{d\mu} + (n-m)(n+m+1)\, u = 0. \tag{5.93}$$

Let us compare it with the equation obtained by differentiating[15] m times the Legendre equation (5.91):

$$(1-\mu^2)\frac{d^{m+2}y}{d\mu^{m+2}} - 2(m+1)\mu\frac{d^{m+1}y}{d\mu^{m+1}} + (n-m)(n+m+1)\frac{d^m y}{d\mu^m} = 0. \tag{5.94}$$

It is immediate to see that, if y_n is a solution of (5.94), the function:

$$u_n^m(\mu) = \frac{d^m}{d\mu^m} y_n(\mu), \tag{5.95}$$

is solution of (5.93) and thus, through (5.92), the function:

$$y_n^m(\mu) = (\mu^2 - 1)^{m/2}\frac{d^m}{d\mu^m} y_n(\mu), \tag{5.96}$$

---

[14] For $|\mu| \le 1$, the function $y(\mu)$ takes complex values for m odd.

[15] Recall that, given two functions $f(x)$ and $g(x)$, the following formula due to Newton holds:

$$\frac{d^n}{dx^n}(fg) = \sum_{k=0}^{n}\binom{n}{k} f^{n-k-1} g^{k+1} =$$

$$= g\frac{d^n f}{dx^n} + n\frac{dg}{dx}\frac{d^{n-1}f}{dx^{n-1}} + \cdots + \frac{n(n-1)\cdots 3 \cdot 2}{(n-1)!}\frac{d^{n-1}g}{dx^{n-1}}\frac{df}{dx} + \frac{d^n g}{dx^n}f.$$

Also note that $\dfrac{d^k(1-\mu^2)}{d\mu^k} = 0$ for $k \ge 3$.

is a solution of the associated Legendre equation. Since, due to our hypothesis on the parameter n, Legendre polynomials are solutions of (5.93) in the interval $-1 < \mu < +1$, the functions:

$$P_n^m(\mu) = (\mu^2 - 1)^{m/2} \frac{d^m}{d\mu^m} P_n(\mu), \qquad (5.97)$$

form, in the same interval, a family of solutions of the associated Legendre equation. They are named associated Legendre functions.[16]

The nature of these functions becomes clear when the Legendre polynomial (5.97) is represented by the Rodrigues[17] formula:

$$P_n^m(\mu) = \frac{(\mu^2 - 1)^{m/2}}{2^n n!} \frac{d^{m+n}}{d\mu^{m+n}} (\mu^2 - 1)^n. \qquad (5.98)$$

It shows that, apart from the factor $(\mu^2 - 1)^{m/2}$, $P_n^m(\mu)$ is a polynomial of degree $(n - m)$. It also helps to prove that the associated functions too satisfy orthogonality conditions (similar to (O.45)):

$$\int_{-1}^{+1} P_n^m(\mu) P_{n'}^m \, d\mu = 0 \qquad\qquad n \neq n'. \qquad (5.99)$$

For a demonstration see the Appendix O.

Following the literature, we define the following functions:

$$T_n^m(\mu) = (-1)^{m/2} P_n^m(\mu) = (-1)^{m/2} \frac{(\mu^2 - 1)^{m/2}}{2^n n!} \frac{d^{m+n}}{d\mu^{m+n}} (\mu^2 - 1)^n, \qquad (5.100)$$

named after Ferrers.[18] These are real functions that retain all the considerations made so far for the associated Legendre functions.

5.5.4 The Spherical Harmonics of Integer Degree

If n and $0 \le m \le n$ are two integers and the variable $\mu = \cos\theta$ spans the interval $-1 < \mu < +1$, the function:

---

[16] Note that, for $m = 0$, the associated Legendre polynomial $P_n^m(\mu)$ becomes a simple Legendre polynomial $P_n(\mu)$. Under these hypotheses, in fact, the (5.82) simplifies into (5.91).

[17] Benjamin Olinde Rodrigues (1795–1851): French banker born from a Jewish-Portuguese family, he was also a mathematician and a social reformer (he was closely associated to the Comte de Saint-Simon).

[18] Norman Macleod Ferrers (1829–1903): English mathematician who held high office at Cambridge University.

$$U_n^m(\theta, \phi) = \Big[A \, \cos(m\phi) + B \, \sin(m\phi) \Big] T_n^m(\cos\theta), \qquad (5.101)$$

is a surface spherical harmonic of integer degree n. For each value of n we distinguish three cases according to value assumed by the parameter m.

(a) **Zonal harmonic**: $m = 0$. In this case the function (5.101) becomes a constant multiple of the Legendre coefficient $P_n(\cos\theta)$. We show in Appendix O.3 that $P_n(\mu)$ has n distinct roots between -1 and $+1$, i.e., $P_n(\cos\theta)$ has n distinct roots between 0 and π, arranged symmetrically around $\theta = \pi/2$. Consequently, on a unit sphere centered in the origin of the reference system, the integer degree surface spherical harmonic:

$$U_n^0(\theta, \phi) = A \, P_n(\cos\theta), \qquad (5.102)$$

cancels on n circles with poles at the points $\theta = 0$ and $\theta = \pi$, arranged symmetrically with respect to the great circle $\theta = \pi/2$ (Fig. 5.7, panel a). For this reason, $U_n^0(\theta, \phi)$ is called a zonal harmonic. The point $\theta = 0$ is named the pole and the diameter through it is named the axis of the zonal harmonic.

(b) **Tesseral harmonic**: $0 < m < n$. In this case, by replacing the expression (5.100) of the Ferres functions in (5.101), we have:

$$U_n^m(\theta, \phi) = \Big[A \, \cos(m\phi) + B \, \sin(m\phi) \Big] \times \sin^m\theta \, \frac{d^{m+n}}{d\mu^{m+n}} (\mu^2 - 1)^n. \qquad (5.103)$$

Let us search for the roots of this function. The first multiplier vanishes when $\tan(m\phi) = -A/B$, i.e., for m equally spaced great circles through the poles of the unit sphere (Fig. 5.7); the planes of two consecutive circles form the constant angle π/m. The second factor cancels m times at the poles. The third behaves like a zonal harmonic of degree $(n-m)$: it cancel on $(n-m)$ minor circles with pole in $\theta = 0$. The name of tesseral[19] harmonics for these functions descends from the appearance of the areas on the surface of the unitary sphere framed by the zeros (Fig. 5.7b).

(c) **Sectoral harmonic**: $m = n$. Since the k-th derivative of a polynomial of degree k is a constant, for $m = n$ the expression (5.101) becomes:

$$U_n^m(\theta, \phi) = \Big[A \, \cos(n\phi) + B \, \sin(n\phi) \Big] \sin^n\theta. \qquad (5.104)$$

This function is canceled on n equispaced great circles through the poles of the unit spherical surface (Fig. 5.7c) and for this reason takes the name of sectoral harmonic.

In summary, for every non-negative integer n, we have $n + 1$ surface spherical harmonics of the type (5.101) broken down into:

---

[19] From Latin "*tessera*", meaning cubic body or die.

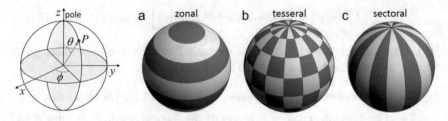

Fig. 5.7 Roots of the spherical harmonics, **a** zonal, **b** tesseral, **c** sectoral

a. one zonal harmonic ($m = 0$),
b. $n-1$ tesseral harmonics ($0 < m < n$),
c. one sectoral harmonic ($m = n$).

The rationale of this distinction concerns the way in which each harmonic modulates the surface of the unit sphere.

Following the current usage, we will sometimes use for (5.101) the compact notation below:

$$U_n^m(\theta, \phi) = \begin{array}{c} \cos m\phi \\ \sin m\phi \end{array} T_n^m(\cos \theta). \qquad (5.105)$$

Appendix O.1 contains some useful theorems concerning harmonic functions. For the integral representation, see p. 130 of [1].

We will use two important theorems presented in full in Appendix P. The first states that each solution of the associated Legendre equation for an integer and non-negative value of n and for $|\mu| < 1$ (i.e., for any spherical harmonic of integer degree) can always be written through a linear combination of the corresponding zonal, tesseral, and sectoral harmonics.

The second theorem, known also as addition theorem (cf. Appendix O.4.3), states that, if the expansion:

$$P_n(\cos \gamma) = A_\circ P_n(\cos \theta) + \sum_{k=1}^{n} \left[A_k \cos(k\phi) + B_k \sin(k\phi) \right] T_n^k(\cos \theta), \quad (5.106)$$

is true, due to the previous theorem the coefficients A_m and B_m are given by the following expressions:

$$\left.\begin{array}{c} A_m \\ B_m \end{array}\right\} = 2 \frac{(n-m)!}{(n+m)!} \times \left\{ \begin{array}{c} \cos(m\phi') \\ \sin(m\phi') \end{array} \right. T_n^m(\cos \theta') . \qquad (5.107)$$

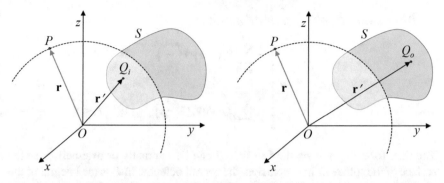

Fig. 5.8 Potential through spherical harmonics. The body of surface S can be either internal to the spherical surface of radius r centered on the origin O of the reference system and containing the point P on which the potential is to be calculated, or external, or partly internal (*left*) and partly external (*right*). It is obvious that this division depends on the choice of the reference system

5.6 Expansion of the Potential through Spherical Harmonics

We now prove that, whatever the mass distribution $\rho(r', \theta', \phi')$, the potential $\mathcal{U}$ in the outer point $P(r, \theta, \phi)$ can be always written in the form:

$$\mathcal{U}(P) = \sum_{i=0}^{\infty} r^i \, Y_i(\theta, \phi) + \sum_{j=0}^{\infty} \frac{1}{r^{j+1}} Z_j(\theta, \phi), \qquad (5.108)$$

where Y_i and Z_j are surface spherical harmonics of integer degree i and j respectively.

Let us introduce the surface S of a sphere centered in the origin O of the reference system and containing P (Fig. 5.8). The potential in P, produced by the mass element:

$$dm = \rho(r', \theta', \phi') \, r'^2 \sin\theta' \, dr' \, d\theta' \, d\phi', \qquad (5.109)$$

placed in $Q(r', \theta', \phi')$, is:

$$d\mathcal{U} = G \frac{dm}{\delta} = G \frac{\rho(r', \theta', \phi') \, r'^2 \sin\theta' \, dr' \, d\theta' \, d\phi'}{(r^2 - 2rr' \cos\gamma + r'^2)^{1/2}}, \qquad (5.110)$$

where, as usual, γ is the angle between the vector radii of the points Q and P. Through (5.41a) we readily have:

$$d\mathcal{U} = G\,\rho(r',\theta',\phi')\,r'^2\,\sin\theta'\,dr'\,d\theta'\,d\phi'\times$$

$$\times \begin{cases} \dfrac{1}{r}\displaystyle\sum_{k=0}^{\infty} P_k(\cos\gamma)\left(\dfrac{r'}{r}\right)^k & \text{if} \quad r' < r, \\[2em] \dfrac{1}{r'}\displaystyle\sum_{k=0}^{\infty} P_k(\cos\gamma)\left(\dfrac{r}{r'}\right)^k & \text{if} \quad r < r'. \end{cases} \qquad (5.111)$$

The dual form depends on the fact that P can lie internally or externally[20] to the surface of the sphere S. In conclusion, the overall potential in P is the integral of the elementary contributions (5.111) extended to the whole volume Σ occupied by the material body. Remembering that $\mathcal{U}$ is expressed by an absolutely convergent series, we have:

$$\mathcal{U}(r,\theta,\phi) = G\sum_{k=0}^{\infty} r^k \iiint_{\Sigma} \frac{\rho(r',\theta',\phi')}{r'^{k-1}}\,P_k(\cos\gamma)\,\sin\theta'\,dr'\,d\theta'\,d\phi' +$$

$$+ G\sum_{k=0}^{\infty} \frac{1}{r^{k+1}} \iiint_{\Sigma} \rho(r',\theta',\phi')\,r'^{k+2}\,P_k(\cos\gamma)\,\sin\theta'\,dr'\,d\theta'\,d\phi'. \quad (5.112)$$

If $\rho \equiv 0$ outside the spherical surface S centered at the origin and containing P $(r > r')$, the first term at the the right-hand side of (5.112) vanishes; if instead $\rho \equiv 0$ inside S (i.e., $r < r'$), it is the second term to vanish. Substituting to the Legendre coefficients, $P_k(\cos,\gamma)$, the expressions given by the addition theorem (O.90), we obtain:

$$\mathcal{U}(r,\theta,\phi) = G\sum_{k=0}^{\infty} r^k \left[A_{\circ,k} P_k(\cos\theta) + \right.$$

$$+ \sum_{m=1}^{k} \Big(A_{m,k}\cos(m\phi) + B_{m,k}\sin(m\phi) \Big)\, T_k^m(\cos\theta) \bigg] +$$

$$+ G\sum_{k=0}^{\infty} \frac{1}{r^{k+1}} \left[A'_{\circ,k} P_k(\cos\theta) + \right.$$

$$+ \sum_{m=1}^{k} \Big(A'_{m,k}\cos(m\phi) + B'_{m,k}\sin(m\phi) \Big)\, T_k^m(\cos\theta) \bigg]. \qquad (5.113)$$

The constants are:

---

[20] Note that, since the origin of the reference system is arbitrarily chosen, the spherical surface discriminating between the two cases can also be changed at will.

$$A_{o,k} = \iiint_\Sigma \frac{1}{r'^{k-1}} P_k(\cos\theta')\, dv', \tag{5.114a}$$

$$A'_{o,k} = \iiint_\Sigma r'^{k+2} P_k(\cos\theta')\, dv', \tag{5.114b}$$

$$A_{m,k} = \iiint_\Sigma 2\frac{(k-m)!}{(k+m)!} \cos(m\phi')\frac{1}{r'^{k-1}} T_k^m(\cos\theta')\, dv', \tag{5.114c}$$

$$A'_{m,k} = \iiint_\Sigma 2\frac{(k-m)!}{(k+m)!} \cos(m\phi')\, r'^{k+2} T_k^m(\cos\theta')\, dv', \tag{5.114d}$$

$$B_{m,k} = \iiint_\Sigma 2\frac{(k-m)!}{(k+m)!} \sin(m\phi')\frac{1}{r'^{k-1}} T_k^m(\cos\theta')\, dv', \tag{5.114e}$$

$$B'_{m,k} = \iiint_\Sigma 2\frac{(k-m)!}{(k+m)!} \sin(m\phi')\, r'^{k+2} T_k^m(\cos\theta')\, dv', \tag{5.114f}$$

where we have placed:

$$dv' = \rho(r',\theta',\phi')\sin\theta'\, dr'\, d\theta'\, d\phi'. \tag{5.115}$$

Using the expression (O.76) of a surface spherical harmonic of integer degree, it is immediate to verify that (5.113) coincides with (5.108), which is what we wanted to prove.

Note 1. The constant $A'_{o,o}$ is proportional to the total mass m of the body generating the potential. Indeed, from the (5.114a) and (5.99), remembering that $P_0(\mu) = 1$, it is:

$$A'_{o,o} = \iiint_\Sigma \rho(r'\theta',\phi')\, r'^2 \sin\theta'\, dr'\, d\theta'\, d\phi' = \iiint_\Sigma dm = m. \tag{5.116}$$

Note 2. We want to emphasize again the fact that if all the mass m producing the potential is enclosed within the surface of the sphere concentric to the origin of the reference system and containing P, in other words if $\rho(r',\theta',\phi') = 0$ for $r' > r$, the potential in P is given by:

$$\mathcal{U}(r,\theta,\phi) = G \sum_{k=0}^{\infty} \frac{1}{r^{k+1}} \Big[A'_{o,k} P_k(\cos\theta) +$$

$$+ \sum_{m=1}^{k} \Big(A'_{m,k} \cos(m\phi) + B'_{m,k} \sin(m\phi) \Big) T_k^m(\cos\theta) \Big]. \tag{5.117}$$

If the contrary is true, i.e., if $\rho(r',\theta',\phi') = 0$ for $r' < r$, we have instead:

$$\mathcal{U}(r, \theta, \phi) = G \sum_{k=0}^{\infty} r^k \left[A_{\circ,k} P_k(\cos\theta) + \right.$$

$$\left. + \sum_{m=1}^{k} \left(A_{m,k} \cos(m\phi) + B_{m,k} \sin(m\phi) \right) T_k^m(\cos\theta) \right]. \qquad (5.118)$$

Note 3. If the mass responsible of the potential is distributed with rotational symmetry around the z-axis, that is, if the density ρ does not depend on ϕ', the expression (5.113) simplifies in:

$$\mathcal{U}(r, \theta) = G \sum_{k=0}^{\infty} \left[A_{\circ,k} r^k + \frac{A'_{\circ,k}}{r^{k+1}} \right] P_k(\cos\theta). \qquad (5.119)$$

In fact, the coefficients $A_{m,k}$, $A'_{m,k}$, $B_{m,k}$, and $B'_{m,k}$, of the tesseral and sectoral harmonics, both in Y_k and in Z_k, are of the type:

$$\text{const} \times \iint \left[\int_0^{2\pi} \rho(r', \theta', \phi') \begin{Bmatrix} \sin(m\phi') \\ \cos(m\phi') \end{Bmatrix} d\phi' \right] \times$$

$$\times \begin{Bmatrix} r'^{k+2} \\ r'^{1-k} \end{Bmatrix} T_k^m(\cos\theta') \sin\theta' \, dr' \, d\theta'. \qquad (5.120)$$

If ρ is independent of ϕ', the integrals:

$$\int_0^{2\pi} \begin{Bmatrix} \sin(m\phi') \\ \cos(m\phi') \end{Bmatrix} d\phi' = \frac{1}{m} \begin{Bmatrix} -\cos(m\phi') \\ \sin(m\phi') \end{Bmatrix} \Bigg|_0^{2\pi}, \qquad (5.121)$$

vanish when $m \neq 0$. Thus, the coefficients of all harmonics but the zonal ones are zero.

Note 4. If, in addition to the hypothesis of *Note 3*, the mass distribution is also symmetrical with respect to the plane containing the x and y-axes, i.e., if:

$$\rho(r', \theta', \phi') = \rho(r', \theta') = \rho(r', \pi - \theta'), \qquad (5.122)$$

the coefficients with odd index in (5.119) are zero, and the expression of the potential reduces further to:

$$\mathcal{U}(r, \theta) = G \sum_{k=0}^{\infty} \left[A_{\circ,2k} r^{2k} + \frac{A'_{\circ,2k}}{r^{2k+1}} \right] P_{2k}(\cos\theta). \qquad (5.123)$$

The proof is immediate. Just remember that the Legendre coefficients of odd index are antisymmetric relative to $\theta = \pi/2$.

Once again we want to draw the attention to the fact that, in order to exploit the properties of this and the previous *Note*, it is not enough that symmetries exist with respect to an axis and a plane orthogonal to it. An appropriate choice of the coordinate system is also required. The symmetry axis must contain the z-axis, and the symmetry plane both the x and y-axes. It is obvious that, under these hypotheses, the origin of the system coincides with the barycenter of mass distribution, since the density ρ ends up depending exclusively on the modulus of z.

Note 5. It can be proved that the operator:

$$O = \ell \frac{\partial}{\partial x} + m \frac{\partial}{\partial y} + n \frac{\partial}{\partial z}, \tag{5.124}$$

where ℓ, m, and n, are constants, allows us to write:

$$O\mathcal{U} = Y_\circ + r\, Y_1 + r^2 Y_2 + \cdots + \frac{Z_1}{r^2} + \frac{Z_2}{r^3} + \cdots \tag{5.125}$$

5.7 Potential of a Thin Spherical Layer

As an application of the results of the previous section, we calculate here the potential due to a thin spherical layer Σ of radius $r_\circ$ and surface density $\sigma(\theta', \phi')$. We introduce a spherical coordinate system with origin in the geometric center O of Σ (which in this case does not necessarily coincide with the barycenter as the surface density is not required to be constant). Based on (5.108), the potential at an external point $P(r, \theta, \phi)$ is given by:

$$\mathcal{U} = G \sum_{k=0}^{\infty} \frac{Z_k(\theta, \phi)}{r^{k+1}} \qquad (r > r_\circ), \tag{5.126}$$

where Z_k is an integer degree spherical harmonic of degree k:

$$Z_k(\theta, \phi) = A_{\circ,k} P_k(\cos\theta) + \sum_{m=1}^{k} \frac{A_{m,k}\cos(m\phi)}{B_{m,k}\sin(m\phi)} T_k^m(\cos\theta). \tag{5.127}$$

The constants $A_{m,k}$ and $B_{m,k}$ are of the type:

$$\left.\begin{matrix} A_{m,k} \\ B_{m,k} \end{matrix}\right\} = \text{const} \times r_\circ^{k+2} \iint_\Sigma \sigma(\theta', \phi') \times$$

$$\times \begin{cases} \cos(m\phi') \\ \sin(m\phi') \end{cases} T_k^m(\cos\theta')\sin\theta'\, d\theta'\, d\phi', \tag{5.128}$$

since the integration with respect to r', constant and equal to $r_\circ$, is suppressed. By factoring $r_\circ^{k+2}$, the (5.126) becomes:

$$\mathcal{U} = G \sum_{k=0}^{\infty} \frac{r_\circ^{k+2}}{r^{k+1}} Y_k(\theta, \phi) \qquad (r > r_\circ), \qquad (5.129)$$

with $Z_k(\theta, \phi) = r_\circ^{k+2} Y_k(\theta, \phi)$. By the same procedure it is easy to demonstrate that, if P is an internal point, then:

$$\mathcal{U} = G \sum_{k=0}^{\infty} \frac{r^k}{r_\circ^{k-1}} Y_k(\theta, \phi) \qquad (r < r_\circ). \qquad (5.130)$$

We now verify that, if the layer is homogeneous ($\sigma = \sigma_\circ$), (5.129) and (5.130) are identical to (5.50) and (5.51) respectively (outer and inner potentials of a homogeneous spherical layer). For the hypothesis made, the present case falls into those contemplated by the *Note 4* at Sect. 5.6, so:

$$\mathcal{U} = G \sum_{k=0}^{\infty} \frac{r_\circ^{2k+2}}{r^{2k+1}} A_{\circ,2k} P_{2k}(\cos\theta) \qquad \text{if } r > r_\circ,$$

$$\qquad\qquad\qquad\qquad\qquad\qquad\qquad\qquad\qquad (5.131)$$

$$\mathcal{U} = G \sum_{k=0}^{\infty} \frac{r^{2k}}{r_\circ^{2k-1}} A_{\circ,2k} P_{2k}(\cos\theta) \qquad \text{if } r < r_\circ,$$

where:

$$A_{\circ,2k} = \sigma_\circ \iint_\Sigma P_{2k}(\cos\theta') \sin\theta'\, d\theta'\, d\phi' = 2\pi\, \sigma_\circ \int_{-1}^{+1} P_{2k}(\mu')\, d\mu', \qquad (5.132)$$

and $\mu' = \cos\theta'$. Remembering that $P_0(\mu') \equiv 1$, for the orthogonality of the Legendre polynomials it is:

$$\int_{-1}^{+1} P_{2k}(\mu')\, d\mu' = \int_{-1}^{+1} P_{2k}(\mu')\, P_0(\mu')\, d\mu' = \begin{cases} 2 & \text{if } k = 0, \\ 0 & \text{if } k \neq 0, \end{cases} \qquad (5.133)$$

and thus all the Legendre coefficients $A_{\circ,k}$ are zero with the exception of $A_{\circ,\circ} = 4\pi\sigma_\circ$. This proves the thesis for the external points. The procedure for the internal points is analogous.

5.8 Potential of a Homogeneous Spheroid in an External Point

Here we call spheroid a body that differs little from a sphere (Fig. 5.9; note that this body shares the same name of the revolution or rotational ellipsoids of Fig. 5.13). The equation of its surface can be put in the form:

Fig. 5.9 Potential of a
homogeneous spheroid. Note
the O is the center of the
sphere of radius r_o defined
by the (5.137) and it is not
necessarily the barycenter of
the spheroid

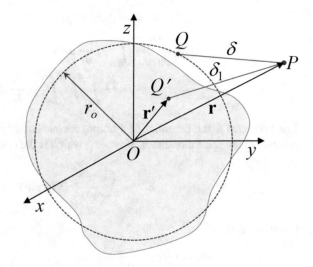

$$r = r_o \left[1 + \alpha\, f(\theta, \phi)\right], \tag{5.134}$$

where r_o is an average radius and α is a parameter whose magnitude establishes
the deviation from the sphere. In the following we assume that α is so small that
its square can be neglected. We will show that, if the spheroid is homogeneous
($\rho(r, \theta, \phi) = \rho_o$), its external potential is given by the sum of that of the solid sphere
of radius r_o with the potential of the thin spherical layer of equal radius and surface
density:

$$\sigma = \rho_o r_o \alpha f(\theta, \phi). \tag{5.135}$$

The potential in $P(r, \theta, \phi)$ is:

$$\mathcal{U}_e(r, \theta, \phi) = G \int_0^{2\pi} \int_0^{\pi} \int_0^{r_1} \rho_o \frac{r'^2}{\delta_1} \sin\theta'\, dr'\, d\theta'\, d\phi', \tag{5.136}$$

where δ_1 is the distance between P and the volume element of the spheroid centered
in $Q'(r', \theta', \phi')$. The upper limit of the integral in r', according to (5.134), is:

$$r_1 = r_o \left[1 + \alpha\, f(\theta', \phi')\right]. \tag{5.137}$$

Placing $\mu' = \cos\theta'$, we break down the integral (5.136) in the sum of two terms:

$$\mathcal{U}_e(r, \theta, \phi) = \mathcal{U}_1(r, \theta, \phi) + \mathcal{U}_2(r, \theta, \phi) =$$

$$= G \int_0^{2\pi} \int_{-1}^{+1} \int_0^{r_o} \rho_o \frac{r'^2}{\delta_1} \, dr' \, d\mu' \, d\phi' +$$

$$+ G \int_0^{2\pi} \int_{-1}^{+1} \int_{r_o}^{r_1} \rho_o \frac{r'^2}{\delta_1} \, dr' \, d\mu' \, d\phi'. \tag{5.138}$$

The first term is the potential of a homogeneous solid sphere of radius r_o. If P is external to the spheroid, that is, if $r > r_1$, which is the case of (5.41b), then:

$$\delta_1^{-1} = \sum_{k=0}^{\infty} \frac{r'^k}{r^{k+1}} P_k(\cos \gamma). \tag{5.139}$$

It follows that:

$$\mathcal{U}_2 = G \int_0^{2\pi} \int_{-1}^{+1} \int_{r_o}^{r_1} \rho_o \, r'^2 \left[\sum_{k=0}^{\infty} \frac{r'^k}{r^{k+1}} P_k(\cos \gamma) \right] dr' \, d\mu' \, d\phi', \tag{5.140}$$

which can be integrated with respect to r':

$$\mathcal{U}_2 = G \int_0^{2\pi} \int_{-1}^{+1} \rho_o \left[\sum_{k=0}^{\infty} \frac{1}{k+3} \frac{r_1^{k+3} - r_o^{k+3}}{r^{k+1}} P_k(\cos \gamma) \right] d\mu' \, d\phi'. \tag{5.141}$$

Expanding $r_1^{k+3} = r_o^{k+3} \left[1 + \alpha \, f(\theta', \phi') \right]^{k+3}$ in powers of α and truncating at the first order (which is allowed by our assumption that α is so small that α^2 can be neglected), it is:

$$r_1^{k+3} = r_o^{k+3} \left[1 + \alpha \, (k+3) \, f(\theta', \phi') \right], \tag{5.142}$$

with which:

$$\mathcal{U}_2 = G \int_0^{2\pi} \int_{-1}^{+1} \rho_o \left[\sum_{k=0}^{\infty} \frac{r_o^{k+3}}{r^{k+1}} \alpha \, f(\theta', \phi') \, P_k(\cos \gamma) \right] d\mu' \, d\phi'. \tag{5.143}$$

Then, by placing $\sigma = \rho_o \, r_o \, \alpha \, f(\theta', \phi')$, we obtain:

$$\mathcal{U}_2 = G \int_0^{2\pi} \int_{-1}^{+1} \sigma \left[\sum_{k=0}^{\infty} \frac{r_o^k}{r^{k+1}} P_k(\cos \gamma) \right] r_o^2 \, d\mu' \, d\phi' =$$

$$= G \int_0^{2\pi} \int_{-1}^{+1} \frac{\sigma \, r_o^2 \, d\mu' \, d\phi'}{\delta}, \tag{5.144}$$

where δ is the distance of P from the point $Q(r_o, \theta', \phi')$ belonging to the surface of the sphere of radius r_o (see Fig. 5.9). The expression of $\mathcal{U}_2$ is that of the external potential of a thin spherical layer of radius r_o and surface density σ, which is what we wanted to prove for an external point. Its extension to an internal point is now obvious.

The integral appearing in (5.144) is not trivial. However, if $f(\theta, \phi)$ in (5.135) is expandable in a series of spherical harmonics:

$$f(\theta', \phi') = \sum_{n=0}^{\infty} Y_n(\theta', \phi'), \tag{5.145}$$

the integration follows in a simpler way owing to the orthogonality of spherical harmonics themselves.[21] Using the addition theorem (Sect. O.4.3) and (O.88) in particular, we can then prove that the external potential of the spheroid is given by following expression:

$$\mathcal{U}_e(r, \theta, \phi) = \frac{4\pi}{3} G \rho_o \frac{r_o^3}{r} +$$

$$+ 4\pi G \rho_o r_o^2 \alpha \sum_{n=0}^{\infty} \frac{1}{2n+1} \left(\frac{r_o}{r}\right)^{n+1} Y_n(\theta, \phi). \tag{5.146}$$

We want to prove now that the spheroid delimited by:

$$r = r_o\left[1 + \alpha \sum_{n=0}^{\infty} Y_n(\theta, \phi)\right], \tag{5.147}$$

has a mass:

$$M = \frac{4\pi}{3} \rho_o r_o^3\left[1 + 3\alpha Y_o\right]. \tag{5.148}$$

In fact, for the orthogonality of surface harmonics (see footnote 21 in this chapter), it is:

$$M = \int_0^{2\pi}\int_{-1}^{+1}\int_0^{r} \rho_o r'^2 \, dr' \, d\mu' \, d\phi' = \int_0^{2\pi}\int_{-1}^{+1} \frac{1}{3} \rho_o r^3 \, d\mu' \, d\phi' =$$

$$\approx \int_0^{2\pi}\int_{-1}^{+1} \frac{1}{3} \rho_o r_o^3\left[1 + 3\alpha \sum_{n=0}^{\infty} Y_n(\theta, \phi)\right] d\mu' \, d\phi' =$$

$$= \frac{4\pi}{3} \rho_o r_o^3\left[1 + 3\alpha Y_o\right], \tag{5.149}$$

---

[21] As usual, to exploit this property it is enough to introduce $P_0(\mu) = 1$ in the integral.

to the first order in α; that is: $\left[1 + \alpha \sum\limits_{n=0}^{\infty} Y_n(\theta, \phi)\right]^3 \approx \left[1 + 3\alpha \sum\limits_{n=0}^{\infty} Y_n(\theta, \phi)\right]$. This result allows us to rewrite the expression (5.146) in the following way[22]:

$$\mathcal{U}_e(r, \theta, \phi) = \frac{4\pi}{3} G \rho_o \frac{r_o^3}{r} + 4\pi G \rho_o r_o^2 \left(\frac{r_o}{r}\right) \alpha Y_0 +$$

$$+ 4\pi G \rho_o r_o^2 \alpha \sum_{n=1}^{\infty} \frac{1}{2n+1} \left(\frac{r_o}{r}\right)^{n+1} Y_n(\theta, \phi) =$$

$$= G\left[\frac{M}{r} + 4\pi \rho_o r_o^2 \alpha \sum_{n=1}^{\infty} \left(\frac{r_o}{r}\right)^{n+1} \frac{Y_n(\theta, \phi)}{2n+1}\right]. \qquad (5.150)$$

It can be proven that the previous expressions lose the term $Y_1(\theta, \phi)$ if the origin of the reference system contains the center of gravity (see also Sect. 5.11).

As an application of the previous formula, we consider an ellipsoid of revolution with axes $a = b$ and c:

$$\frac{x^2 + y^2}{a^2} + \frac{z^2}{c^2} = 1. \qquad (5.151)$$

We place $a = c(1 + \alpha)$. If α is such that its powers higher than the first are negligible, it can be proven[23] that:

$$r_1 = c\,(1 + \alpha \sin^2 \theta). \qquad (5.152)$$

Recalling the expression (5.40c) of the Legendre coefficient $P_2(\cos \theta)$, it results:

$$r_1 = c\left[1 + \alpha\left(\frac{2}{3} - \frac{2}{3} P_2(\cos \theta)\right)\right]. \qquad (5.153)$$

The comparison of this with the (5.147) gives as $r_o = c$, $Y_o = \frac{2}{3}$, $Y_1 = 0$, $Y_2 = -\frac{2}{3} P_2(\cos \theta)$. Substituting in (5.150), it is:

---

[22] Note that at the first order it is: $\dfrac{\alpha}{1 + 2\alpha} = \dfrac{\alpha(1 - 2\alpha)}{1 - 4\alpha^2} = \dfrac{\alpha - 2\alpha^2}{1 - 4\alpha^2} = \alpha$.

[23] From (5.151), with $a = c(1 + \alpha)$:

$$1 = r^2 \left(\frac{\sin^2 \theta}{c^2(1 + \alpha)^2} + \frac{\cos^2 \theta}{c^2}\right) \approx r^2 \left(\frac{1 + 2\alpha \cos^2 \theta}{c^2(1 + 2\alpha)}\right),$$

by ignoring the terms in α^2, and thus:

$$r^2 \approx \frac{c^2(1 + 2\alpha)}{1 + 2\alpha \cos^2 \theta} \simeq \frac{c^2(1 + 2\alpha)(1 - 2\alpha \cos^2 \theta)}{1 - 4\alpha^2 \cos^4 \theta} \simeq c^2(1 + 2\alpha \sin^2 \theta) \simeq$$
$$\approx c^2(1 + 2\alpha \sin^2 \theta + \alpha^2 \sin^4 \theta) \approx c^2(1 + \alpha \sin^2 \theta)^2,$$

where the addition of the term $(\alpha^2 \sin^4 \theta)$ is justified by the negligibility of α^2.

$$\mathcal{U}_e(r,\theta) = G\left[\frac{M}{r} - \frac{8\pi}{15}\rho_o\,c^2\,\alpha\left(\frac{c}{r}\right)^3 P_2(\cos\theta)\right] =$$

$$\approx G\left[\frac{M}{r} - \frac{2}{5}\alpha\,c^2\,\frac{M}{r^3} P_2(\cos\theta)\right] =$$

$$= G\frac{M}{r}\left[1 - \frac{2}{5}\alpha\,c^2\,\frac{1}{r^2} P_2(\cos\theta)\right], \qquad (5.154)$$

where $M = \dfrac{4\pi}{3}\rho_o c^3(1+2\alpha)$ is the total mass of the ellipsoid approximated by (5.152).

5.9 Potential of a Homogeneous Ellipsoid in an Inner Point

We now calculate the potential generated by a homogeneous ellipsoid. In the general case it cannot be expressed through elementary functions. We start by calculating the potential at a point internal to a homogeneous ellipsoidal shell with constant density ρ (Fig. 5.10). This is a figure limited by two triaxial ellipsoids of semi-axes a, b, c (outer surface) and pa, pb, pc (inner surface as $p < 1$). Since the internal potential is constant (Newton's theorem, Sect. 5.4, we simply calculate its value at the central point O.

Consider a cone with the apex in O, which subtends a solid angle $d\omega$ and whose axis intercepts the surfaces of the figure at distances pr and r from O (Fig. 5.11). The potential in O due to the volume element of the homogeneous ellipsoidal shell cut by the cone is:

$$d\mathcal{U}_i(O) = G\int_{pr}^{r}\rho\,r^2\,d\omega\frac{dr}{r} = \frac{1}{2}G\,\rho\,(1-p^2)\,r^2\,d\omega, \qquad (5.155)$$

and thus the total potential is:

Fig. 5.10 Homogeneous triaxial ellipsoid

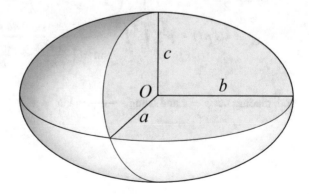

Fig. 5.11 Ellipsoidal shell

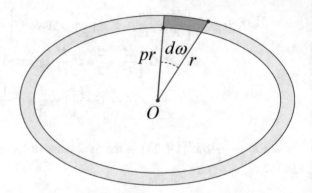

$$\mathcal{U}_i(O) = \frac{1}{2} G \rho \left(1 - p^2\right) \int_{sup} r^2 \, d\omega. \tag{5.156}$$

The integration must be carried out on the external surface of the figure, whose equation is:

$$\frac{x^2}{a^2} + \frac{y^2}{b^2} + \frac{z^2}{c^2} = 1. \tag{5.157}$$

Let $(r\ell, rm, rn)$ be the coordinates of a point of the external surface in an generic reference system centered in O; ℓ, m, n represent, as usual, direction cosines: $\ell = \sin\theta \cos\phi$, $m = \sin\theta \sin\phi$, and $n = \cos\theta$. Then:

$$r^2 \left(\frac{\ell^2}{a^2} + \frac{m^2}{b^2} + \frac{n^2}{c^2} \right) = 1. \tag{5.158}$$

Furthermore, since the integrand in (5.156) is symmetric with respect to each of the axes, its integral will be equal to eight times that calculated on a single octant. Therefore:

$$\mathcal{U}_i(O) = \frac{1}{2} G \rho \left(1 - p^2\right) \int_{sup} \left(\frac{\ell^2}{a^2} + \frac{m^2}{b^2} + \frac{n^2}{c^2} \right)^{-1} d\omega =$$

$$= 4G \rho \left(1 - p^2\right) \int_0^{\frac{\pi}{2}} \int_0^{\frac{\pi}{2}} \frac{\sin\theta \, d\theta \, d\phi}{\sin^2\theta \left(\dfrac{\cos^2\phi}{a^2} + \dfrac{\sin^2\phi}{b^2} \right) + \dfrac{\cos^2\theta}{c^2}}, \tag{5.159}$$

and, placing $\tan\phi = t$ and using $\dfrac{1}{\cos^2\phi} = \tan^2\phi + 1$:

$$\mathcal{U}_i(O) = 4G\,\rho\,(1 - p^2) \times$$

$$\times \int_0^{\frac{\pi}{2}} \int_0^\infty \frac{\sin\theta\,d\theta\,dt}{\dfrac{\sin^2\theta}{a^2} + \dfrac{\cos^2\theta}{c^2} + \left(\dfrac{\sin^2\theta}{b^2} + \dfrac{\cos^2\theta}{c^2}\right)t^2}. \tag{5.160}$$

Let us place:

$$A' = \sqrt{\frac{\sin^2\theta}{a^2} + \frac{\cos^2\theta}{c^2}}, \tag{5.161a}$$

$$B' = \sqrt{\frac{\sin^2\theta}{b^2} + \frac{\cos^2\theta}{c^2}}. \tag{5.161b}$$

The inner integral in (5.160) now is:

$$\int_0^\infty \frac{dt}{A'^2 + B'^2 t^2} = \frac{1}{A'B'}\arctan\left(\frac{B'}{A'}t\right)\bigg|_0^\infty = \frac{1}{2}\frac{\pi}{A'B'}, \tag{5.162}$$

then:

$$\mathcal{U}_i(O) = 2\pi\,G\,\rho\,(1 - p^2) \times$$

$$\times \int_0^{\pi/2} \left\{\left(\frac{\sin^2\theta}{a^2} + \frac{\cos^2\theta}{c^2}\right)\left(\frac{\sin^2\theta}{b^2} + \frac{\cos^2\theta}{c^2}\right)\right\}^{-1/2}\sin\theta\,d\theta =$$

$$= 2\pi\,G\,\rho\,(1 - p^2)\int_0^{\pi/2} \frac{\sin\theta\,d\theta}{A'B'}. \tag{5.163}$$

By setting $u = c^2\tan^2\theta$ we have:

$$\cos^2\theta = \frac{c^2}{c^2 + u}, \quad \sin^2\theta = \frac{u}{(c^2 + u)}, \tag{5.164a}$$

$$\sin\theta\,d\theta = \frac{\cos\theta\sin^2\theta}{2u}\,du = \frac{c}{2(c^2 + u)^{3/2}}\,du, \tag{5.164b}$$

and the functions A' and B' are:

$$A' = \frac{\cos\theta}{ac}\sqrt{a^2 + u}, \tag{5.165a}$$

$$B' = \frac{\cos\theta}{bc}\sqrt{b^2 + u}. \tag{5.165b}$$

Substituting it in (5.163) we finally obtain the potential of the ellipsoidal layer in the central point O:

$$\mathcal{U}_i(O) = \pi\,G\,\rho\,(1 - p^2)\,abc \times I, \tag{5.166}$$

Fig. 5.12 Homogeneous
solid ellipsoid

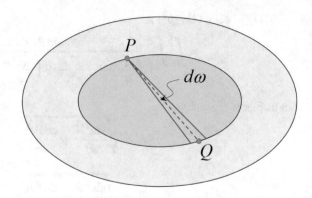

where:

$$I = \int_0^\infty \frac{du}{\Delta}, \tag{5.167a}$$

$$\Delta^2 = (a^2 + u)(b^2 + u)(c^2 + u). \tag{5.167b}$$

Placing $p = 0$, we obtain the potential at the center of a solid and homogeneous ellipsoid.

Now we want to calculate the force exerted at a point $P(x, y, z)$ inside a homogeneous ellipsoid. Since the shell having the internal surface passing through P provides a null contribution to the force in P (since its contribution to the potential is constant), it is sufficient to calculate the force exerted by the ellipsoid containing P. To this end we adopt the following strategy. We consider an infinitesimal cone that has the vertex in P and its axis directed towards a point Q also belonging to the surface of the ellipsoid (Fig. 5.12). To calculate the total potential in P we will just need to integrate the elementary contributions by the part of the ellipsoid contained within the cone, allowing Q to vary across the entire outer surface.

Let (ℓ, m, n) be the direction cosines of $\overrightarrow{PQ}$, with which the coordinates of Q are $x + \ell r$, $y + mr$, and $z + nr$. The requirement that P and Q are on the outer surface of the ellipsoid is accomplished by replacing the corresponding coordinates in the (5.157). From these, by subtraction, we have:

$$r^2 \left(\frac{\ell^2}{a^2} + \frac{m^2}{b^2} + \frac{n^2}{c^2} \right) + 2r \left(\frac{\ell x}{a^2} + \frac{my}{b^2} + \frac{nz}{c^2} \right) = 0. \tag{5.168}$$

The force in P due to the mass within the conical element of vertex P, axis $\overrightarrow{PQ}$, and solid angle $d\omega$, has the modulus:

$$|df(P)| = G \rho \, d\omega \int_0^r dr = G \rho r \, d\omega, \tag{5.169}$$

and components:

$$G \rho \, \ell r \, d\omega, \qquad G \rho \, mr \, d\omega, \qquad G \rho \, nr \, d\omega. \qquad (5.170)$$

Each component of the total force will therefore result from the integration of the elementary contributions for all the possible values of ω on the half-space limited by the plane tangent in P and containing the ellipsoid (in fact, each line through P has only one real intersection with the boundary surface of the ellipsoid).

Let us obtain the component of the total force along the z-axis:

$$f_z = \frac{1}{2} G \rho \int_{sup} nr \, d\omega =$$

$$= -\frac{1}{2} G \rho \int_{sup} 2n \left(\frac{\ell x}{a^2} + \frac{my}{b^2} + \frac{nz}{c^2} \right) \left(\frac{\ell^2}{a^2} + \frac{m^2}{b^2} + \frac{n^2}{c^2} \right)^{-1} d\omega, \qquad (5.171)$$

where we used the relation (5.168). For reasons of symmetry, the terms in nl and nm give a null contribution to the integral in (5.171); for instance, for each value of nl there will be a corresponding value with opposite sign. So, the (5.171) reduces to:

$$f_z = -G \rho \int_{sup} \frac{n^2 z}{c^2} \left(\frac{\ell^2}{a^2} + \frac{m^2}{b^2} + \frac{n^2}{c^2} \right)^{-1} d\omega = -G \rho C z, \qquad (5.172)$$

where we used the result of integration by ϕ given by (5.162) and the expressions (5.164a) and (5.164b):

$$C = \frac{1}{c^2} \int_{sup} n^2 \left(\frac{\ell^2}{a^2} + \frac{m^2}{b^2} + \frac{n^2}{c^2} \right)^{-1} d\omega = \frac{4\pi}{c^2} \int_0^{\pi/2} \frac{\sin \theta \cos^2 \theta \, d\theta}{A' B'} =$$

$$= 2\pi abc \int_0^\infty \frac{du}{(c^2 + u)\sqrt{(a^2 + u)(b^2 + u)(c^2 + u)}} =$$

$$= 2\pi abc \int_0^\infty \frac{du}{(c^2 + u)\Delta}. \qquad (5.173)$$

Analogous expressions are obtained for the other two components along x and y, in which the integrals A and B will appear similar to C. Given the linear dependence of each component of the force from the corresponding coordinate:

$$f_x = -2x \tilde{A},$$
$$f_y = -2y \tilde{B}, \qquad (5.174)$$
$$f_z = -2z \tilde{C},$$

where:

$$\begin{pmatrix} \tilde{A} \\ \tilde{B} \\ \tilde{C} \end{pmatrix} = \pi G \rho \, a \, b \, c \int_0^\infty \frac{du}{\Delta} \times \begin{pmatrix} (a^2 + u)^{-1} \\ (b^2 + u)^{-1} \\ (c^2 + u)^{-1} \end{pmatrix}, \qquad (5.175)$$

we are then able to write the the element of the potential as:

$$d\mathcal{U} = f_x dx + f_y dy + f_z dz = -2(\tilde{A} x dx + \tilde{B} y dy + \tilde{C} z dz). \qquad (5.176)$$

After integrating, the total potential of the ellipsoid in the inner point P is:

$$\mathcal{U}_i(x, y, z) = \mathcal{U}(O) - \left(\tilde{A} x^2 + \tilde{B} y^2 + \tilde{C} z^2 \right), \qquad (5.177)$$

where, in correspondence to the limits of the integral, the integration constant $\mathcal{U}_i(O)$ is equal to the potential at the center ($x = y = z = 0$) of the homogeneous ellipsoid, defined by (5.166) for $p = 0$.

In conclusion, the potential of the homogeneous ellipsoid in any inner point is:

$$\mathcal{U}_i(x, y, z) = G \pi \, \rho \, abc \int_0^\infty \left\{ 1 - \frac{x^2}{a^2 + u} - \frac{y^2}{b^2 + u} - \frac{z^2}{c^2 + u} \right\} \frac{du}{\Delta}, \qquad (5.178)$$

or, remembering (5.167a) and (5.167b):

$$\mathcal{U}_i(x, y, z) = G \pi \, \rho \, abc \left\{ I + \frac{x^2}{a} \frac{\partial I}{\partial a} + \frac{y^2}{b} \frac{\partial I}{\partial b} + \frac{z^2}{c} \frac{\partial I}{\partial c} \right\}. \qquad (5.179)$$

From (5.172) we easily deduce two relations among A, B, and C:

$$A + B + C = \int_{sup} d\omega = 4\pi, \qquad (5.180a)$$

$$A a^2 + B b^2 + C c^2 = \int_{sup} \left(\frac{\ell^2}{a^2} + \frac{m^2}{b^2} + \frac{n^2}{c^2} \right)^{-1} d\omega. \qquad (5.180b)$$

Note that A, B, and C depend only on the axial ratios.

In the general case, the integrals (5.180a) and (5.180b) can be solved in terms of elliptical integrals. If, however, two of the axes are equal, that is, if the ellipsoid reduces to a spheroid (a figure of revolution in this case), the integration becomes possible in terms of simple functions. Suppose that $a = b$, with which $A = B$. From (5.180b) we obtain:

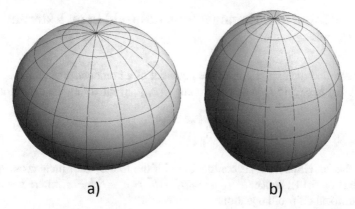

a)   b)

Fig. 5.13 Oblate (**a**) and prolate (**b**) spheroids. The oblate spheroid looks like a sphere equally flattened at the poles; its extreme shape is that of a disc or lentil. The prolate spheroid looks like a rugby ball or, in the extreme case, a cigar

$$2A\,a^2 + C\,c^2 = \int_{sup} \left(\frac{\ell^2 + m^2}{a^2} + \frac{n^2}{c^2}\right)^{-1} d\omega =$$

$$= \int_0^\pi \int_0^{\pi/2} \left(\frac{\sin^2\theta}{a^2} + \frac{\cos^2\theta}{c^2}\right)^{-1} \sin\theta\, d\theta\, d\phi =$$

$$= 2\pi \int_{-1}^{+1} \left\{\frac{1}{a^2} + \left(\frac{1}{c^2} - \frac{1}{a^2}\right)u^2\right\}^{-1} du, \qquad (5.181)$$

where we have placed $u = \cos\theta$. We distinguish two cases (Fig. 5.13). Oblate spheroid, or $a > c$ (since we chose to put $a = b$):

$$2Aa^2 + Cc^2 = \frac{4\pi a^2 c}{\sqrt{a^2 - c^2}} \arctan\sqrt{\frac{a^2}{c^2} - 1} =$$

$$= 4\pi a^2 \frac{\sqrt{1 - e^2}}{e} \arctan\frac{e}{\sqrt{1 - e^2}}, \qquad (5.182)$$

where e is the ellipticity $\left(e^2 = 1 - c^2/a^2\right)$.
Prolate spheroid, or $a < c$:

$$2A\,a^2 + C\,c^2 = \frac{\pi a^2 c^2}{c^2 - a^2} \ln\left\{\frac{c + \sqrt{c^2 - a^2}}{c - \sqrt{c^2 - a^2}}\right\} = \frac{\pi a^2}{e^2} \ln\frac{1 + e}{1 - e}, \quad (5.183)$$

where $e^2 = 1 - a^2/c^2$.

These solutions combine with (5.180a) to provide the values of A and C, that is, what is needed to obtain the potential.

5.10   Potential of a Homogeneous Ellipsoid in an External Point

We want to find the expression of the potential of a homogeneous ellipsoid E with semi-axes a, b, c (5.157) in an outer point $P(x, y, z)$ satisfying the condition:

$$\frac{x^2}{a^2} + \frac{y^2}{b^2} + \frac{z^2}{c^2} > 1. \qquad (5.184)$$

We say that an ellipsoid E' is confocal to E if the foci of their main cross-sections coincide (Fig. 5.14); hence: $a'^2 - a^2 = b'^2 - b^2 = c'^2 - c^2 = s$, where s is the root of the confocal ellipsoid equation:

$$\frac{x^2}{a^2 + s} + \frac{y^2}{b^2 + s} + \frac{z^2}{c^2 + s} = 1. \qquad (5.185)$$

We take E' in such a way that it contains the point P (therefore $s > 0$) and assume that it has the same constant density ρ as E. Denote the projections of the gravitational forces exerted by E' on P as f_x', f_y', f_z'.

To obtain the outer potential of the ellipsoid we can use the Laplace-Maclaurin theorem[24]: homogeneous confocal ellipsoids attract an outer point with identical forces proportional to their masses:

Fig. 5.14 Cross sections of confocal ellipsoids by a plane containing two principal axes

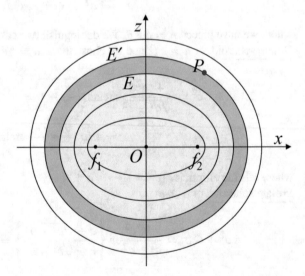

$$\frac{f_x}{f'_x} = \frac{M}{M'} = \frac{abc}{a'b'c'}. \tag{5.186}$$

Since the point P for the ellipsoid E' is located at its surface, we can use the expression for the inner potential (because P is the limiting cases of the inner point for E'). So, expressing the projection f'_x of the force using (5.172), we obtain:

$$f_x = \frac{abc}{a'b'c'} \left(-2\pi G\rho a'b'c' \, x \int_0^\infty \frac{du}{(a'^2 + u)\Delta'} \right), \tag{5.187}$$

where:

$$\Delta'(u) = \sqrt{(a'^2 + u)(b'^2 + u)(c'^2 + u)} =$$
$$= \sqrt{(a^2 + s + u)(b^2 + s + u)(c^2 + s + u)} = \Delta(s + u). \tag{5.188}$$

The (5.187) can be rewritten by introducing the new variable $u' = s + u$:

$$f_x = -2\pi G\rho abc \, x \int_s^\infty \frac{du'}{(a^2 + u')\Delta(u')}. \tag{5.189}$$

By repeating the same considerations for the inner potential, by analogy with (5.178) and omitting the primes, we obtain the expression for the outer potential of the ellipsoid:

$$\mathcal{U}_e(x, y, z) = G\pi \, \rho \, abc \int_s^\infty \left\{ 1 - \frac{x^2}{a^2 + u} - \frac{y^2}{b^2 + u} - \frac{z^2}{c^2 + u} \right\} \frac{du}{\Delta}, \tag{5.190}$$

where Δ is defined by the expression (5.167b). Thus, the projection of the gravitational force and the potential of the ellipsoid in an outer point differ from the inner case in the lower limit of the integral only.

5.11 The Explicit Form of the Potential in an External Point

The potential of a generic body in an external point $P(r, \theta, \phi)$ is (cf. (5.111)):

$$\mathcal{U}_e(r, \theta, \phi) = G \iiint_M \frac{dM}{r} \left[\sum_{k=0}^\infty \left(\frac{r'}{r} \right)^k P_k(\cos \gamma) \right], \tag{5.191}$$

with $dM = \rho r'^2 \sin\theta' \, dr' \, d\theta' \, d\phi'$. By expliciting the Legendre coefficients up to the second degree, we obtain:

$$\mathcal{U}_e(r, \theta, \phi) = G\frac{M}{r} + \frac{G}{r^2} \iiint_M r' \cos\gamma \, dM +$$

$$+ \frac{G}{2r^3} \iiint_M r'^2 \left(3\cos^2\gamma - 1\right) dM +$$

$$+ G \iiint_M \frac{dM}{r} \left[\sum_{k=3}^{\infty} \left(\frac{r'}{r}\right)^k P_k(\cos\gamma)\right]. \quad (5.192)$$

We now assume that the center of gravity of the body contains the origin of the coordinate system O, and indicate with ℓ, m, and n, the direction cosines of $\overrightarrow{OP}$. Hence:

$$r' \cos\gamma = \ell x' + my' + nz', \quad (5.193)$$

where x', y', and z', are the Cartesian coordinates of the point $Q(r', \theta', \phi')$, and thus, by definition, of the barycenter:

$$\iiint_M r' \cos\gamma \, dM = \iiint_M \left(\ell x' + my' + nz'\right) dM =$$

$$= \ell \iiint_M x' \, dM + m \iiint_M y' \, dM + n \iiint_M z' \, dM = 0. \quad (5.194)$$

Therefore, the second term at the second member of (5.192) is zero.[25] The integral of the third term is:

$$r'^2(3\cos^2\gamma - 1) = 3(\ell x' + my' + nz')^2 - (x'^2 + y'^2 + z'^2) =$$

$$= 3(\ell^2 x'^2 + m^2 y'^2 + n^2 z'^2) - (x'^2 + y'^2 + z'^2) +$$

$$+ 6(\ell m \, x'y' + \ell n \, x'z' + mn \, y'z'), \quad (5.195)$$

with the usual meaning of the symbols. We now observe that the axes of the reference system are main axes of inertia for the body of mass M. Then, by indicating with I_x, I_y, and I_z, the components of the inertia moment vector:

$$I_x = \iiint_M (y'^2 + z'^2) \, dM,$$

$$I_y = \iiint_M (z'^2 + x'^2) \, dM, \quad (5.196)$$

$$I_z = \iiint_M (x'^2 + y'^2) \, dM,$$

it is easily verified that:

---

[25] Only because the center of gravity is at the origin of the axes, however!

$$\iiint_M x'^2 \, dM = \frac{I_x + I_y + I_z}{2} - I_x,$$

$$\iiint_M y'^2 \, dM = \frac{I_x + I_y + I_z}{2} - I_y, \tag{5.197}$$

$$\iiint_M z'^2 \, dM = \frac{I_x + I_y + I_z}{2} - I_z.$$

Thus, since the mixed inertia products vanish, the integral of the third term of (5.192) is $\left[I_x + I_y + I_z - 3(\ell^2 I_x + m^2 I_y + n^2 I_z) \right]$, and the external potential takes the expression:

$$\mathcal{U}_e(r, \theta, \phi) = G\frac{M}{r} + \frac{G}{2r^3}\left[I_x + I_y + I_z - 3(\ell^2 I_x + m^2 I_y + n^2 I_z) \right] +$$

$$+ G\sum_{k=3}^{\infty} \frac{1}{r^{k+1}} \iiint_M r'^k \, P_k\left(\frac{\ell x' + my' + nz'}{r'} \right) dM. \tag{5.198}$$

Note that, according to the Pythagorean theorem, the quantity:

$$I_{OP} = \iiint_M \left[r'^2 - (\ell x' + my' + nz')^2 \right] dM, \tag{5.199}$$

is the moment of inertia of the body with respect to the axis $\overrightarrow{OP}$.

We now suppose that the ellipsoid of inertia has two axes (that is, it is either oblate or prolate). Making the z-axis coincide with the symmetry axis, we set $I_x = I_y$. It is also $n = \cos\theta$. The coefficient of the term of the potential in $1/r^3$ undergoes a further reduction, becoming:

$$2I_x + I_z - 3(\ell^2 + m^2)I_x - 3n^2 I_z =$$
$$= 2I_x - 3(1 - n^2)I_x - (3n^2 - 1)I_z =$$
$$= (3n^2 - 1)(I_x - I_z) = (3\cos^2\theta - 1)(I_x - I_z). \tag{5.200}$$

In conclusion, neglecting the terms of higher order than the second in the expansion (5.192), the external potential of a rotationally symmetric body reduces to:

$$\mathcal{U}_e(r, \theta) \approx G\frac{M}{r} - G\frac{I_z - I_x}{2r^3}(3\cos^2\theta - 1), \tag{5.201}$$

if the reference system is barycentric, with the z-axis coincident with the axis of symmetry.

To calculate the force at an external point P, it is convenient to rotate the reference system so that P contains the plane xz, so that $r = (x^2 + z^2)^{1/2}$ and $\cos\theta = z/r$, with which the (5.201) becomes:

$$\mathcal{U}_e(x, 0, z) \approx \frac{G\,M}{(x^2 + z^2)^{1/2}} - G\,\frac{(I_z - I_x)}{2(x^2 + z^2)^{3/2}} \left(\frac{3z^2}{x^2 + z^2} - 1\right) =$$

$$= G\,\frac{M}{(x^2 + z^2)^{1/2}} - G\,\frac{(I_z - I_x)}{2}\,\frac{2z^2 - x^2}{(x^2 + z^2)^{5/2}}. \qquad (5.202)$$

Let us now calculate the force in P as the gradient of $\mathcal{U}_e(P)$. The first partial derivatives of (5.202) are:

$$\frac{\partial \mathcal{U}_e}{\partial x} = -G\,\frac{M}{r^3}\,x + G\,\frac{(I_z - I_x)}{2r^5}\,x\,(15\cos^2\theta - 3),$$

$$\frac{\partial \mathcal{U}_e}{\partial y} = 0, \qquad (5.203)$$

$$\frac{\partial \mathcal{U}_e}{\partial z} = -G\,\frac{M}{r^3}\,z + G\,\frac{(I_z - I_x)}{2r^5}\,z\,(15\cos^2\theta - 9).$$

In conclusion (Fig. 5.15), the acceleration exerted in P results:

$$\mathbf{a} \approx -G\left[\frac{M}{r^3} - \frac{I_z - I_x}{2r^5}\left(15\cos^2\theta - 3\right)\right]\mathbf{r} - 3\,G\,(I_z - I_x)\frac{\cos\theta}{r^4}\,\mathbf{k}, \qquad (5.204)$$

where $\mathbf{k}$ indicates the unit vector of the z-axis. The (5.204) shows that the acceleration is not purely radial due to the presence of the term oriented as the z-axis. For an oblate ellipsoid ($I_z > I_x$), this term produces an acceleration directed towards the xy-plane.

Figure 5.15 allows us to guess the physical reason of precession, i.e., for the change in direction of the orbital axis of a body revolving in an axisymmetric gravitation field. Suppose that P is an artificial satellite on an inclined orbit relative to the

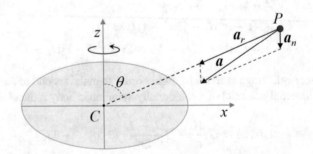

Fig. 5.15 External potential of an ellipsoid. The off-centering of the acceleration $\mathbf{a}$ is exaggerated for graphical reasons. Note that, at a point P at latitude $0 < \frac{\pi}{2} - \theta < \frac{\pi}{2}$, the acceleration $\mathbf{a}$ is not radial and can be decomposed into two components, radial ($\mathbf{a_r}$) and normal to the equatorial plane ($\mathbf{a_n}$)

equatorial plane of the Earth, and assume that the planet can be modeled as an oblate spheroid. According to (5.204), the satellite is subject to an acceleration that has both radial and normal components. If the radial component were alone, being a central force it would give rise to a purely plane motion even if not Keplerian (hence the orbit would not be closed). The presence of a second component perpendicular to the radius vector, $|\mathbf{a_n}| = -3\,G\,(I_z - I_x)\dfrac{\cos\theta}{r^4}$, produces a torque:

$$\tau = |\mathbf{r} \times \mathbf{a}_n| = r|\mathbf{a}_n| \sin\theta \propto \cos\theta \, \sin\theta. \tag{5.205}$$

The phenomenon resembles that of a spinning top subject to Earth's gravity. Obviously, as P moves along its orbit, the variation of θ affects the torque applied by a_n, which is zero on the equator and pole, and maximal for $\theta = \pi/4$. This qualitative consideration complements the quantitative treatment of Sect. 4.13.

5.12 Rotational Distortion of the Earth

In addition to gravity $\mathbf{g}$, each point of the Earth's surface feels the centrifugal force induced by the diurnal solid-body rotation of the planet. In a reference frame centered at the barycenter with the z-axis oriented as the angular velocity $\overrightarrow{\omega}_\oplus$ of the Earth's rotation,[26] the potential of the centrifugal force per unit mass, $\omega^2 r \sin\theta$, perpendicular to the z-axis and directed outwards, is (Fig. 5.16):

$$\mathcal{U}_c = \int_0^r r'\omega^2 \sin^2\theta \, dr' = \frac{1}{2}\omega^2 r^2 \sin^2\theta. \tag{5.206}$$

Adding to this term the gravitational potential in the approximation of the third order (5.201), we obtain the total potential for unit mass at a point $P(r,\theta)$ of the Earth surface:

$$\mathcal{U}(r,\theta) = G\frac{M_\oplus}{r} - G\frac{I_z - I_x}{2r^3}\left(3\cos^2\theta - 1\right) + \frac{1}{2}\omega^2 r^2 \left(1 - \cos^2\theta\right). \tag{5.207}$$

Let us ask what the shape of the Earth's equipotential surfaces should be with reference only to the centrifugal and gravitational forces.[27] Rotational symmetry of the total potential ensures that these surfaces cannot be functions of the azimuthal

---

[26] The angular velocity of the Earth is $\omega = \dfrac{2\pi}{86164}$ rad s^{-1}, as it must be computed using the sidereal day, which is ~4 min shorter that the solar day.

[27] The reader will appreciate the tasty representation of the debate on the shape of the Earth (prolate versus oblate) between Jacques Cassini (1677-1756), son of Giovanni Domenico, forefather of the family, and second director of the Paris Observatory, and Pierre-Louis Moreau de Maupertuis, in M. Terrall, *Representing the Earth's Shape: The Polemics Surrounding Maupertuis's Expedition to Lapland*, ISIS, 83, No. 2, 218–237, 1992.

Fig. 5.16 The effective gravity of the Earth, $\mathbf{g}_{eff}$, is the vector sum of the gravity $\mathbf{g}$ (approximately) directed towards the center of the Earth C and the opposite of the centripetal force $\mathbf{a}_c = \vec{\omega}_\oplus \times (\vec{\omega}_\oplus \times \mathbf{s})$, where $|\mathbf{s}| = r_\oplus \sin\theta$

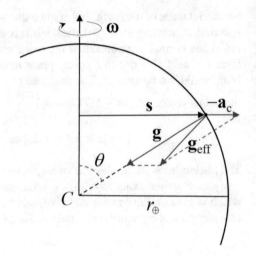

coordinate ϕ. Furthermore, measurements and observations of the Earth, and analogy with the shape of the other Solar System bodies, suggest that the equipotential surfaces must be nearly spherical. We therefore assume an analytic expression of the type (5.134):

$$r = r_\oplus \left[1 - Y(\cos\theta)\right], \tag{5.208}$$

where $r_\oplus$ is the equatorial radius of the Earth, and $Y(\cos\theta)$ a function such that $|Y| \ll 1$ for each value of the variable (actually Y does not exceed 0.3%). With this function and with the condition of equipotentiality, we obtain the equation:

$$\mathfrak{U} = \frac{1}{2}\omega^2 r_\oplus^2 (1-Y)^2(1-\cos^2\theta) + \frac{G\,M_\oplus}{r_\oplus(1-Y)} +$$
$$- \frac{G\,(I_z - I_x)}{2\,r_\oplus^3\,(1-Y)^3}\left(3\cos^2\theta - 1\right) = \mathfrak{U}_o, \tag{5.209}$$

where $\mathfrak{U}_o$ is a constant. We can simplify this expression by approximating $(1-Y)^2$ and $(1-Y)^3$ to unity. We obtain:

$$\mathfrak{U} = \frac{1}{2}\omega^2 r_\oplus^2 (1-\cos^2\theta) + \frac{G\,M_\oplus}{r_\oplus(1-Y)} - \frac{G\,(I_z - I_x)}{2\,r_\oplus^3}\left(3\cos^2\theta - 1\right). \tag{5.210}$$

Moreover, since the surface is equipotential, the constant can be calculated at any point. Choosing $r = r_\oplus$ (equatorial radius) at $\theta = \pi/2$, that is, $Y(\cos\theta) = 0$ for $\theta = \pi/2$, the (5.210) becomes:

$$\mathfrak{U} = \frac{1}{2}\omega^2 r_\oplus^2 + \frac{G\,M_\oplus}{r_\oplus} + \frac{G\,(I_z - I_x)}{2\,r_\oplus^3}. \tag{5.211}$$

By equating (5.210) to (5.211), being both equal to $\mathcal{U}_o$, we obtain:

$$Y(\cos\theta) \approx \frac{Y}{1-Y} = \left[\frac{3(I_z - I_x)}{2M_\oplus r_\oplus^2} + \frac{\omega^2 r_\oplus^3}{2G\,M_\oplus}\right]\cos^2\theta. \tag{5.212}$$

We note that the second term in square brackets is the ratio between the centrifugal and the gravitational potentials at the equator:

$$\phi = \frac{\omega^2\,r_\oplus^3}{G\,M_\oplus} = 2\,\frac{\mathcal{U}_c}{\mathcal{U}_g}. \tag{5.213}$$

In conclusion, within the limits of the approximations we have adopted, the equipotential surface of the Earth is given by the equation:

$$r = r_\oplus\left(1 - f\,\cos^2\theta\right), \tag{5.214a}$$

with:

$$f = \frac{3}{2}\frac{I_z - I_x}{M_\oplus r_\oplus^2} + \frac{1}{2}\phi. \tag{5.214b}$$

By analogy with (5.152), we deduce that this surface is a spheroid with flattening $f = \dfrac{a-c}{a}$, where $a = r_\oplus$ and c are the semi-axes.[28] Note that, from knowledge of f and of ϕ, through (5.214b) we could calculate the difference between the polar and equatorial inertia moments. However it is almost impossible to measure the constant f with sufficient accuracy using triangulations. It is possible to infer it from the study of the precession of the equinoxes or from the measures of acceleration of gravity, as the latter is normal to equipotential surfaces, $\mathbf{g}_{eff} = \nabla\mathcal{U} = \left(\dfrac{\partial\mathcal{U}}{\partial r}, \dfrac{1}{r}\dfrac{\partial\mathcal{U}}{\partial\theta}\right)$.

From (5.207) we calculate the gravitational acceleration:

$$|\mathbf{g}_{eff}| = \left[\left(\frac{\partial\mathcal{U}}{\partial r}\right)^2 + \left(\frac{1}{r}\frac{\partial\mathcal{U}}{\partial\theta}\right)^2\right]^{1/2} \approx \left|\frac{\partial\mathcal{U}}{\partial r}\right| =$$

$$= G\frac{M_\oplus}{r^2} - 3G\frac{I_z - I_x}{2r^4}\left(3\cos^2\theta - 1\right) - \omega^2 r\left(1 - \cos^2\theta\right), \tag{5.215}$$

since, under our assumptions about the moderate distortion of the Earth, $-\dfrac{1}{r}\dfrac{\partial\mathcal{U}}{\partial\theta}$ can be neglected. Using $r = r_\oplus$ together with (5.213) and (5.214b), and eliminating $\left(I_z - I_x\right)$, after some simple reductions we have:

---

[28] Note the difference with (5.152), where we modulated c instead of a.

$$g_{eff} = G \frac{M_\oplus}{r_\oplus^2}\left(1 + f - \frac{3}{2}\phi\right) + \left(\frac{5}{2}r_\oplus\omega^2 - 3f\, G\frac{M_\oplus}{r_\oplus^2}\right)\cos^2\theta =$$

$$= g_\circ\left(1 + \frac{\frac{5}{2}\phi - 3f}{1 + f - \frac{3}{2}\phi}\right)\cos^2\theta, \tag{5.216}$$

where the term:

$$g_\circ = G\frac{M_\oplus}{r_\oplus^2}\left(1 + f - \frac{3}{2}\phi\right), \tag{5.217}$$

indicates the gravity acceleration at the equator. The (5.217) is known as Clairaut[29] equation. It shows that, at the first order, the acceleration of gravity varies with the square of the latitude ($\ell = (\pi/2) - \theta$).

5.13 Potential of a Homogeneous Circular Torus

We now want to obtain the potential of a torus, a figure which is a solid of revolution generated by the complete rotation of a circle around a coplanar and non-intersecting axis. Its name is borrowed from Latin, where it means "a round, swelling, elevation, protuberance". Toroidal structures have acquired increasing importance in astrophysics as they are, for instance, key components of Active Galactic Nuclei (AGN's).

A torus is more complex than a solid sphere or ellipsoid because it is not a simply connected body (in short, we can draw circles on its surface that cannot contract to a point without leaving the torus figure, as is instead the case with a solid sphere). Moreover, the surface of a torus has a curvature[30] which depends on the angle in the torus meridian cross-section (angle ψ in the upper panel of Fig. 5.18).

The easiest way to get the torus potential is by direct integration over the volume but this procedure does not allow us to simplify the final expression and to explore the gravitational properties of the this body. Here we address the problem by modeling the torus through a set of infinitely thin rings, i.e., massive circles characterized by one radius only.

---

[29] Alexis Claude Clairaut (1713–1765) was a French mathematician, astronomer, and geophysicist. Child prodigy and prominent supporter of Newtonianism, he was a leading scientist in the expedition to Lapland (1736–37) led by Maupertuis: an adventure that helped confirming Newton's theory for the figure of the Earth, opposite to the prolate hypothesis formulated by Jacques Cassini, faithful Cartesian.

[30] The Gaussian curvature of the torus surface is: $K = \dfrac{\cos\psi}{R_0(R + R_0\cos\psi)}$, where ψ is the angle in the torus meridian cross-section (see Fig. 5.18). This implies $K > 1$ in the external part of the surface, $K < 1$ in the inner part (tracing the hole). The curvature is null for two angles, $\psi = \pi/2$, $3\pi/2$.

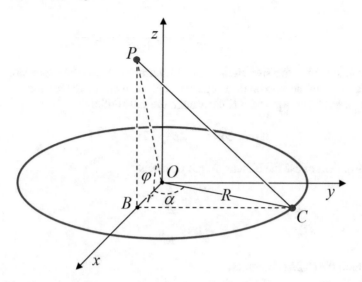

Fig. 5.17 Scheme of the attraction of a test particle by an infinitely thin ring (massive circle)

5.13.1 Potential of an Infinitely Thin Ring

We will start by considering the potential generated by a massive circle of total mass M and radius R in a generic point P:

$$\mathcal{U} = G \int_M \frac{dm}{|PC|}, \qquad (5.218)$$

where the mass of the ring element in C (Fig. 5.17) is:

$$dm = \frac{M}{2\pi R} dl = \frac{M}{2\pi} d\alpha, \qquad (5.219)$$

and PC is the distance of P from C. In the following we will use a cylindrical coordinate system. The distance PC can be found from the triangle PBC of Fig. 5.17:

$$PC^2 = z^2 + BC^2, \qquad (5.220)$$

where BC is given by the triangle BOC:

$$BC^2 = r^2 + R^2 - 2Rr \cos \alpha. \qquad (5.221)$$

Taking into account (5.218) and (5.219), we obtain:

$$\mathcal{U}(r, z) = \frac{GM}{\pi} \int_0^\pi \frac{d\alpha}{\sqrt{r^2 + z^2 + R^2 - 2Rr \cos \alpha}}. \tag{5.222}$$

Note that, since the massive circle is an axisymmetric body, the expression for the potential does not depend on the azimuthal angle α. It can be rewritten in form of an elliptic integral (see Appendix N.4) through the substitution:

$$\cos \alpha = \cos(\pi - 2\beta) = 2 \sin^2 \beta - 1. \tag{5.223}$$

It is convenient to use the dimensionless variables:

$$\begin{aligned} \rho &= \frac{r}{R}, \\ \zeta &= \frac{z}{R}, \end{aligned} \tag{5.224}$$

with which the (5.222) becomes:

$$\mathcal{U}(\rho, \zeta) = \frac{GM}{\pi R} \sqrt{\frac{m}{\rho}} K(m) = \frac{GM}{\pi R} \phi(\rho, \zeta), \tag{5.225}$$

where:

$$\phi(\rho, \zeta) = \sqrt{\frac{m}{\rho}} K(m), \tag{5.226}$$

is the dimensionless potential of a massive circle, and:

$$K(m) = \int_0^{\pi/2} \frac{d\beta}{\sqrt{1 - m \sin^2 \beta}}, \tag{5.227}$$

is the complete elliptic integral of the first kind with module:

$$m = \frac{4\rho}{(\rho + 1)^2 + \zeta^2}. \tag{5.228}$$

We can use the same simple considerations applied to the thin spherical shell (cf. Sect. 5.4) to prove here that a massive circle is an attractor. Since it is a one-dimensional figure, the mass element is $dm \propto dl$ and the corresponding element of force $dF \propto dl/r^2$. So, there is no compensation of the forces acting on the particle and coming from the opposite side. Equilibrium is possible only at the center: a characteristic shared also by the torus. But the torus has also a second set of weightless points forming a circle inside its volume.

5.13.2 Potential of a Homogeneous Circular Torus

We now want to find the potential of a homogeneous circular torus modeling it by a set of the massive circles [2]. The approach is similar to that we have adopted in Sect. 5.4 for the potential of the solid sphere consisting of infinitely thin spherical layers. Before getting started, it is useful to remember two theorems on solids of revolution named after Pappus and Guldinus.[31] The first theorem states that the area A of a surface obtained by rotating a plane curve of length ℓ by an angle of 2π around a coplanar axis is equal to:

$$A = 2\pi R\ell, \tag{5.229a}$$

where R is the distance of the center of the curve from the rotation axis. The second theorem asserts that the volume V of a solid obtained by rotating a plane figure of area S around a coplanar axis, without intersecting it, is equal to:

$$V = 2\pi RS, \tag{5.229b}$$

where again R is the distance of the center of the plane figure from the rotational axis.

That said, we take into consideration a torus with mass M, major radius R, and minor (cross-section) radius R_0 (Fig. 5.18). According to the (5.229), it has an area $S = (2\pi R)(2\pi R_0)$ and a volume $V = (2\pi R)(\pi R_0^2)$. In order to evaluate its potential $\mathcal{U}(P)$ at any point P, including both internal and external regions (that is, inside and outside the volume of the figure), we decompose the torus into a set of massive circles (infinitely thin rings) coplanar to the symmetry plane. Note again that, due to the symmetry of the figure about the z-axis of the coordinate system with origin at the torus barycenter (Fig. 5.18), P can always be chosen to lie on the rz-plane. We first compute the contribution by a generic massive circle of mass M_r, using the dimensionless coordinates $\rho = r/R$, $\zeta = z/R$ of $P(r, z)$ (see (5.224)). It is identified by its intersect on the cross-section of the torus with the rz-plane, whose dimensionless coordinates (hereafter circle coordinates) are:

$$\eta' = \frac{x'}{R},$$

$$\zeta' = \frac{z'}{R}. \tag{5.230}$$

---

[31] These theorems were originally obtained by the Greek mathematicians Pappus of Alexandria (290–350 AD), but the first known demonstrations belong to the Swiss Jesuit Paul Guldin (1577–1643). Kepler knew these theorems (1615). There is a story about a barrel of wine that motivated the German astronomer to study how to calculate areas and volumes. It happened when he had to argue with a merchant who was selling him the wine to celebrate his second marriage in Linz. The man was likely trying to cheat about the quantity of the product. Kepler reacted by writing a book entitled *Nova stereometria doliorum vinariorum* (New solid geometry of wine barrels).

Fig. 5.18 *Up*: a 3D scheme
of a torus configuration.
Down: schematic
cross-section of a torus
showing the position of the
central massive circle and of
another circle of the set
composing the torus itself

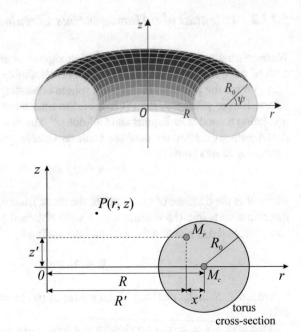

Owing to the results of Sect. 5.13.1, the potential $\mathcal{U}_r$ at $P(\rho, \zeta)$ generated by the massive circle with coordinates (η', ζ') is:

$$\mathcal{U}_r(\rho, \zeta; M_r, \eta', \zeta') = \frac{GM_r}{\pi R'} \, \phi_r(\rho, \zeta; \eta', \zeta'), \qquad (5.231)$$

where R' is the radius of the circle and the expression for ϕ_r is obtained by replacing r/R with r/R' and z/R with $(z - z')/R'$ in (5.225) and (5.226). We now want to refer the coordinates of the circle to the center $C(R, 0)$ of the torus cross-section (Fig. 5.18): $x' = R' - R$, from where $R' = R + x'$. Consequently, the parameter of the elliptic integral m in (5.228) transforms into:

$$m_r = \frac{4r/R'}{\left(1 + \dfrac{r}{R'}\right)^2 + \dfrac{(z - z')^2}{R'^2}} = \frac{4r/(1 + x')}{\left(1 + \dfrac{r}{1 + x'}\right)^2 + \dfrac{(z - z')^2}{(1 + x')^2}} =$$

$$= \frac{4r(1 + x')}{(1 + x' + r)^2 + (z - z')^2} = \frac{4\rho(1 + \eta')}{(1 + \eta' + \rho)^2 + (\zeta - \zeta')^2}. \qquad (5.232)$$

In conclusion, the expression for the dimensionless potential of the component massive circle takes the form:

$$\phi_r(\rho, \zeta; \eta', \zeta') = \sqrt{\frac{(1+\eta')m_r}{\rho}}\, K(m_r). \tag{5.233}$$

The homogeneity of the torus implies the equality of the reduced masses $\kappa_c = \kappa_r$, where $\kappa_c = M_c/(2\pi R)$ is the reduced mass of the central circle and $\kappa_r = M_r/(2\pi R')$ that of the component circle. Then: $M_r = M_c R'/R$. Using this expression for (5.231), we obtain the potential of the component circle:

$$\mathcal{U}_r(\rho, \zeta; \eta', \zeta') = \frac{GM_c}{\pi R}\, \phi_r(\rho, \zeta; \eta', \zeta'). \tag{5.234}$$

Due to the additive property, the potential of the torus is just the sum of all the contributions by the component circles. To perform this integration, we replace in (5.234) the finite mass of the circle, M_c, with the differential dM, which for the circular torus of total mass M equals:

$$dM = \frac{M}{\pi r_0^2}\, d\eta' d\zeta', \tag{5.235}$$

where $r_0 = R_0/R$ is the dimensionless minor radius of the torus (a geometrical parameter telling us the size of the torus cross-section in units of the torus width; cf. Fig. 5.18). In conclusion, the potential of the homogeneous circular torus takes the form:

$$\mathcal{U}_{tor}(\rho, \zeta) = \frac{GM}{\pi^2 R r_0^2} \int_{-r_0}^{r_0} \int_{-\sqrt{r_0^2 - \eta'^2}}^{\sqrt{r_0^2 - \eta'^2}} \sqrt{\frac{(1+\eta')m_r}{\rho}}\, K(m_r)d\eta' d\zeta' =$$

$$= \frac{GM}{\pi^2 R r_0^2} \int_{-r_0}^{r_0} \int_{-\sqrt{r_0^2 - \eta'^2}}^{\sqrt{r_0^2 - \eta'^2}} \phi_r(\rho, \zeta; \eta', \zeta')d\eta' d\zeta', \tag{5.236}$$

where $K(m_r)$ is the complete elliptic integral of the first kind (5.227) with the parameter (5.232). This integral expression gives the torus potential for points belonging either to the inner volume bounded by the torus surface (inside the torus body) and to the region outside it.

Figure 5.19 plots the trend of the torus potential with the radial coordinate ρ in the equatorial plane $\zeta = 0$, computed numerically from (5.236) for different values of the geometrical parameter r_0 (here and for the next figure we use the unity system $G = 1, M = 1, R = 1$). The curves for all values of r_0 are seen to be inscribed within the potential curve of the central massive circle (infinitely thin ring) of the same mass M and radius R, located at the torus symmetry plane. The potential curve to the right of the torus surface ($\rho > 1 + r_0$) virtually coincides with the potential curve of the circle, while to the left ($\rho < 1 - r_0$) it stays lower and differs by a quantity that depends on r_0. We will investigate this in the next section.

Fig. 5.19 Dependence of the potential on the radial coordinate ρ for tori with different values of the geometrical parameter r_0. The potential of a massive circle (5.225) with the same mass M of the torus is shown by the dashed line

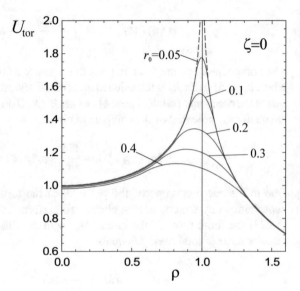

In conclusion, Fig. 5.19 tells us that the outer potential of the torus can be approximately represented by that of a massive circle of the same mass up to torus surface. For $\rho \to 0$, the values of the torus potential differ from those of the massive circle by a quantity that depends on the geometric parameter r_0, which is especially evident for a thick torus ($r_0 > 0.5$).

5.13.3   Approximation of the Homogeneous Torus Potential in the Outer Region

We now justify analytically the result obtained in the previous section where we showed that the outer potential of the torus can be represented by the potential of a central massive circle of equal mass.[32]

The outer region implies that $(\rho - 1)^2 + \zeta^2 \geq r_0^2$. Within this region, the integrand $\phi_r(\rho, \zeta; \eta', \zeta')$ in (5.236) has no singularities for all η', ζ'; therefore, we can expand in a Maclaurin[33] power series of η', ζ' in the vicinity of the center of the torus cross-section, $\eta' = \zeta' = 0$. Since the integrals in symmetrical limits of the series terms containing cross derivatives and derivatives of the odd orders are equal to zero, only the terms with the even orders remain in the expansion. By restricting to the quadratic terms of the series, the potential of the component circle is:

---

[32] The approximate expression for the torus potential in the inner region can be represented by the sum of the cylinder potential and a term consisting of the curvature of the torus surface [2].

[33] Colin Maclaurin (1698–1746): Scottish mathematician who gave important contributions to geometry and algebra.

$$\phi_r(\rho, \zeta; \eta', \zeta') \approx \phi_c(\rho, \zeta) + \frac{1}{2} \frac{\partial^2 \phi_r}{\partial \eta'^2}\bigg|_{\substack{\eta'=0 \\ \zeta'=0}} \eta'^2 + \frac{1}{2} \frac{\partial^2 \phi_r}{\partial \zeta'^2}\bigg|_{\substack{\eta'=0 \\ \zeta'=0}} \zeta'^2, \qquad (5.237)$$

where $\phi_c = \sqrt{\dfrac{m}{\rho}} K(m)$ is the dimensionless potential of the central circle (5.226)
with the module of the elliptic integral given by (5.228). The coordinates of this
circle are $(R, 0)$, that is, the circle is located on the equatorial plane and its radius is
equal to the major radius of the torus, R.

The substitution of (5.237) into (5.236) gives us:

$$\mathcal{U}_{tor}(\rho, \zeta) \approx \frac{GM}{\pi^2 R r_0^2} \int_{-r_0}^{r_0} \int_{-\sqrt{r_0^2-\eta'^2}}^{\sqrt{r_0^2-\eta'^2}} \bigg[\phi_c(\rho, \zeta) +$$

$$+ \frac{1}{2} \frac{\partial^2 \phi_r}{\partial \eta'^2}\bigg|_{\substack{\eta'=0 \\ \zeta'=0}} \eta'^2 + \frac{1}{2} \frac{\partial^2 \phi_r}{\partial \zeta'^2}\bigg|_{\substack{\eta'=0 \\ \zeta'=0}} \zeta'^2 \bigg] d\eta' d\zeta'. \qquad (5.238)$$

The first integral is solved immediately; it is the area of the torus cross-section:

$$\int_{-r_0}^{r_0} \int_{-\sqrt{r_0^2-\eta'^2}}^{\sqrt{r_0^2-\eta'^2}} d\eta' d\zeta' = \pi r_0^2. \qquad (5.239)$$

The second and the third integrals are:

$$\int_{-r_0}^{r_0} \int_{-\sqrt{r_0^2-\eta'^2}}^{\sqrt{r_0^2-\eta'^2}} \eta'^2 d\eta' d\zeta' = \int_{-r_0}^{r_0} \int_{-\sqrt{r_0^2-\eta'^2}}^{\sqrt{r_0^2-\eta'^2}} \zeta'^2 d\eta' d\zeta' = \frac{\pi r_0^4}{4}. \qquad (5.240)$$

Using these expression in (5.238), we obtain:

$$\mathcal{U}_{tor}(\rho, \zeta) \approx \frac{GM}{\pi R} \phi_c \times \left(1 + \frac{r_0^2}{8\phi_c} \left[\frac{\partial^2 \phi_r}{\partial \eta'^2}\bigg|_{\substack{\eta'=0 \\ \zeta'=0}} + \frac{\partial^2 \phi_r}{\partial \zeta'^2}\bigg|_{\substack{\eta'=0 \\ \zeta'=0}} \right] \right). \qquad (5.241)$$

Although each term in the square brackets is cumbersome, their sum can be reduced
to a compact form. Finally, after some transformations, the approximate expression
for the outer region torus potential takes the form:

$$\mathcal{U}_{tor}(\rho, \zeta) \approx \frac{GM}{\pi R} \phi_c(\rho, \zeta) \left(1 - \frac{r_0^2}{16} + \frac{r_0^2}{16} S(\rho, \zeta) \right). \qquad (5.242)$$

The function:

$$S(\rho, \zeta) = \frac{\rho^2 + \zeta^2 - 1}{(\rho - 1)^2 + \zeta^2} \frac{E(m)}{K(m)}, \qquad (5.243)$$

contains the complete elliptic integral of the second kind (see Appendix N.4):

$$E(m) = \int_0^{\pi/2} d\beta \sqrt{1 - m \sin^2 \beta}, \qquad (5.244)$$

with m defined by (5.228). We may conveniently adopt the new variable $\eta = \rho - 1$, which is counted from the torus cross-section. It allows the expression (5.242) to be written as:

$$S(\eta, \zeta) = \frac{\eta^2 + \zeta^2 + 2\eta}{\eta^2 + \zeta^2} \frac{E(m)}{K(m)}, \qquad (5.245)$$

where:

$$m = \frac{4(\eta + 1)}{(\eta + 2)^2 + \zeta^2}. \qquad (5.246)$$

Expression (5.242) (we call it the S-approximation), coupled with (5.243) or (5.245), represents the torus potential quite accurately in the outer region (Fig. 5.19). Since $|S| \leq 1$, the second multiplier in (5.242) is a slowly varying function of ρ and ζ. Let us simplify the expression (5.242) replacing the second multiplier by its asymptotic approximations.

When $\rho \to 0$, corresponding to $\eta \to -1$, that is, near the symmetry axis of the torus, the parameter $m \to 0$ and $E(m)/K(m) \to 1$; therefore, $S \to (\zeta^2 - 1)/(\zeta^2 + 1)$. The expression for the torus potential is then:

$$\mathcal{U}_{tor}(\rho, \zeta) \approx \frac{GM}{\pi R} \phi_c(\rho, \zeta) \left(1 - \frac{r_0^2}{16} + \frac{r_0^2}{16} \frac{\zeta^2 - 1}{\zeta^2 + 1} \right). \qquad (5.247)$$

Since the dimensionless potential of the massive circle (5.225) at the symmetry axis is $\phi_c = \pi/\sqrt{1 + \zeta^2}$, we obtain for the torus:

$$\mathcal{U}_{tor}(0, \zeta) \approx \frac{GM}{R} \frac{1}{\sqrt{1 + \zeta^2}} \left(1 - \frac{r_0^2}{16} + \frac{r_0^2}{16} \frac{\zeta^2 - 1}{\zeta^2 + 1} \right), \qquad (5.248)$$

and for $\zeta = 0$ in the symmetry center of the torus:

$$\mathcal{U}_{tor}(0, 0) \approx \frac{GM}{R} \left(1 - \frac{r_0^2}{8} \right). \qquad (5.249)$$

The second term $r_0^2/8$ in (5.249) describes the difference of the torus potential at the symmetry axis with respect to the potential of an infinitely thin ring.

At large η, the parameter $m \to 0$ and $S \to 1$ (5.249). In this second case:

$$\mathcal{U}_{tor}(\rho, \zeta) \approx \frac{GM}{\pi R} \phi_c(\rho, \zeta), \qquad (5.250)$$

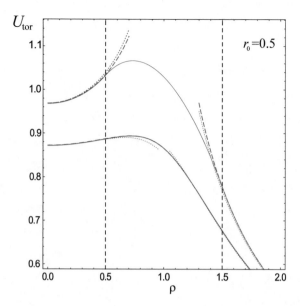

Fig. 5.20 Dependence on ρ of the torus potential with $r_0 = 0.5$ for $\zeta = 0$ (upper curves) and $\zeta = r_0 = 0.5$ (lower curves). Solid lines show the potential calculated with the exact formula (5.236). The S-approximation (5.242) of the potential is shown by the dashed lines, while the dotted lines represent the limiting cases of the S-approximation: the potential curve for the infinitely thin ring (5.250) is to the right of the torus cross-section, and the curve representing a *"shifted"* potential of the infinitely thin ring (5.247) is to the left. The boundaries of the torus cross-section are shown by the dotted vertical lines

that is, the torus potential is equal to the potential of the massive circle with the same mass M and radius R. It is seen from Fig. 5.20 that the S-approximation for the torus outer potential (5.242) is applicable up to the torus surface (upper curves).

In summary, the outer potential of the torus can be represented with good accuracy by a potential of the massive circle of the same mass. The dependence of the geometrical parameter r_0 appears only in the torus hole; it is taken into account in the "shifted" potential of the massive circle given by (5.247). These approximations are valid up to the surface of the torus and are a simple and useful tool. There is some analogy to the known result concerning the external potential of a solid sphere of mass M, which is the same as that generated by a point mass M located at the center (see Sect. 5.4).

References

1. A. Sommerfeld, *Partial Differential Equations* (Academic Press, New York, 1949)
2. E. Yu. Bannikova, V.G. Vakulik, V.M. Shulga, Gravitational potential of a homogeneous circular torus: a new approach. Monthly Notices R. Astron. Soc. **411**, 557 (2011)

Appendix A
Fundamental Formulas of Spherical Trigonometry

Plato, phrase engraved at the Academy's door
Let no one ignorant of geometry enter here.

We start with some definitions and some intuitive properties. The intersection of a spherical surface with a plane is called either maximum or minor circle depending on whether the plane contains the center of the sphere or not. Only one maximum circle passes for any two distinct points of the spherical surface. Three maximum circles pass for three distinct points provided that these points do not all lie on a same maximum circle. Each of the three distinct maximum circles are divided into four arcs. Each closed surface on the sphere generated by three such arcs is called spherical triangle.

Three points identify a maximum of eight spherical triangles, one of which has its three arcs, called sides, not exceeding a semicircle. In each spherical triangle three angles are also defined by the tangents to the corresponding maximum circles in the three crossing points, called vertexes of the angles. It is then clear that ABC in Fig. A.1 is a spherical triangle, as opposed to the triangle DEC, which is not since it is composed by a minor circle too (the circle making the DE side).

Consider[1] now the spherical triangle ABC in Fig. A.2 and trace the straight lines through OB and OC, then the tangents in A to the maximum circles passing through BA and CA. Being on the same planes, these lines give rise to two intersection points E and D that we join. We call a, b, and c the sides of the triangle, which correspond to the top angles as in the figure. It is clear then that the angle $\widehat{EAD}$ of the plane triangle AED corresponds to the vertex angle in A of the spherical triangle ABC.

Let us take a look at the two plane triangles OED and AED. The length of their common side $\overline{ED}$ can be written through the Carnot[2] cosine theorem:

---

[1] This section is largely inspired by W.M. Smart's classical book [1].

© The Editor(s) (if applicable) and The Author(s), under exclusive license to Springer Nature Switzerland AG 2022
E. Bannikova and M. Capaccioli, *Foundations of Celestial Mechanics*, Graduate Texts in Physics, https://doi.org/10.1007/978-3-031-04576-9

Fig. A.1 Triangles on the sphere

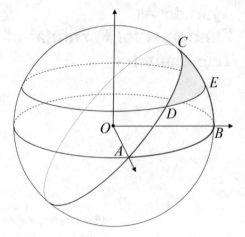

Fig. A.2 Projection of a spherical triangle on a plane

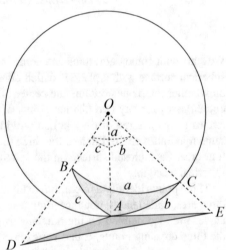

$$\overline{ED}^2 = \overline{OE}^2 + \overline{OD}^2 - 2\,\overline{OE} \cdot \overline{OD} \cos a,$$
$$\overline{ED}^2 = \overline{AE}^2 + \overline{AD}^2 - 2\,\overline{AE} \cdot \overline{AD} \cos A, \tag{A.1}$$

where A is the angle corresponding to the corner A of the triangle ABC (the angle $\widehat{BAC} = \widehat{DAE} = A$). Thus:

$$\overline{OE}^2 + \overline{OD}^2 - 2\,\overline{OE} \cdot \overline{OD} \cos a = \overline{AE}^2 + \overline{AD}^2 - 2\overline{AE} \cdot \overline{AD} \cos A. \tag{A.2}$$

---

[2] Lazare Nicolas Marguérite Carnot (1753–1823): French general, mathematician, physicist, and politician. Extraordinary personality, he lived at the turn of the French Revolution into the Napoleonic age. He was also the father of the physicist and engineer Nicolas Léonard Sadi Carnot, one of the founders of thermodynamics.

From the plane right triangles AOE and AOD, we have:

$$\overline{OA} = \overline{OE} \cos b,$$
$$\overline{OA} = \overline{OD} \cos c, \tag{A.3}$$

from which:

$$\overline{OE} = \frac{\overline{OA}}{\cos b},$$
$$\overline{OD} = \frac{\overline{OA}}{\cos c}, \tag{A.4}$$

and also:

$$\overline{AE} = \overline{OA} \tan b = \overline{OA} \frac{\sin b}{\cos b},$$
$$\overline{AD} = \overline{OA} \tan c = \overline{OA} \frac{\sin c}{\cos c}. \tag{A.5}$$

By replacing in (A.2), we obtain:

$$\frac{\overline{OA}^2}{\cos^2 b} + \frac{\overline{OA}^2}{\cos^2 c} - 2\frac{\overline{OA}^2 \cos a}{\cos b \cos c} =$$
$$= \overline{OA}^2 \frac{\sin^2 b}{\cos^2 b} + \overline{OA}^2 \frac{\sin^2 c}{\cos^2 c} - 2\overline{OA}^2 \frac{\sin b \sin c \cos A}{\cos b \cos c}, \tag{A.6}$$

which, after simplification and proper grouping, gives:

$$\frac{1 - \sin^2 b}{\cos^2 b} + \frac{1 - \sin^2 c}{\cos^2 c} = 2 \left(\frac{\cos a}{\cos b \cos c} - \frac{\sin b \sin c \cos A}{\cos b \cos c} \right), \tag{A.7}$$

and, by simplifying further:

$$\cos a = \cos b \cos c + \sin b \sin c \cos A. \tag{A.8}$$

In the same way we obtain the cosine formulas that constitute the group:

$$\begin{cases} \cos a = \cos b \cos c + \sin b \sin c \cos A, \\ \cos b = \cos a \cos c + \sin a \sin c \cos B, \\ \cos c = \cos a \cos b + \sin a \sin b \cos C, \end{cases} \tag{A.9}$$

where B and C are the angles corresponding to the corners B and C of the triangle ABC. Let us now write the first of them in the form:

$$\cos a - \cos b \cos c = \sin b \sin c \cos A, \tag{A.10}$$

and square it:

$$\cos^2 a + \cos^2 b \, \cos^2 c - 2 \, \cos a \, \cos b \, \cos c = \sin^2 b \, \sin^2 c \, \cos^2 A. \qquad \text{(A.11)}$$

The right part of (A.11) can be written as:

$$\sin^2 b \, \sin^2 c \, \cos^2 A = \sin^2 b \, \sin^2 c \, (1 - \sin^2 A) =$$
$$= \sin^2 b \, \sin^2 c - \sin^2 b \, \sin^2 c \, \sin^2 A =$$
$$= (1 - \cos^2 b)(1 - \cos^2 c) - \sin^2 b \, \sin^2 c \, \sin^2 A =$$
$$= 1 - \cos^2 b - \cos^2 c + \cos^2 b \, \cos^2 c - \sin^2 b \, \sin^2 c \, \sin^2 A. \qquad \text{(A.12)}$$

Replacing (A.12) in (A.11), after simplification we have:

$$\sin^2 b \, \sin^2 c \, \sin^2 A = 1 - \cos^2 a - \cos^2 b - \cos^2 c + 2 \cos a \, \cos b \, \cos c. \quad \text{(A.13)}$$

The symmetry of the right-hand side ensures that the left term takes the same value whatever the choice of the group equation (A.9). So we can write:

$$\sin^2 b \, \sin^2 c \, \sin^2 A = \sin^2 a \, \sin^2 b \, \sin^2 B = \sin^2 a \, \sin^2 c \, \sin^2 C. \qquad \text{(A.14)}$$

Since the implied angles are all lower than π, the latter can be put in the usual form:

$$\frac{\sin a}{\sin A} = \frac{\sin b}{\sin B} = \frac{\sin c}{\sin C}, \qquad \text{(A.15)}$$

which are the so-called sinus formulas for a spherical triangle. Let us now write the second of the (A.9) in the form:

$$\cos b - \cos a \, \cos c = \sin a \, \sin c \, \cos B, \qquad \text{(A.16)}$$

and replace $\cos a$ with the first of them. It is:

$$\cos b - \cos c \, (\cos b \, \cos c + \sin b \, \sin c \, \cos A) =$$
$$= \cos b - \cos^2 c \, \cos b - \sin b \, \sin c \, \cos c \, \cos A =$$
$$= \cos b - \cos b \, (1 - \sin^2 c) - \sin b \, \sin c \, \cos c \, \cos A =$$
$$= \cos b - \cos b + \cos b \, \sin^2 c - \sin b \, \sin c \, \cos c \, \cos A =$$
$$= \cos b \, \sin^2 c - \sin b \, \sin c \, \cos c \, \cos A. \qquad \text{(A.17)}$$

Finally we have:

$$\cos b \, \sin^2 c - \sin b \, \sin c \, \cos c \, \cos A = \sin a \, \sin c \, \cos B, \qquad \text{(A.18)}$$

which, divided by $\sin c$, gives:

$$\sin a \, \cos B = \cos b \, \sin c - \sin b \, \cos c \, \cos A. \qquad (\text{A.19})$$

Together with similar expressions, this latter forms the group:

$$\begin{cases} \sin a \, \cos B = \cos b \, \sin c - \sin b \, \cos c \, \cos A, \\ \sin b \, \cos C = \cos c \, \sin a - \sin c \, \cos a \, \cos B, \\ \sin c \, \cos A = \cos a \, \sin b - \sin a \, \cos b \, \cos C. \end{cases} \qquad (\text{A.20})$$

Equations (A.9), (A.15), and (A.20) form first spherical group, also known as the first Gauss group.

The vertex angles can be also expressed with similar formulas that we give without demonstration:

$$\begin{cases} \cos A = -\cos B \, \cos C + \sin B \, \sin C \, \cos a, \\ \cos B = -\cos A \, \cos C + \sin A \, \sin C \, \cos b, \\ \cos C = -\cos A \, \cos B + \sin A \, \sin B \, \cos c, \end{cases} \qquad (\text{A.21})$$

$$\begin{cases} \sin A \, \cos b = \cos B \, \sin C + \sin B \, \cos C \, \cos a, \\ \sin B \, \cos c = \cos C \, \sin A + \sin C \, \cos A \, \cos b, \\ \sin C \, \cos a = \cos A \, \sin B + \sin A \, \cos B \, \cos c. \end{cases} \qquad (\text{A.22})$$

Let us now consider the second of the (A.9), where we replace $\cos c$ with the expression given by the third of the same group:

$$\cos b = \cos a \, (\cos a \, \cos b + \sin a \, \sin b \, \cos C) + \sin a \, \sin c \, \cos B =$$
$$= \cos^2 a \, \cos b + \cos a \, \sin a \, \sin b \, \cos C + \sin a \, \sin c \, \cos B. \qquad (\text{A.23})$$

After some simplification, it is:

$$\cos b \, (1 - \cos^2 a) = \cos b \, \sin^2 a =$$
$$= \cos a \, \sin a \, \sin b \, \cos C + \sin a \, \sin c \, \cos B. \qquad (\text{A.24})$$

We now divide the left and right sides by $\sin a \, \sin b$ and obtain:

$$\frac{\cos b \, \sin a}{\sin b} = \cos a \, \cos C + \frac{\sin c}{\sin b} \cos B. \qquad (\text{A.25})$$

But, from (A.15) we have that:

$$\frac{\sin c}{\sin b} = \frac{\sin C}{\sin B}, \qquad (\text{A.26})$$

which, substituted in (A.25), gives in the end:

$$\sin a \, \cot b = \cos a \, \cos C + \sin C \, \cot B. \qquad \text{(A.27)}$$

Together with the other two analogous formulas, it forms the group:

$$\begin{cases} \sin a \, \cot b = \cos a \, \cos C + \sin C \, \cot B, \\ \sin b \, \cot c = \cos b \, \cos A + \sin A \, \cot C, \\ \sin c \, \cot a = \cos c \, \cos B + \sin B \, \cot A. \end{cases} \qquad \text{(A.28)}$$

Another group is:

$$\begin{cases} \sin A \, \cot B = -\cos A \, \cos c + \sin c \, \cot b, \\ \sin B \, \cot C = -\cos B \, \cos a + \sin a \, \cot c, \\ \sin C \, \cot A = -\cos C \, \cos b + \sin b \, \cot a. \end{cases} \qquad \text{(A.29)}$$

The (A.28), (A.29) are called cotangent formulas.

Appendix B
Formulas of Transformation of Coordinate Systems

Giuseppe Tomasi di Lampedusa, *The Leopard*

Everything must change so that everything can stay the same.

Let $S_1[x, y, z]$ and $S_2[x', y', z']$ be two orthogonal Cartesian reference systems having the same origin O. As usual, we denote with $\mathbf{i}, \mathbf{j}, \mathbf{k}$, and $\mathbf{i}', \mathbf{j}', \mathbf{k}'$, the unit vectors of the two systems. The coordinates of a point $P(x', y', z')$ relative to the second system are linked to those in S_1, $P(x, y, z)$, by means of the transformation formulas[3]:

$$
\begin{aligned}
x' &= a_{11}\, x + a_{12}\, y + a_{13}\, z, \\
y' &= a_{21}\, x + a_{22}\, y + a_{23}\, z, \\
z' &= a_{31}\, x + a_{32}\, y + a_{33}\, z,
\end{aligned}
\tag{B.2}
$$

where each coefficient a_{ij} represents the cosine of the angle between the i-th axis of the system S_2 and the j-th axis of the system S_1:

$$
\begin{aligned}
a_{11} &= \cos(\widehat{x'x}) & a_{12} &= \cos(\widehat{x'y}) & a_{13} &= \cos(\widehat{x'z}), \\
a_{21} &= \cos(\widehat{y'x}) & a_{22} &= \cos(\widehat{y'y}) & a_{23} &= \cos(\widehat{y'z}), \\
a_{31} &= \cos(\widehat{z'x}) & a_{32} &= \cos(\widehat{z'y}) & a_{33} &= \cos(\widehat{z'z}).
\end{aligned}
\tag{B.3}
$$

In fact, if $\mathbf{r} = \overrightarrow{OP}$, it is:

---

[3] Matrix notation can be also used:

$$
\begin{vmatrix} x' \\ y' \\ z' \end{vmatrix} = \begin{vmatrix} a_{11} & a_{12} & a_{13} \\ a_{21} & a_{22} & a_{23} \\ a_{31} & a_{32} & a_{33} \end{vmatrix} \times \begin{vmatrix} x \\ y \\ z \end{vmatrix}.
\tag{B.1}
$$

© The Editor(s) (if applicable) and The Author(s), under exclusive license to Springer 279
Nature Switzerland AG 2022
E. Bannikova and M. Capaccioli, *Foundations of Celestial Mechanics*, Graduate Texts in Physics, https://doi.org/10.1007/978-3-031-04576-9

$$\mathbf{r} = x\mathbf{i} + y\mathbf{j} + z\mathbf{k} = x'\mathbf{i}' + y'\mathbf{j}' + z'\mathbf{k}', \tag{B.4}$$

which, multiplied scalarly by $\mathbf{i}'$, $\mathbf{j}'$, and $\mathbf{k}'$, shows us that the (B.2) follow from the orthogonality of the axes. In the same way we prove that the system (B.2) can be inverted in:

$$\begin{aligned} x &= a_{11}\, x' + a_{21}\, y' + a_{31}\, z', \\ y &= a_{12}\, x' + a_{22}\, y' + a_{32}\, z', \\ z &= a_{13}\, x' + a_{23}\, y' + a_{33}\, z'. \end{aligned} \tag{B.5}$$

The transformation between the two coordinate systems therefore depends, via (B.2) and (B.5), on the nine direction cosines a_{lm} ($l, m = 1, 2, 3$). However, they are not all independent from each other, as there are the following six relations:

$$a_{l1}a_{m1} + a_{l2}a_{m2} + a_{l3}a_{m3} = \delta_{lm} \begin{cases} = 0 & \text{if } l \neq m, \\ = 1 & \text{if } l = m, \end{cases} \tag{B.6}$$

with $m \geq l = 1, 2, 3$. They are easily obtained from (B.4) by replacing the vector $\mathbf{r}$ with the three unit vectors of each of the two reference systems:

$$\begin{aligned} \mathbf{i}' &= a_{11}\mathbf{i} + a_{12}\mathbf{j} + a_{13}\mathbf{k}, \\ \mathbf{j}' &= a_{21}\mathbf{i} + a_{22}\mathbf{j} + a_{23}\mathbf{k}, \\ \mathbf{k}' &= a_{31}\mathbf{i} + a_{32}\mathbf{j} + a_{33}\mathbf{k}, \end{aligned} \tag{B.7}$$

and:

$$\begin{aligned} \mathbf{i} &= a_{11}\mathbf{i}' + a_{21}\mathbf{j}' + a_{31}\mathbf{k}', \\ \mathbf{j} &= a_{12}\mathbf{i}' + a_{22}\mathbf{j}' + a_{32}\mathbf{k}', \\ \mathbf{k} &= a_{13}\mathbf{i}' + a_{23}\mathbf{j}' + a_{33}\mathbf{k}'. \end{aligned} \tag{B.8}$$

The six relations (B.6) are obtained by internally multiplying each l-th identity of (B.7) with any other m-th identity of (B.8), respecting the condition $m > l$. In addition, by replacing into the identities:

$$\begin{aligned} \mathbf{i}' &= \mathbf{j}' \times \mathbf{k}', \\ \mathbf{j}' &= \mathbf{k}' \times \mathbf{i}', \\ \mathbf{k}' &= \mathbf{i}' \times \mathbf{j}', \end{aligned} \tag{B.9}$$

the expressions of the unit vectors for the system S_2 given by (B.7), further nine relations are obtained that, while not independent, will be of some use to us:

Fig. B.1 Relations between concentric orthogonal reference systems

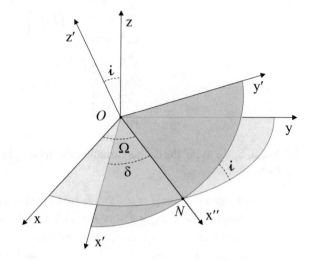

$$a_{11} = a_{22} a_{33} - a_{23} a_{32},$$
$$a_{21} = a_{13} a_{32} - a_{12} a_{33}, \qquad\qquad \text{(B.10a)}$$
$$a_{31} = a_{12} a_{33} - a_{13} a_{23},$$

$$a_{12} = a_{23} a_{31} - a_{21} a_{33},$$
$$a_{22} = a_{11} a_{33} - a_{13} a_{31}, \qquad\qquad \text{(B.10b)}$$
$$a_{32} = a_{13} a_{21} - a_{11} a_{23},$$

$$a_{13} = a_{21} a_{32} - a_{22} a_{31},$$
$$a_{23} = a_{12} a_{31} - a_{11} a_{32}, \qquad\qquad \text{(B.10c)}$$
$$a_{33} = a_{11} a_{22} - a_{12} a_{21}.$$

The six identities (B.6) show that three angles are sufficient to identify all nine direction cosines involved in the transformation from the system S_1 to S_2 and vice versa. In the applications of interest for us we choose (see Fig. B.1):

- the angle $0 \leq i \leq \pi$ between the axes z and z',
- the angle $0 \leq \Omega \leq 2\pi$ between the x axis and the intersection x'' between the coordinate planes xy and $x'y'$ (line of nodes), and
- the angle $0 \leq \delta \leq 2\pi$ between x'' and x'.

We now compute the direction cosines a_{lm}. From Fig. B.1 it is apparent that:

$$\widehat{xz} = \widehat{x''z} = \widehat{yz} = \frac{\pi}{2},$$
$$\widehat{xx''} = \Omega,$$
$$\widehat{x''y} = \frac{\pi}{2} - \Omega,$$
$$\widehat{x'x''} = \delta,$$
$$\widehat{x''y'} = \frac{\pi}{2} - \delta,$$
$$\widehat{zz'} = i.$$

(B.11)

With that, by applying the cosine formula to suitable spherical triangles of Fig. B.1, we have:

triangle $xx''x'$: $a_{11} = \cos(\widehat{xx'}) = \cos\Omega \cos\delta - \sin\Omega \sin\delta \cos i,$ (B.12a)

triangle $yx''x'$: $a_{12} = \cos(\widehat{yx'}) = \sin\Omega \cos\delta + \cos\Omega \sin\delta \cos i,$ (B.12b)

triangle $zx''x'$: $a_{13} = \cos(\widehat{zx'}) = \sin\delta \sin i,$ (B.12c)

triangle $xx''y$: $a_{21} = \cos(\widehat{xy'}) = -\cos\Omega \sin\delta - \sin\Omega \cos\delta \cos i,$ (B.12d)

triangle $yx''y'$: $a_{22} = \cos(\widehat{yy'}) = -\sin\Omega \sin\delta + \cos\Omega \cos\delta \cos i,$ (B.12e)

triangle $zx''y'$: $a_{23} = \cos(\widehat{zy'}) = \cos\delta \sin i,$ (B.12f)

triangle xzz' : $a_{31} = \cos(\widehat{xz'}) = \sin\Omega \sin i,$ (B.12g)

triangle yzz' : $a_{32} = \cos(\widehat{yz'}) = -\cos\Omega \sin i,$ (B.12h)

and, directly:

angle zz' : $a_{33} = \cos(\widehat{zz'}) = \cos i.$ (B.12i)

Appendix C
About Vector Operators

Lao Tzu

If you do not change direction, you may end up where you are heading.

By the vector differential operator:

$$\nabla = \mathbf{i}\frac{\partial}{\partial x} + \mathbf{j}\frac{\partial}{\partial y} + \mathbf{k}\frac{\partial}{\partial z}, \tag{C.1}$$

commonly known as nabla,[4] the gradient of a derivable scalar field $f(x, y, z)$ writes as:

$$\text{grad } f = \nabla f = \frac{\partial f}{\partial x}\mathbf{i} + \frac{\partial f}{\partial y}\mathbf{j} + \frac{\partial f}{\partial z}\mathbf{k}. \tag{C.2}$$

The gradient of f is the vector providing direction and value of the maximal variation per unit length of f. The divergence is the application of ∇ to a derivable vector field $\mathbf{V}(x, y, z) = V_x\mathbf{i} + V_y\mathbf{j} + V_z\mathbf{k}$:

$$\text{div}\mathbf{V} = \nabla\mathbf{V} = \frac{\partial V_x}{\partial x} + \frac{\partial V_y}{\partial y} + \frac{\partial V_z}{\partial z}. \tag{C.3}$$

It measures the local tendency of a vector field to diverge or converge. The meaning of this operator becomes clear by considering the continuity equation that expresses

---

[4] The term comes from the name of a traditional stringed musical instrument of the ancient Middle East peoples (Hebrews and Phoenicians), the 'nebel' or 'nabla'. It is similar to a lyre and a harp, but with a sound box with a triangular profile, which recalls that of an inverted Δ. The symbol, originally adopted by Hamilton in the form of the lying Δ, came in use at the middle of the 19-th century.

© The Editor(s) (if applicable) and The Author(s), under exclusive license to Springer
Nature Switzerland AG 2022
E. Bannikova and M. Capaccioli, *Foundations of Celestial Mechanics*, Graduate Texts
in Physics, https://doi.org/10.1007/978-3-031-04576-9

the conservation law, in a local form, for a generic physical quantity using its flow through a closed surface. For instance, in fluid dynamics (cf. [2]):

$$\frac{\partial \rho}{\partial t} + \nabla(\rho \mathbf{v}) = 0. \tag{C.4}$$

It shows that, if the divergence of the matter current $\nabla(\rho \mathbf{v})$ is positive, then there is a local depletion as $\partial \rho/\partial t$ is negative; if on the contrary $\nabla(\rho \mathbf{v}) < 0$, there is an enrichment as $\partial \rho/\partial t > 0$. If the divergence is null, than $\partial \rho/\partial t = 0$ and the flow is stationary.

The curl of the vector field $\mathbf{V}$ is defined as:

$$\mathrm{rot}\,\mathbf{V} = \nabla \times \mathbf{V} =$$
$$= \left(\frac{\partial V_z}{\partial y} - \frac{\partial V_y}{\partial z}\right)\mathbf{i} + \left(\frac{\partial V_x}{\partial z} - \frac{\partial V_z}{\partial x}\right)\mathbf{j} + \left(\frac{\partial V_y}{\partial x} - \frac{\partial V_x}{\partial y}\right)\mathbf{k}. \tag{C.5}$$

This formula can be memorized by applying the rules of determinants to the form:

$$\nabla \times \mathbf{V} = \begin{vmatrix} \mathbf{i} & \mathbf{j} & \mathbf{k} \\ \dfrac{\partial}{\partial x} & \dfrac{\partial}{\partial y} & \dfrac{\partial}{\partial z} \\ V_x & V_y & V_z \end{vmatrix}. \tag{C.6}$$

It measures the tendency of a vector field to generate a rotation, giving both the direction and the orientation of the rotation axis and, with its module, the amount of rotation at each point.

The divergence of the gradient of a scalar field f is:

$$\nabla \cdot (\nabla f) = \nabla^2 f = \frac{\partial^2 f}{\partial x^2} + \frac{\partial^2 f}{\partial x^2} + \frac{\partial^2 f}{\partial z^2}, \tag{C.7}$$

where $\nabla^2 = \triangle = \nabla \cdot \nabla$ is also called Laplace operator.

Gradient, divergence, and curl satisfy the following properties:

$$\nabla \times (\nabla f) = 0, \tag{C.8a}$$
$$\nabla \cdot (\nabla \times \mathbf{V}) = 0, \tag{C.8b}$$
$$\nabla (f + g) = \nabla f + \nabla g, \tag{C.8c}$$
$$\nabla \cdot (\mathbf{V} + \mathbf{U}) = \nabla \cdot \mathbf{V} + \nabla \cdot \mathbf{U}, \tag{C.8d}$$
$$\nabla \times (\mathbf{V} + \mathbf{U}) = \nabla \times \mathbf{V} + \nabla \times \mathbf{U}, \tag{C.8e}$$
$$\nabla (fg) = f\nabla g + g\nabla f, \tag{C.8f}$$
$$\nabla \cdot (f\mathbf{V}) = f\nabla \cdot \mathbf{V} + \mathbf{V} \cdot \nabla f, \tag{C.8g}$$
$$\nabla \times (f\mathbf{V}) = f\nabla \times \mathbf{V} + \mathbf{V} \times \nabla f, \tag{C.8h}$$

$$\nabla \cdot (\mathbf{V} \times \mathbf{U}) = \mathbf{U} \cdot \nabla \times \mathbf{V} - \mathbf{V} \cdot \nabla \times \mathbf{U}, \tag{C.8i}$$

$$\nabla \times (\mathbf{V} \times \mathbf{U}) = (\mathbf{U} \cdot \nabla)\mathbf{V} - \mathbf{U}(\nabla \cdot \mathbf{V}) - (\mathbf{V} \cdot \nabla)\mathbf{U} + \mathbf{V}(\nabla \cdot \mathbf{U}), \tag{C.8j}$$

$$\nabla \times (\nabla \times \mathbf{V}) = \nabla(\nabla \cdot \mathbf{V}) - \nabla^2 \mathbf{V}, \tag{C.8k}$$

$$\nabla(\mathbf{V} \cdot \mathbf{U}) = (\mathbf{U} \cdot \nabla)\mathbf{V} + (\mathbf{V} \cdot \nabla)\mathbf{U} +$$
$$+ \mathbf{U} \times (\nabla \times \mathbf{V}) + \mathbf{V} \times (\nabla \times \mathbf{U}). \tag{C.8l}$$

It is useful to express the vector operators in spherical coordinates $O[r, \theta, \phi]$:

$$\nabla f = \mathbf{i}_r \frac{\partial f}{\partial r} + \mathbf{i}_\theta \frac{1}{r} \frac{\partial f}{\partial \theta} + \mathbf{i}_\phi \frac{1}{r \sin \theta} \frac{\partial f}{\partial \phi}, \tag{C.9a}$$

$$\nabla \cdot \mathbf{V} = \frac{1}{r^2} \frac{\partial}{\partial r}(r^2 V_r) + \frac{1}{r \sin \theta} \frac{\partial}{\partial \theta}(\sin \theta \, V_\theta) + \frac{1}{r \sin \theta} \frac{\partial V_\phi}{\partial \phi}, \tag{C.9b}$$

$$\nabla \times \mathbf{V} = \mathbf{i}_r \frac{1}{r \sin \theta} \left\{ \frac{\partial}{\partial \theta}(\sin \theta V_\phi) - \frac{\partial V_\theta}{\partial \phi} \right\} +$$
$$+ \mathbf{i}_\theta \frac{1}{r} \left\{ \frac{1}{\sin \theta} \frac{\partial V_r}{\partial \phi} - \frac{\partial}{\partial r}(r V_\phi) \right\} + \mathbf{i}_\phi \frac{1}{r} \left\{ \frac{\partial}{\partial r}(r V_\theta) - \frac{\partial V_r}{\partial \theta} \right\}, \tag{C.9c}$$

where $\mathbf{i}_r$, $\mathbf{i}_\theta$, $\mathbf{i}_\phi$ are the unit vectors of the spherical coordinate axes and V_r, V_θ, V_ϕ are the components of the vector $\mathbf{V}$ along these axes.

The Laplace operator can be expressed in spherical coordinates in the two equivalent forms:

$$\nabla^2 f = \frac{1}{r^2} \frac{\partial}{\partial r}\left(r^2 \frac{\partial f}{\partial r}\right) + \frac{1}{r^2 \sin \theta} \frac{\partial}{\partial \theta}\left(\sin \theta \frac{\partial f}{\partial \theta}\right) + \frac{1}{r^2 \sin^2 \theta} \frac{\partial^2 f}{\partial \phi^2}, \tag{C.10a}$$

$$\nabla^2 f = \frac{1}{r} \frac{\partial^2}{\partial r^2}(rf) + \frac{1}{r^2 \sin \theta} \frac{\partial}{\partial \theta}\left(\sin \theta \frac{\partial f}{\partial \theta}\right) + \frac{1}{r^2 \sin^2 \theta} \frac{\partial^2 f}{\partial \phi^2}. \tag{C.10b}$$

Finally, we want to mention two theorems without proving them.

Divergence theorem[5] the flow of a regular vector field $\mathbf{V}$ through a closed surface S is equal to the integral of the divergence of the field extended to the volume Σ enclosed by this surface:

$$\iint_S \mathbf{V} \cdot \mathbf{n} \, dS = \iint_S \mathbf{V} \cdot d\mathbf{S} = \iiint_\Sigma \nabla \cdot \mathbf{V} \, d\Sigma, \tag{C.11}$$

where $\mathbf{n}$ is the vector normal to the surface element dS (oriented outwards), and $d\mathbf{S} = dS \, \mathbf{n}$. We can visualize the theorem by saying that the flow through S is given by the sum of the contributions of all the sources present in V, counted with their sign (a negative divergence is equivalent to a drain). It is useful to note that this theorem

---

[5] Known also as Gauss's or Ostrogradsky's theorem, after the name of the Russian Empire mathematician and physicist Michail Vasilyevich Ostrogradsky (1801–1862).

generalizes integration by parts which holds at one dimension and Green's theorem which holds at two dimensions.

Stokes[6] or curl theorem: in 3D space the circulation of a regular vector field **V** (i.e., its line integral) along a closed contour ℓ is equal to the flow of the field curl of **V** through the surface encircled by ℓ:

$$\oint_\ell \mathbf{V}\,d\ell = \iint_S \nabla\times\mathbf{V} \cdot \mathbf{n}\,dS. \tag{C.12}$$

---

[6] George Gabriel Stokes (1819–1903): Irish mathematician and physicist, made important contributions to fluid dynamics, optics, and mathematical physics.

Appendix D
Relative Motion

Blaise Pascal

Our nature consists in motion; complete rest is death.

We want to review the formula providing the acceleration in a reference frame $S'[x', y', z']$, which is in accelerated motion with respect to an inertial frame $S[x, y, z]$. For sake of simplicity, we first distinguish two cases, when:

1. S' moves of rectilinear motion (but not uniform, otherwise nothing would happen according to Galilei's principle) with a velocity $\mathbf{v}_r$ and an acceleration $\mathbf{a}_r$ with respect to S;
2. O' is fixed but the system S' rotates around an axis with an angular velocity $\vec{\omega}$ which can be a function of time.

Obviously the complete formula will contemplate both cases.

Pure Translation

We consider a point P and its vector radii from O and O' respectively: $\mathbf{r} = \overrightarrow{OP}$ and $\mathbf{r}' = \overrightarrow{O'P}$ (Fig. D.1). It is;

$$\overrightarrow{OP} = \overrightarrow{OO'} + \overrightarrow{O'P}, \tag{D.1}$$

and, placing $\mathbf{R} = \overrightarrow{OO'}$, by time derivatives:

$$\frac{d\mathbf{r}}{dt} = \frac{d\mathbf{R}}{dt} + \frac{d\mathbf{r}'}{dt}, \tag{D.2a}$$

$$\frac{d^2\mathbf{r}}{dt^2} = \frac{d^2\mathbf{R}}{dt^2} + \frac{d^2\mathbf{r}'}{dt^2}, \tag{D.2b}$$

© The Editor(s) (if applicable) and The Author(s), under exclusive license to Springer Nature Switzerland AG 2022
E. Bannikova and M. Capaccioli, *Foundations of Celestial Mechanics*, Graduate Texts in Physics, https://doi.org/10.1007/978-3-031-04576-9

Fig. D.1 Relative coordinate systems

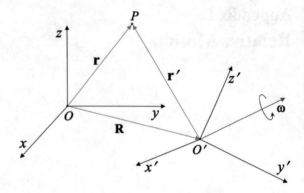

where dr/dt and d^2r/dt^2 are velocity and acceleration in S, dr'/dt and d^2r'/dt^2 the relative velocity and acceleration in S', and dR/dt and d^2R/dt^2 the velocity and acceleration of O' as seen from an observer in S, i.e., from the inertial reference frame.

Pure Rotation

We now exclude translation and suppose instead that S' rotates around an axis passing through O' with an angular velocity $\vec{\omega}$. With no loss of generality, we may assume that O and O' are coincident. Remembering that any vector $\mathbf{V}$ of constant modulus and rigidly connected with S' has a velocity $d\mathbf{V}/dt = \vec{\omega} \times \mathbf{V}$, and therefore the derivatives with respect to time of the unit vectors of the coordinate axes of S' are:

$$\frac{d\mathbf{i}'}{dt} = \vec{\omega} \times \mathbf{i}', \qquad \frac{d\mathbf{j}'}{dt} = \vec{\omega} \times \mathbf{j}', \qquad \frac{d\mathbf{k}'}{dt} = \vec{\omega} \times \mathbf{k}', \qquad (D.3)$$

the velocity of a vector $\mathbf{X} = X_x\mathbf{i} + X_y\mathbf{j} + X_z\mathbf{k} = X_{x'}\mathbf{i}' + X_{y'}\mathbf{j}' + X_{z'}\mathbf{k}'$ is:

$$\frac{d\mathbf{X}}{dt} = \frac{dX_{x'}}{dt}\mathbf{i}' + \frac{dX_{y'}}{dt}\mathbf{j}' + \frac{dX_{z'}}{dt}\mathbf{k}' + X_{x'}\frac{d\mathbf{i}'}{dt} + X_{y'}\frac{d\mathbf{j}'}{dt} + X_{z'}\frac{d\mathbf{k}'}{dt}, \qquad (D.4)$$

that is, using the (D.3):

$$\frac{d\mathbf{X}}{dt} = \frac{d\mathbf{X}'}{dt} + \vec{\omega} \times \mathbf{X}', \qquad (D.5)$$

where $\dfrac{d\mathbf{X}'}{dt}$ is the velocity relative to the rotating system S' (the velocity that an observer rigidly connected to S' would measure). A further derivation with respect to time gives us the acceleration:

$$\frac{d^2\mathbf{X}}{dt^2} = \frac{d}{dt}\Big(\frac{d\mathbf{X}'}{dt} + \big(\vec{\omega} \times \mathbf{X}'\big)\Big) + \vec{\omega} \times \frac{d\mathbf{X}}{dt} =$$

$$= \frac{d^2\mathbf{X}'}{dt^2} + \frac{d}{dt}\left(\vec{\omega} \times \mathbf{X}'\right) + \vec{\omega} \times \left(\frac{d\mathbf{X}'}{dt} + \vec{\omega} \times \mathbf{X}'\right) =$$

$$= \frac{d^2\mathbf{X}'}{dt^2} + \frac{d\vec{\omega}}{dt} \times \mathbf{X}' + \vec{\omega} \times \frac{d\mathbf{X}'}{dt} + \vec{\omega} \times \frac{d\mathbf{X}'}{dt} + \vec{\omega} \times \left(\vec{\omega} \times \mathbf{X}'\right) =$$

$$= \frac{d^2\mathbf{X}'}{dt^2} + \frac{d\vec{\omega}}{dt} \times \mathbf{X}' + 2\vec{\omega} \times \frac{d\mathbf{X}'}{dt} + \vec{\omega} \times \left(\vec{\omega} \times \mathbf{X}'\right). \tag{D.6}$$

The first term at the right-hand side of (D.6) is the acceleration as seen sitting on S'. The second is the contribution given by the changing with time of $\vec{\omega}$ (Euler's acceleration). The third is the so-called Coriolis[7] acceleration, and the forth is the centripetal acceleration, always directed toward the rotational axis and orthogonal to $\vec{\omega}$. In fact, using (C.8k):

$$\vec{\omega} \times \left(\vec{\omega} \times \mathbf{X}'\right) = \left(\vec{\omega} \cdot \mathbf{X}'\right)\vec{\omega} - \left(\vec{\omega} \cdot \vec{\omega}\right)\mathbf{X}', \tag{D.7}$$

with which $\vec{\omega} \times \left(\vec{\omega} \times \mathbf{X}'\right) \cdot \vec{\omega} = 0$.

Translation + Rotation

Finally, placing $\mathbf{X} = \mathbf{r}'$ and summing up the (D.2b) with the (D.6), we obtain the complete expression for the relative acceleration:

$$\frac{d^2\mathbf{r}}{dt^2} = \frac{d^2\mathbf{R}}{dt^2} + \frac{d^2\mathbf{r}'}{dt^2} + \frac{d\vec{\omega}}{dt} \times \mathbf{r}' + 2\vec{\omega} \times \frac{d\mathbf{r}'}{dt} + \vec{\omega} \times \left(\vec{\omega} \times \mathbf{r}'\right). \tag{D.8}$$

---

[7] Gaspard-Gustave de Coriolis (1792–1843) was a French mathematician, physicist, and mechanical engineer. He lived in the period of the industrial revolution, whose problems led him to study the rotation of bodies. His name is among those of the 72 French scientists, engineers, and mathematicians engraved on the sides of the Eiffel Tower under the first balcony.

Appendix E
The Mirror Theorem

The form of the equations of motion of the N-body problem allows us to formulate an interesting theorem, known as mirror theorem, developed by the British astronomer Archie R. Roy and the Canadian astronomer Michael W. Ovenden in the 1950s along the lines traced by Poicaré [3], which is useful in studying non-integrable systems: if the N masses are subject only to mutual gravitational attraction and if at some time $\bar{t}$ each radius vector from the gravity center, $\mathbf{r}_i$, is orthogonal to each velocity vector, $\dot{\mathbf{r}}_j$ $(i \neq j = 1, \ldots, N)$, the orbit described by each of the masses for $t \geq \bar{t}$ is the mirror image of the orbit prior to $\bar{t}$, symmetrically with respect to $\bar{t}$.

The rigorous proof of this theorem is straightforward. We start observing that there can be only two configurations (at the time $\bar{t}$) capable of satisfying the hypothesis (Fig. E.1):

1. all the radius vectors $\mathbf{r}_i$ are contained on the same plane and all the velocities $\dot{\mathbf{r}}_i$ are orthogonal to that plane (and therefore parallel or antiparallel to each other);
2. all the radius vectors $\mathbf{r}_i$ are contained on the same straight line where the velocities are orthogonal (without necessarily being parallel to each other, though).

Let us consider the first of these configurations, which we call C. By changing directions of all velocities, we obtain a configuration C' that mirrors the previous one. In fact, it is sufficient to change the direction of the coordinate axis perpendicular to the plane containing the radius vectors to get back a configuration C'' of positions and velocities identical to the starting one. The identity of the initial conditions in C and C'' at the time $t = \bar{t}$ guarantees that the orbits of each of the N-bodies will be the same in both cases. On the other hand, the reference systems of C and C'' are mirror-like; so, if we refer the individual orbits of both of them to the same coordinate system, they will be mirror images.

© The Editor(s) (if applicable) and The Author(s), under exclusive license to Springer Nature Switzerland AG 2022

E. Bannikova and M. Capacciolo, *Foundations of Celestial Mechanics*, Graduate Texts in Physics, https://doi.org/10.1007/978-3-031-04576-9

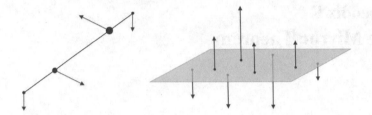

Fig. E.1 Collinear (*left*) and planar (*right*) mirror configurations

Fig. E.2 Collinear mirror
configuration for
Earth-Moon-Sun

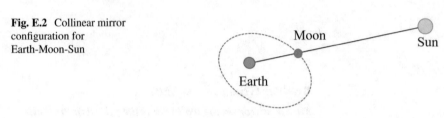

This is exactly what happens if we compare the orbits of C and C'. Now, the change in the direction of velocities from C to C' can be thought of as a reversal of the time axis. In other words, at each instant $t - \bar{t}$, C equals C' at the time $\bar{t} - t$, i.e., the evolution of C' on a timeline originating at $\bar{t}$ and oriented towards the past represents the evolution of C from $\bar{t}$ forward. But, as we have seen, the orbits in the two cases are one the mirror of the other; therefore the theorem remains proved for the first of the two configurations listed above. We leave the reader to repeat, for the collinear configuration, the reasoning just made, reversing the direction of the straight line that contains the vector radii.

We now prove a corollary of the mirror theorem: if a self-gravitating system of N massive points satisfies the mirror theorem in two distinct epochs, separated by the time interval Δt, the orbits of all the bodies of the system are necessarily periodic of the same period $P = 2 \, \Delta t$. Again the proof is quite simple. If the mirror configuration C takes place at the time $t = -\bar{t}$ and the configuration D at time $t = 0$, the first one will occur once more at $t = \bar{t}$, the second one at the time $t = 2\bar{t}$ and so on (in other words, two reflections cancel each other). This means that the orbits must be periodic of period $P = 2\bar{t}$. As an example of the application of this theorem, let us imagine that we have three bodies of which we know, for some reason, that at two successive epochs t_1 and t_2 they produce an eclipse (Fig. E.2). Thanks to the corollary of the mirror theorem we can then say that they describe periodic orbits of period $P = 2|t_2 - t_1|$.

Appendix F
On the Solution of Kepler's Equation

Johannes Kepler

Nature uses as little as possible of anything.

Introduced in Sect. 2.7 and represented in Fig. F.1, Kepler's equation:

$$M = E - e \sin E, \tag{F.1}$$

is a transcendent relation between the mean anomaly M and the eccentric anomaly E, holding for the elliptical motion with eccentricity e. Since Eq. (F.1) surpasses the powers of algebraic methods, appropriate expansions are in order to solve or use it in complex expressions. We consider here the developments made by Lagrange and Fourier, as well as an iterative numerical method.

F.1 Lagrange's Inversion Theorem

Consider the following relation between two variables x and y:

$$y = x + \alpha \, \phi(y), \tag{F.2}$$

where ϕ is a function and α is a parameter. A theorem due to Lagrange[8] states that any function $f(y)$, whose independent variable y satisfies a relation of the type (F.2), can be expanded in a power series of α:

---

[8] A simpler proof of this theorem was given by Laplace in 1780, 10 years after Lagrange's work.

E. Bannikova and M. Capaccioli, *Foundations of Celestial Mechanics*, Graduate Texts in Physics, https://doi.org/10.1007/978-3-031-04576-9

Fig. F.1 Kepler's equation
for a set of values of the
eccentricity

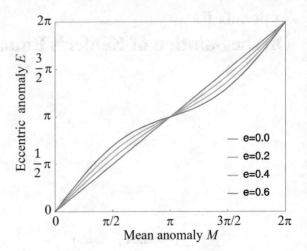

$$f(y) = f(x) + \sum_{n=1}^{\infty} \frac{\alpha^n}{n!} \frac{d^{n-1}}{dx^{n-1}} \left(\phi^n(x) \frac{df(x)}{dx} \right), \tag{F.3}$$

provided that both functions ϕ and f are indefinitely derivable and the α is sufficiently
small to assure convergence [4].

To apply Lagrange's expansion to the Kepler equation (F.1), let's assume $y = E$,
$x = M$, $\phi(y) = \sin E$, $\alpha = e$, and $f(y) = y$, with which the series (F.3) becomes:

$$E = M + \sum_{n=1}^{\infty} \frac{e^n}{n!} \frac{d^{n-1}}{dM^{n-1}} \sin^n M. \tag{F.4}$$

For example, by truncating the expansion at the third order of e, with some manipulations we obtain:

$$E \approx M + \left(e - \frac{e^3}{8} \right) \sin M + \frac{e^2}{2} \sin(2M) + \frac{3e^3}{8} \sin(3M). \tag{F.5}$$

It can be proven that the complete series (F.4) converges absolutely for any value of
M as long as $e < e_l \simeq 0.6627$ [5]. This threshold value of the eccentricity is called
Laplace limit as it was firstly found by the French mathematician. If $e_l \leq e < 1$, the
series does not converge for any value of M but multiples of π. So, the Lagrange
expansion is suitable to approximate the Kepler equation when the eccentricity is not
too high, which is the case for the major planets and satellites of the Solar System
(where it is always $e < 0.25$). For higher eccentricities, another type of development
is required which, via the Fourier theorem, leads to a series of Bessel functions.

F.2   Fourier's Expansion Theorem

The Fourier theorem asserts that, if $f(x)$ is a periodic function of period 2π, it can be expanded in a trigonometric series [6]:

$$f(x) = \frac{a_o}{2} + \sum_{n=1}^{\infty} \left(a_n \cos(nx) + b_n \sin(nx) \right), \qquad (F.6)$$

which converges uniformly[9] in the interval $0 \le x \le 2\pi$. The coefficients a_n and b_n are:

$$a_n = \frac{1}{\pi} \int_0^{2\pi} f(x) \cos(nx)\, dx,$$
$$b_n = \frac{1}{\pi} \int_0^{2\pi} f(x) \sin(nx)\, dx. \qquad (F.7)$$

To prove it, we multiply both terms of (F.6) by $\cos(mx)\, dx$ (and then by $\sin(mx)\, dx$) and integrate between 0 and 2π making use of the orthogonality of the trigonometric functions.

The conditions under which $f(x)$ can be expanded in a Fourier series are extremely large, and will not be analyzed here. It will be enough for us to say that, if $f(x)$ is absolutely integrable in $0 \le x \le 2\pi$, the expansion (F.6) holds at any point of the domain where it is continuous and derivable. It is also easy to prove that, if $f(x)$ is an even function, i.e., $f(x) = f(-x)$, the trigonometric series reduces to a series of only cosines, as the coefficients b_n cancel out. If the function is odd instead, $f(x) = -f(-x)$, the series has only sinuses, because all the coefficients a_n vanish.

If we write the Kepler equation in the form:

$$E - M = e \sin E, \qquad (F.8)$$

the function $E - M$ is periodic of period 2π with respect to M, absolutely integrable and indefinitely derivable in the interval $0 \le M \le 2\pi$, and also odd (as it is a sine function). Therefore:

$$E - M = \sum_{n=1}^{\infty} b_n \sin(nM), \qquad (F.9)$$

with:

$$b_n = \frac{1}{\pi} \int_0^{2\pi} (E - M) \sin(nM)\, dM. \qquad (F.10)$$

---

[9] This means that for every $\epsilon > 0$, it exists an index n such that, for all points x within the interval $0 \le x \le 2\pi$, the partial sum $f_k(x)$, with $k \ge n$, satisfies the condition $|f_k(x) - f(x)| < \epsilon$.

Integration by parts of (F.10) gives:

$$b_n = \frac{1}{n\pi} \int_0^{2\pi} (dE - dM) \cos(nM) = \frac{1}{n\pi} \int_0^{2\pi} \cos(nM) \, dE =$$

$$= \frac{1}{n\pi} \int_0^{2\pi} \cos\left(n(E - e\sin E)\right) dE. \quad \text{(F.11)}$$

We see that b_n is function of n (which is a wave number) and of the eccentricity e only. We can then put:

$$b_n = \frac{2}{n} J_n(ne), \quad \text{(F.12)}$$

where:

$$J_n(y) = \frac{1}{2\pi} \int_0^{2\pi} \cos[nx - y\sin x] \, dx, \quad \text{(F.13)}$$

is a Bessel function of the first order type n (see Appendix N.3). Consequently, Kepler's equation becomes:

$$E = M + \sum_{n=1}^{\infty} \frac{2}{n} J_n(ne) \sin(nM), \quad \text{(F.14)}$$

which has an explicit form for E and converges for any $e < 1$.

F.3 Numerical Solutions: An Iterative Method

The expansions of the Kepler equation considered above have the advantage of the analytical form, but they are not very convenient in purely numerical calculations, for which a Newton iterative method is more friendly. Consider the sequence:

$$
\begin{aligned}
E_1 &= M + e\sin M \\
E_2 &= M + e\sin E_1 \\
E_3 &= M + e\sin E_2 \\
&\cdots\cdots\cdots\cdots\cdots \\
E_{n+1} &= M + e\sin E_n \\
&\cdots\cdots\cdots\cdots\cdots
\end{aligned}
\quad \text{(F.15)}
$$

We will prove that it converges to the value $E = M + e\sin E$. In fact, subtracting from (F.1) the first of (F.15) and dividing both terms by $(E - M)$, we obtain:

$$\frac{E - E_1}{E - M} = e\,\frac{\sin E - \sin M}{E - M} =$$

$$= e\,\frac{2}{E - M}\cos\left(\frac{E + M}{2}\right)\sin\left(\frac{E - M}{2}\right). \quad \text{(F.16)}$$

Since:

$$\left|\cos\frac{E + M}{2}\right| \leq 1, \qquad \text{and} \qquad \left|\frac{\sin\dfrac{E - M}{2}}{\dfrac{E - M}{2}}\right| \leq 1, \quad \text{(F.17)}$$

we have:

$$\left|\frac{E - E_1}{E - M}\right| \leq e. \quad \text{(F.18)}$$

Similarly, it is easy to show that:

$$\left|\frac{E - E_2}{E - E_1}\right| \leq e. \quad \text{(F.19)}$$

The product of (F.18) with (F.19) gives the inequality:

$$\left|\frac{E - E_2}{E - M}\right| \leq e^2, \quad \text{(F.20)}$$

which generalizes to:

$$\left|\frac{E - E_n}{E - M}\right| \leq e^n. \quad \text{(F.21)}$$

But, since $|E - M| = e|\sin E| \leq e$, then:

$$|E - E_n| \leq e^{n+1}. \quad \text{(F.22)}$$

Therefore, for each value of the eccentricity $0 \leq e < 1$, it results:

$$\lim_{n \to \infty} |E - E_n| = 0. \quad \text{(F.23)}$$

The convergence of this iterative procedure is fast for small values of the eccentricity. For example, in order to obtain a precision of 7 digits, the condition $e^{n+1} \leq 10^{-7}$ must be satisfied. This sets the following upper limits to the iterations: $n = 6$ for $e = 0.1$, $n = 22$ for $e = 0.5$, $n = 152$ for $e = 0.9$, and $n = 1.6 \times 10^6$ if $e = 0.99999$. We then conclude that 11 iterations are sufficient to obtain the required accuracy for any planet in the Solar System ($e < 0.25$). Note however that this estimate of the

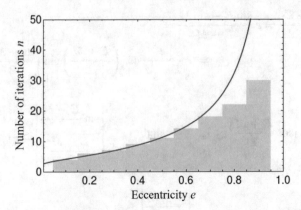

Fig. F.2 Number n of iterations needed to solve Kepler's equation (F.1) with an accuracy of 7 digits as a function of the ellipticity e, for $M = \pi/2$. The continuous curve plots the equation: $10^{-7} = e^{n-1}$ (here e is the eccentricity), providing an upper limit to the number of iterations required. The histogram shows the value of n for a discrete set of values of e, computed by direct numerical iteration of the sequence (F.15) for the same level of accuracy and for $M = \pi/2$

upper limit of the numbers of iterations diverges with increasing ellipticity, as it is shown in Fig. F.2, where the red points correspond to direct estimates by numerical calculations.

Appendix G
Bohr's Model for the Hydrogen Atom

George Wald

A physicist is an atom's way of knowing about atoms.

This appendix presents an application of the two-body problem to a context outside celestial mechanics. We consider here the first attempts to model the structure of the atom as it emerged from Thomson's[10] and Rutherford's[11] discoveries. Obviously, the interpretative effort of the numerous evidences collected in laboratory by spectroscopists initially focused on the simplest physical case, that of hydrogen, in which a single negatively charged particle, the electron, was imagined to circulate around a positively charged nucleus, consisting of a single proton. The first model capable of accounting for appearances was developed in 1913 by Niels Bohr.[12]

Based on ad hoc hypotheses inserted in a substrate of classical physics, Bohr's model was able to reproduce the empirical formula of Rydberg.[13] It was only the first step of a path that in a few years would contribute to the birth of a new, revolutionary vision of the physical world, quantum mechanics [7, 8].

---

[10] Joseph Thomson (1856–1940): British physicist of Scottish origin, known for having discovered the electron.

[11] Ernest Rutherford, (1871–1937): New Zealand physicist, student of Thomson, known as the father of nuclear physics. He was the precursor of the orbital theory of the atom. He won the Nobel Prize for chemistry in 1908.

[12] Niels Bohr (1885–1962): Danish physicists and mathematicians, one of the most influential scientists of the XX century, he gave essential contributions to knowledge the atomic structure and to quantum mechanics. Nobel Prize in physics in 1922.

[13] Johannes Rydberg (1854–1919): Swedish physicist and mathematician, best known for the generalization, in 1888, of the formula found by the Swiss teacher Johann Balmer (1825–1898). The latter was used to predict the wavelengths λ_H of the lines of the hydrogen spectrum appearing in the visible domain: $\lambda_m = B \dfrac{m^2}{m^2 - n^2}$, with $B = 3645.6\,\text{Å}$. The two integers are $n = 2$ and $m > n$.

© The Editor(s) (if applicable) and The Author(s), under exclusive license to Springer Nature Switzerland AG 2022
E. Bannikova and M. Capaccioli, *Foundations of Celestial Mechanics*, Graduate Texts in Physics, https://doi.org/10.1007/978-3-031-04576-9

G.1 Bohr's Model of the Hydrogen Atom

Modelled on an intuition of Rutherford, Bohr's atom is seen as a two-body system in which gravitational interaction is replaced by the preponderant[14] electric interaction between the nuclear proton and the electron, also attractive because it is exerted between two charges of opposite sign. Bohr introduced three fundamental postulates:

1. The electron revolves around the central nucleus on circular orbits (the generalization to the elliptical orbits is due to Sommerfeld[15] to take into account some failures of the theory regarding, for example, the reproduction of helium spectra). These orbits can only take quantized angular momentum values: $L = n\hbar = nh/2\pi$, where h is the Planck[16] constant and n the principal quantum number. Consequently, the electron energy is also quantized, and the allowed electronic orbits have predetermined radii.
2. Contrary to what is expected from Maxwell's equations for the accelerated motion of an electric charge,[17] electrons do not radiate energy in their orbital motion. In other words, the allowed orbits are also stable.
3. When an electron jumps from one 'permitted' orbit to another, the difference in energy between the two, ΔE, is managed by the emission or absorption (depending on the sign of ΔE) of a quantum of light (photon) with a frequency $\nu = \frac{|\Delta E|}{h}$.

Three different equations derive from the previous hypotheses.

1. The energy of the electron in orbit is the sum of its kinetic E_{kin} and potential the E_{pot} energy:

$$E = E_{kin} + E_{pot} = \frac{1}{2}m_e v^2 - \frac{ke^2}{r}, \tag{G.1}$$

where m_e and e are the mass and charge of the electron, v is its velocity, r the radius of the orbit, and $k = 1/(4\pi\epsilon_0)$, with ϵ_0 the dielectric constant of vacuum.
2. The angular momentum is quantized, that is:

$$L = m_e v r = n\frac{h}{2\pi} = n\hbar. \tag{G.2}$$

3. The Coulombian force that keeps the electron in the orbit must be the same as the centripetal one:

---

[14] If we compare the gravitational force exerted by a proton on an electron to the electrostatic force (Coulombian) at equal distance, we have: $\frac{Gm_p m_e}{ke^2} = 4 \times 10^{-40}$.

[15] Arnold Sommerfeld (1868–1951): German theoretical physicist. Great master, he had among his students Werner Heisenberg, Wolfgang Pauli, Peter Debye, and Hans Bethe.

[16] Max Planck (1858–1947). German physicist, he gave substantial contributions to theoretical physics. He also originated the quantum theory. Nobel prize in physics in 1918.

[17] Classical electrodynamics predicts that, because of the centripetal acceleration, an electron on a circular orbit with period P radiates light at a frequency $2\pi/P$. Since its total energy is proportional to $-1/r$ (cf. (2.66)), the the orbital radius should progressively decrease.

Fig. G.1 Bohr's model of hydrogen atom. The electronic orbits are not in scale

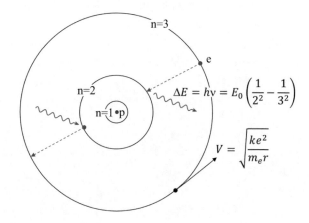

$$\frac{ke^2}{r^2} = \frac{m_e v^2}{r}.\tag{G.3}$$

After some manipulation of the above formulas, we obtain the following expression for the energy of the n-th level:

$$E_n = -\frac{m_e e^4}{8h^2\epsilon_0^2}\frac{1}{n^2} = \frac{E_0}{n^2},\tag{G.4}$$

where:

$$E_0 = -\frac{m_e e^4}{8h^2\epsilon_0^2} = -13.6 \text{ eV},\tag{G.5}$$

($1 \text{ eV} = 1.60218 \times 10^{-19}$ J). So, the lowest energy level ($n = 1$) is around -13.6 eV, the next higher level ($n = 2$) is at -3.4 eV, and so on. Since the system is bound, these energies are negative. In the case of positive energies, we speak of an ionized atom as in this case the electron is made free, in formal analogy with the results obtained in Sect. 2.7. The change of the energy levels produces the emission ($m > n$) or the absorption ($m < n$) of a photon with frequency ν according to the formula:

$$\Delta E_{nm} = |E_n - E_m| = h\nu = E_0\left(\frac{1}{n^2} - \frac{1}{m^2}\right);\tag{G.6}$$

see Fig. G.1.

Appendix H
The Variation of Constants

Joseph Louis Lagrange

Newton was the greatest genius that ever existed, and the most fortunate, for we cannot find more than once a system of the world to establish.

We present here the method of the variation of constants,[18] also called variation of parameters, in the context of the theory of differential equations. Given the ordinary differential equation:

$$\frac{d^2 y}{dx^2} + p(x)\frac{dy}{dx} + q(x)\,y = r(x), \tag{H.1}$$

let us assume that we know two independent solutions, $y_1(x)$ and $y_2(x)$, of the associated homogeneous equation:

$$\frac{d^2 y}{dx^2} + p(x)\frac{dy}{dx} + q(x)\,y = 0. \tag{H.2}$$

The postulated independence ensures that the Wronskian determinant[19] differs from zero, that is:

$$\begin{vmatrix} y_1(x) & y_2(x) \\ \dfrac{dy_1}{dx} & \dfrac{dy_2}{dx} \end{vmatrix} \neq 0. \tag{H.4}$$

---

[18] This method is also known as variation of parameters and it was invented by Euler in 1748 with applications to planetary perturbations and, starting 18 years later, it was perfected by Lagrange.

[19] If n functions $g_1, \ldots, g_n$ of real of complex variable admit $n-1$ derivatives in a given interval, there we may define the Wronskian function, after the Polish mathematician Josef Maria Hoene-Wroński (1776–1853):

© The Editor(s) (if applicable) and The Author(s), under exclusive license to Springer 303
Nature Switzerland AG 2022
E. Bannikova and M. Capaccioli, *Foundations of Celestial Mechanics*, Graduate Texts in Physics, https://doi.org/10.1007/978-3-031-04576-9

In this case, the general integral of the homogeneous equation (H.2) is given by a linear combination of y_1 and y_2:

$$y(x) = c_1 \, y_1(x) + c_2 \, y_2(x), \tag{H.5}$$

where c_1 and c_2 are arbitrary constants.[20]

The strategy of the method of the variation of constant consists in finding a solution of the equation (H.1) in the form (H.5), but assuming that the quantities c_i, instead of being constants, are functions of x:

$$y(x) = c_1(x) \, y_1(x) + c_2(x) \, y_2(x). \tag{H.6}$$

The first derivative of (H.6):

$$\frac{dy}{dx} = c_1 \frac{dy_1}{dx} + c_2 \frac{dy_2}{dx} + \frac{dc_1}{dx} y_1 + \frac{dc_2}{dx} y_2, \tag{H.7}$$

contains 4 terms, but we can simplify it by noting that there are only one unknown function, y, and two conditions. For example, we can arbitrarily impose that:

$$\frac{dc_1}{dx} y_1 + \frac{dc_2}{dx} y_2 = 0, \tag{H.8}$$

with which Eq. (H.7) simplifies in:

$$\frac{dy}{dx} = c_1 \frac{dy_1}{dx} + c_2 \frac{dy_2}{dx}. \tag{H.9}$$

We now calculate the second derivative of y from (H.9) and replace in the original Eq. (H.1) together with the first derivative (H.9). After proper groupings we obtain:

$$W(g, \ldots, g_n) = \begin{vmatrix} g_1(x) & g_2(x) & g_3(x) & \cdots & g_n(x) \\ g_1'(x) & g_2'(x) & g_3'(x) & \cdots & g_n'(x) \\ \cdots & \cdots & \cdots & \cdots & \cdots \\ g_1^{(n-1)}(x) & g_2^{(n-1)}(x) & g_3^{(n-1)}(x) & \cdots & g_n^{(n-1)} \end{vmatrix}. \tag{H.3}$$

The Wronskian is useful for determining whether a set of functions is linearly independent in a given interval. If it does not vanish identically, then the associated functions are linearly independent; if, on the other hand, they are linearly dependent, then their Wronskian is equal to zero (but the reverse is not true).

[20] Consider, for instance, the forced harmonic motion: $\ddot{x}(t) - \omega^2 x(t) = f(t)$, where $f(t)$ is some external periodic force. It is well known that $\sin(\omega t)$ and $\cos(\omega t)$ are independent solutions of the associated homogeneous equation: $\ddot{x}(t) - \omega^2 x(t) = 0$, as it can be proven by direct substitution. The general solution is then: $x(t) = c_1 \sin(\omega t) + c_2 \cos(\omega t)$.

$$c_1 \left(\frac{d^2 y_1}{dx^2} + p \frac{dy_1}{dx} + q\, y_1 \right) + c_2 \left(\frac{d^2 y_2}{dx^2} + p \frac{dy_2}{dx} + q\, y_2 \right) +$$

$$+ \frac{dc_1}{dx} \frac{dy_1}{dx} + \frac{dc_2}{dx} \frac{dy_2}{dx} = -r(x). \qquad \text{(H.10)}$$

Since y_1 and y_2 are solutions of (H.2), it is evident that the expressions in the brackets of (H.10) are zero. Therefore, the equation reduces to:

$$\frac{dc_1}{dx} \frac{dy_1}{dx} + \frac{dc_2}{dx} \frac{dy_2}{dx} = r(x), \qquad \text{(H.11)}$$

with which the condition (H.8) we imposed is associated. Thus we have a system of two equations (H.8) and (H.11) in the unknowns dc_1/dx and dc_2/dx. It admits a solution since the determinant of the system coincides with the Wronskian (H.4), which is different from zero by hypothesis. Thus, the solution of (H.1) reduces to finding $c_1(x)$ and $c_2(x)$ by means of two quadratures.

Now consider a system of differential equations of order n:

$$\frac{dx_k}{dt} = f_k(x_1, x_2, \dots, x_n, t) \qquad (k = 1, \dots, n), \qquad \text{(H.12)}$$

which defines n functions $x_k(t)$ of the independent variable t. Let us assume that we know the general solution of this system. It will take the form:

$$x_k = x_k(t, c_1, c_2, \dots, c_n) = x_k(t, c) \qquad (k = 1, \dots, n), \qquad \text{(H.13)}$$

with n arbitrary constants c_i ($i = 1, \dots, n$). We now modify the second terms of (H.12) by adding n functions $r_k(x_1, x_2, \dots, x_n, t)$, which in practice procure a modest contribution collectively, when compared to the functions f. In other words, let us consider the system of n equations:

$$\frac{dx_k}{dt} = f_k(x_1, x_2, \dots, x_n, t) + r_k(x_1, x_2, \dots, x_n, t). \qquad \text{(H.14)}$$

Once again we look for the solution of (H.14) through the solution (H.13) of (H.12), but thinking of the c_i as functions of the variable t. The derivative of x_k becomes the sum of $n + 1$ terms:

$$\frac{dx_k}{dt} = \frac{\partial x_k}{\partial t} + \sum_{i=1}^{n} \frac{\partial x_k}{\partial c_i} \frac{dc_i}{dt}. \qquad \text{(H.15)}$$

Replaced in (H.14), we have:

$$\frac{\partial x_k}{\partial t} + \sum_{i=1}^{n} \frac{\partial x_k}{\partial c_i} \frac{dc_i}{dt} = f_k\big(x_1(t, c), x_2(t, c), \ldots, x_n(t, c), t\big) +$$

$$+ r_k\big(x_1(t, c), x_2(t, c), \ldots, x_n(t, c), t\big) \qquad (k = 1, \ldots, n), \qquad \text{(H.16)}$$

which leads to an obvious reduction. In fact, it must be satisfied if the r_k functions are canceled and if the c_i are reduced to constants; in this case (H.16) is identified with (H.12). In conclusion, we obtain that the c_k functions must satisfy the following n conditions:

$$\sum_{i=1}^{n} \frac{\partial x_k}{\partial c_i} \frac{dc_i}{dt} = r_k\big(x_1(t, c), x_2(t, c), \ldots, x_n(t, c), t\big), \qquad \text{(H.17)}$$

for each value of the index k between 1 and n. We consider the (H.17) as a system in the unknowns $dc_i(t)/dt$, which admit a solution since the functional determinant $\left|\dfrac{\partial x_i}{\partial c_j}\right|$ is necessarily different from zero. If not, the (H.13) would not represent the general integral of the (H.12), contrary to the hypothesis. In conclusion, the n equations (H.17) can be placed in the form of a system of differential equations of order n:

$$\frac{dc_k}{dt} = c_k(c_1, c_2, \ldots, c_n, t) \qquad (k = 1, \ldots, n), \qquad \text{(H.18)}$$

whose solution determines, together with arbitrary initial conditions, the n functions c_k that we were looking for.

It is worth noting that the integration method shown above can also be applied to the Eq. (H.1) rewritten in the form:

$$\frac{dy}{dx} = y',$$
$$\frac{dy'}{dx} = -p\, y' - q\, y + r. \qquad \text{(H.19)}$$

Appendix I
Lagrange Multipliers

Georg Cantor

In mathematics the art of proposing a question must be held of higher value than solving it.

Let $f(x_1, x_2, \ldots, x_n)$ be a derivable function of n free variables x_i. It is known that the extremes (maxima and minima) of f are provided by the solutions of the system constructed by canceling all the first n partial derivatives: $\partial f / \partial x_i = 0$, for $i = 1, \ldots, n$. If we introduce $m < n$ constraints of the type $g_j(x_i) = 0$, the search for the extremes of f requires the elimination of m of the n independent variables. For example, let us look for the maximum of the function: $f = x_1 x_2$, in the presence of the constraint $g = x_1 + x_2 - 2a = 0$. We have:

$$f = x_1 (2a - x_1) = 2ax_1 - x_1^2,$$
$$\frac{\partial f}{\partial x_1} = 2a - 2x_1 = 0, \tag{I.1}$$

hence $x_1 = a$, and then $x_2 = x_1 = a$. The result illustrates the well-known property of the square with the largest area f for a same perimeter length $4a$ (given by g).

But replacements are not always as simple. Sometimes they give rise to very laborious or even impractical calculations; for example, if the constraints are expressed by transcendent functions. However, there is an elegant method, developed by Lagrange, that removes some of the difficulties. It does not require substitutions of variables and therefore does not destroy any symmetries of the function f, if any. The basic idea is to find the points where ∇f is parallel to ∇g_i. Let us have a closer look to it.

Given the function $f(x_i)$ $(i = 1, n)$ constrained by m relations[21] $g_j(x_i) = 0$, we build the auxiliary function, named Lagrangian:

---

[21] Note that, if $g_i(x) = c_i$, where $c_i = \text{const} \neq 0$, we can always rewrite: $g_i'(x) = g_i(x) - c_i = 0$.

© The Editor(s) (if applicable) and The Author(s), under exclusive license to Springer Nature Switzerland AG 2022
E. Bannikova and M. Capaccioli, *Foundations of Celestial Mechanics*, Graduate Texts in Physics, https://doi.org/10.1007/978-3-031-04576-9

$$G(x_i) = f(x_i) + \sum_{j=1}^{m} \lambda_j \, g_j(x_i). \tag{I.2}$$

The auxiliary variables λ_i are called Lagrange multipliers. They must be eliminated from the system:

$$\begin{cases} \dfrac{\partial G}{\partial x_1} = \dfrac{\partial f}{\partial x_1} + \sum_{j=1}^{m} \lambda_j \, \dfrac{\partial g_j}{\partial x_1} = 0, \\ \quad\dotfill \\ \dfrac{\partial G}{\partial x_n} = \dfrac{\partial f}{\partial x_n} + \sum_{j=1}^{m} \lambda_j \, \dfrac{\partial g_j}{\partial x_n} = 0, \end{cases} \tag{I.3}$$

which thus provides the extremes of f.

Let us solve once more the problem of the rectangle of maximum area given above using Lagrange multipliers. We have:

$$G(x_1, x_2) = x_1 x_2 + \lambda(x_1 + x_2 - 2a), \tag{I.4}$$

from where:

$$\frac{dG}{dx_1} = x_2 + \lambda = 0,$$
$$\frac{dG}{dx_2} = x_1 + \lambda = 0. \tag{I.5}$$

The elimination of λ gives $x_1 = x_2$ as above.

Appendix J
The Two-Body Problem Applied to Visual Binary Orbits

Marcus Aurelius, *Meditations*

Dwell on the beauty of life. Watch the stars, and see yourself running with them.

Let us consider an isolated, gravitationally bound system of two stars with a large enough physical separation to prevent any effects due to the finite sizes of the objects (which might require additional terms in the expression of the potential; cf. Chap. 5). With these assumptions the system can be treated as a two-body problem. Each star, assimilated to a material point (Sect. 5.4), describes an elliptical orbit around the other.

If the system is so close to the Earth that the two stars can be seen well apart, we say they form a visual binary system. In this case the observer may trace the apparent orbit of one of the two stars, usually the fainter (S_2 in Fig. J.1), around the other one (S_1). This orbit is the projection of the true orbit on the plane tangent to the celestial sphere at the position of the binary system. It can be sampled by measuring, at different epochs t_i, the angular separation[22] $\rho(t_i)$ and the position angle[23] $\theta(t_i)$ of the vector connecting S_1 to S_2.

In this appendix we will study how to extract graphically the parameters of the true orbit from the projected orbit noting that, since the latter lies on the plane of the sky which is orthogonal to the line of sight, the projection is also orthogonal. The exercise no longer has a practical use but has some educational content.

To begin, let us consider two non-coincident planes, π and π', and call the line of nodes their intersection (Fig. J.2). We say that the orthogonal projection of a point P of π onto the plane π' is the point P_1 intercepted by the perpendicular to π' passing

---

[22] The physical separation requires the knowledge of the distance to the binary system.

[23] The position angle is the angle that a vector on the plane of the sky forms with the direction of north, counted eastwards.

E. Bannikova and M. Capaccioli, *Foundations of Celestial Mechanics*, Graduate Texts in Physics, https://doi.org/10.1007/978-3-031-04576-9

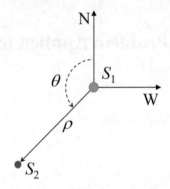

Fig. J.1 Angular separation ρ and position angle θ of the fainter star S_2 of a visual binary system with respect to the brighter one S_1, measured on the plane of the sky at a given epoch

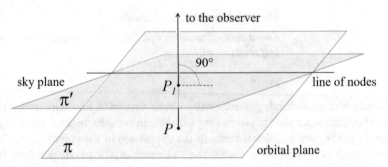

Fig. J.2 Orthogonal projections on the plane of the sky

through P. The orthogonal projection of any figure, plane or solid, is the orthogonal projection of all its points.

We recall some properties of orthogonal projection. The tangents to the curves are conserved (i.e., their projection is tangent to the projected curve) and the parallel lines project into parallel lines. Ratios of the lengths of parallel segments and of the areas of coplanar closed surfaces are also conserved. Every triangle can be positioned in such a way that the relative shadow below an orthogonal projection is equilateral. Furthermore, the medians of a triangle are projected into the medians of the image triangle. The barycenter of a triangle[24] projects into the center of gravity of the triangle of the corresponding projected image. Spheroids project into ellipses (or circles, in the degenerate case). Ellipses project into ellipses, which turn into circles if the line of nodes is parallel to the one of the main axes of the ellipse and the inclination is properly chosen. The center of an ellipse is always conserved in projection, but usually the foci do not project into those of the projected image unless the line of the nodes is parallel to one of the main axes.

---

[24] The barycenter of a triangle is the point of intersection of its medians, i.e., the segments joining each vertex with the midpoint of the opposite side.

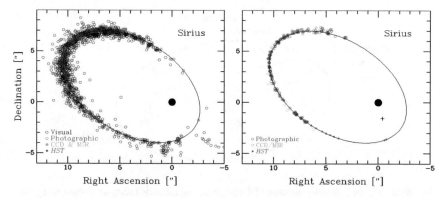

Fig. J.3 Apparent positions of the compact companion Sirius B around the giant star Sirius A (black dot), together with the best fitting apparent elliptical orbit. The shift of the apparent position of primary star relative to the focus of the apparent ellipse (*cross*) is evident. See H.E. Bond et al, The Astrophysical Journal, 840, 70B, 2017. Courtesy of H.E. Bond

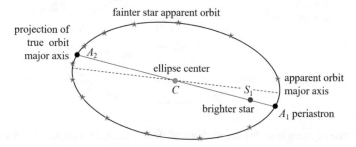

Fig. J.4 Apparent orbit of a binary system. Notice that the primary star does not occupy the focus of the ellipse

Going back to our visual binary system, we assume that the apparent orbit of S_2 has been duly sampled and the positional data best-fitted by an ellipse (cf. Sect. O.1 and see Fig. J.3). In fact, since the apparent orbit is the projection of an ellipse, it must itself be an ellipse on the plane of the sky that also contains the projection of S1 (usually distinct from each of the two foci, although it is itself a focus of the true orbit).

To reconstruct the true orbit graphically, we begin by noting that the chord of the projected ellipse containing the center C and the point S_1 is the projection of the major axis of the true orbit (Fig. J.4). Then, $\overline{(CS_1/CA_1)}$ is the measure of the eccentricity e.

In addition, the bisector of the chords parallel to CA_1 (Fig. J.5) is the projection of the normal to the major axis of the true orbit, i.e., it contains the projection of the minor axis. Tracing the parallel to this latter passing through one of the projections of the true foci (Fig. J.5), we can find a point of the projection of the pedal circumference (see Sect. 2.6.1) using the proportion, valid for the true orbit and therefore

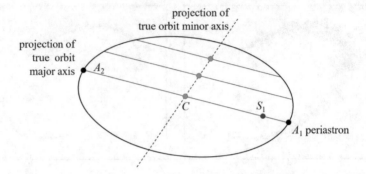

Fig. J.5 Binary system: determination of the projection of the minor semi-axis of the true orbit

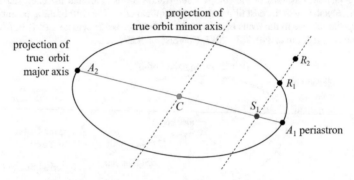

Fig. J.6 Binary system: identification of a point of the projection of the pedal circumference

also for the projected one[25]: $\left(\overline{S_1 R_1}/\overline{S_1 R_2}\right) = \sqrt{1 - e^2}$. Since the ellipse which is a projection of the pedal circle also passes through A_1 and A_2, we have the three distinct points needed to draw it (Fig. J.6); of course, this procedure does not account for measurement errors.

The major axis of this ellipse, $N_1 N_2$, coincides with the line of nodes, i.e., the direction along which the lengths are preserved. In this way we also have the angular value of the semi-major axis of the true orbit $a = \overline{CN_1} = \overline{CN_2}$ and the inclination $i = \arcsin\left(\overline{CB}/\overline{CN_1}\right)$ (Fig. J.7).

Once the period P is known (it is measured directly over a complete rotation, whether this is possible, or computed from the time taken to cover a true arc of orbit) and the distance d of the system from the Earth (which is generally hard to determine), the third Kepler law in the form (2.89) provides the sum of the masses of the two stars. The determination of the individual masses requires some other condition. One can resort, for example, to the estimate of the mass ratio provided by the spectroscopic types of the two stars. But this topic goes beyond our scope.

---

[25] The chord of an ellipse perpendicular to the major axis and passing through one focus is called the '*latus rectum*' (perpendicular side). It is easy to prove that its length is: $\ell = b^2/a = a(1 - e^2)$.

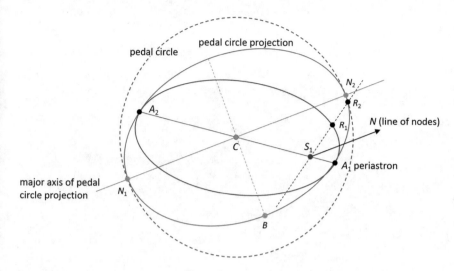

Fig. J.7 Binary system: construction of the pedal circumference and determination of the line of nodes

Appendix K
Planarity of the Stationary Solution of the 3-Body Problem

Lev Tolstoy, *War and peace*
The two most powerful warriors are patience and time.

We want to show that the stationary solutions of the 3-body problem are planar. It will be enough to prove it for the triangular solutions, as the property is obvious for the collinear solutions because, as we already know (Sect. 3.1.1), in this configuration the forces are evidently central.

Let S be the plane containing the three (not aligned) points at the generic epoch t_o. We adopt an orthogonal Cartesian reference system with origin at the (fixed) barycenter B, with axes x and y belonging to S, and assume that it is rigidly connected to the three massive points (for instance, the x-axis contains one of the three points). It is apparent that the co-moving system $B[x, y, z]$ is not inertial. Let $\mathbf{r}_i = \rho(t)\left[x_i^\circ \mathbf{i} + y_i^\circ \mathbf{j}\right]$ be the barycentric radius vector of the i-th point; the superscript indicates that the variable is taken at t_o. As usual, $\mathbf{i}$ and $\mathbf{j}$ are the unit vectors of the axes x and y. Relative velocity and acceleration of P_i in this system are:

$$\mathbf{v}_i = \dot{\rho}(t)\left[x_i^\circ \mathbf{i} + y_i^\circ \mathbf{j}\right] = \dot{\rho}(t)\,\mathbf{r}_i^\circ, \tag{K.1a}$$

$$\frac{d\mathbf{v}_i}{dt} = \ddot{\rho}(t)\,\mathbf{r}_i^\circ; \tag{K.1b}$$

the resulting force is instead:

$$\frac{\mathbf{f}_i}{m_i} = \frac{Gm_j}{r_{ij}^3}(\mathbf{r}_j - \mathbf{r}_i) + \frac{Gm_k}{r_{ik}^3}(\mathbf{r}_k - \mathbf{r}_i) = \frac{1}{\rho^2(t)}\left(f_{ix}^\circ \mathbf{i} + f_{iy}^\circ \mathbf{j}\right), \tag{K.1c}$$

with $i \neq j \neq k = 1, 2, 3$, where f_{ix}° and f_{iy}° are the components of $\mathbf{f}_i/m_i$ at t_o. Finally, calling $\vec{\omega}\,(\omega_x, \omega_y, \omega_z)$ the vector of the angular velocity of the reference system $B[x, y, z]$ relative to a concentric inertial system, the second principle of

E. Bannikova and M. Capaccioli, *Foundations of Celestial Mechanics*, Graduate Texts in Physics, https://doi.org/10.1007/978-3-031-04576-9

dynamics applied to the non-inertial system (Appendix D), using the (C.8k), gives the equation:

$$\frac{d\mathbf{v}_i}{dt} + 2\vec{\omega} \times \mathbf{v}_i + \vec{\omega}\left(\mathbf{r}_i \cdot \vec{\omega}\right) - \mathbf{r}_i\omega^2 + \vec{\omega} \times \mathbf{r}_i = \frac{\mathbf{f}_i}{m_i}. \tag{K.2}$$

With the (K.1), the component along z is:

$$2\dot{\rho}\left(\omega_x y_i^\circ - \omega_y x_i^\circ\right) + \rho\omega_z\left(\omega_x x_i^\circ + \omega_y y_i^\circ\right) + \rho\left(\dot{\omega}_x y_i^\circ - \dot{\omega}_y x_i^\circ\right) = 0. \tag{K.3}$$

Factorizing x_i° and y_i°, it is:

$$\left[-2\dot{\rho}\omega_y + \rho\left(\omega_z\omega_x - \dot{\omega}_y\right)\right]x_i^\circ + \left[2\dot{\rho}\omega_x + \rho\left(\omega_z\omega_y + \dot{\omega}_x\right)\right]y_i^\circ = 0. \tag{K.4}$$

We see that the coefficients in the square brackets are functions of time and independent of the index i. Denoting them as $p(t)$ and $q(t)$ respectively, we have:

$$\frac{p(t)}{q(t)} = -\frac{y_i^\circ}{x_i^\circ}; \tag{K.5}$$

this is the necessary condition for collinearity. But we have assumed that the three points P_i form a triangle. The (K.5) is compatible with this hypothesis only if the functions $p(t)$ and $q(t)$ vanish identically, that is, if:

$$2\dot{\rho}\omega_x = -\rho\left(\omega_z\omega_y + \dot{\omega}_x\right), \tag{K.6a}$$
$$2\dot{\rho}\omega_y = \rho\left(\omega_z\omega_x - \dot{\omega}_y\right). \tag{K.6b}$$

Multiplying by ω_x and ω_y respectively and summing up, we obtain:

$$\frac{2\dot{\rho}}{\rho} = -\frac{\omega_x\dot{\omega}_x + \omega_y\dot{\omega}_y}{\omega_x^2 + \omega_y^2}, \tag{K.7}$$

from which, by a quadrature:

$$\omega_x^2 + \omega_y^2 = \frac{C}{\rho^2}, \tag{K.8}$$

where C is an integration constant.

Similarly, we project the (K.2) on the x and y axes, obtaining:

$$\ddot{x}_i^\circ - 2\omega_z\dot{\rho}y_i^\circ + \omega_x\rho\left(x_i^\circ\omega_x + y_i^\circ\omega_y\right) +$$
$$- \rho x_i^\circ\left(\omega_x^2 + \omega_y^2 + \omega_z^2\right) - \dot{\omega}_z\rho y_i^\circ = \frac{f_{ix}^\circ}{\rho^2}, \tag{K.9a}$$

$$\ddot{p}y_i^\circ + 2\omega_z \dot{p}x_i^\circ + \omega_y \rho\left(x_i^\circ \omega_x + y_i^\circ \omega_y\right) +$$

$$- \rho y_i^\circ \left(\omega_x^2 + \omega_y^2 + \omega_z^2\right) + \dot{\omega}_z \rho x_i^\circ = \frac{f_{iy}^\circ}{\rho^2}, \tag{K.9b}$$

and, by factorizing:

$$\left[\ddot{p} - \rho\left(\omega_y^2 + \omega_z^2\right)\right]x_i^\circ - \left[2\omega_z \dot{p} + \rho\left(\dot{\omega}_z - \omega_x \omega_y\right)\right]y_i^\circ = \frac{f_{ix}^\circ}{\rho^2}, \tag{K.10a}$$

$$\left[2\omega_z \dot{p} + \rho\left(\omega_x \omega_y + \dot{\omega}_z\right)\right]x_i^\circ + \left[\ddot{p} - \rho\left(\omega_x^2 + \omega_z^2\right)\right]y_i^\circ = \frac{f_{iy}^\circ}{\rho^2}. \tag{K.10b}$$

Again, the square brackets contain functions of time. Now, the system of three equations that we obtain from (K.10a) for the three values of the index i is linear and at constant coefficients if both terms are multiplied by ρ^2. Therefore, the expressions within square brackets must be constant too. The same argument applies to (K.10b). In conclusion, the difference between the first term in the square brackets in (K.10a) and the second in (K.10b), once divided by ρ^2, must also be a constant:

$$\omega_x^2 - \omega_y^2 = \frac{A}{\rho^3}, \tag{K.11a}$$

and so the difference between the second of (K.10a) and the first of (K.10b), again divided by ρ^2:

$$2\omega_x \omega_y = \frac{B}{\rho^3}. \tag{K.11b}$$

We can use the complex notation:

$$\omega_x \pm \omega_y = \frac{A \pm iB}{\rho^3}, \tag{K.11c}$$

with modulus:

$$\omega_x^2 + \omega_y^2 = \frac{(A^2 + B^2)^{1/2}}{\rho^3} = \frac{D}{\rho^3}. \tag{K.11d}$$

By the comparison with (K.8):

$$\rho(t) = \frac{C}{D} = \text{const}, \tag{K.12}$$

unless both C and D are zero; and this is precisely the case! In fact, if $\rho = \text{const}$, ω_x and ω_y are also constant and, according to (K.6a) and (K.6b), $\omega_z = 0$. But then, choosing properly the coordinate system, it can be even made that $\omega_y = 0$, with which, from (K.9a), it is also $f_{ix} = 0$, and therefore the three points are aligned, contrary to our starting hypothesis.

In conclusion, $C = D = 0$ implies that $\omega_x = \omega_y = 0$. This proves that the plane S containing the three points rotates on itself (with angular velocity $\vec{\omega} = \omega_z \mathbf{k}$); hence its normal is fixed in space (remember the plane of motion contains the barycenter which by assumption is fixed).

Appendix L
A Simplified Approach to Gravitational Impact

Sun Tzu, *The art of war*

Appear weak when you are strong, and strong when you are weak.

It is known that the heliocentric velocity vector of a minor body in the Solar System can be modified by close gravitational interaction with a major planet to the point that the orbit can change from bound to unbound and vice versa. For instance, these interactions are held responsible for the formation of the families of short period comets ($P < 200$ years).

To understand the mechanism of the phenomenon (three-body impact), which is the basis of the so-called gravity assist commonly used to accelerate spacecrafts,[26] we consider here a simplified model of interaction.

Assume that the small body C of negligible mass m_c revolves around the Sun in an unperturbed elliptical orbit with semi-major axis a_c, ellipticity e_c, and angular momentum h_c, coplanar to the orbit of Jupiter (mass $M_{\u{2643}}$). In other words, we deliberately ignore the gravitational attraction by all massive bodies in the Solar System other then the Sun, and treat the motion of C as a pure two-body problem at all heliocentric distances r_c of C from the Sun except inside a quasi-spherical surface centered at Jupiter that we call gravitational influence sphere (SOI hereafter). We can identify it with the smaller of the Hill surfaces that meet at the Lagrangian point L_1 (see Fig. 3.11). Hence, based on the first of the (3.77), the distance of L_1 from Jupiter, thus equal to the radius of SOI, is $\simeq 5.2 \times 10^7$ km, being $a_{\u{2643}} = 5.2$ AU $= 7.8 \times 10^{13}$ cm,

---

[26] The application of gravity-assist to space travels between planets was considered for the first time in 1918–19 by the Ukrainian engineer and mathematician Yuri Kondratyuk (1897–1942), who also devised the strategy to reach the Moon, then adopted by NASA for the Apollo 11 mission in 1969. The gravity-assist is also called Oberth maneuver in memory of Hermann Oberth, who independently described it in 1927. The first use of gravity assist maneuver was in 1959 by the Soviet probe Luna 3 which photographed the far side of the Moon.

© The Editor(s) (if applicable) and The Author(s), under exclusive license to Springer Nature Switzerland AG 2022
E. Bannikova and M. Capaccioli, *Foundations of Celestial Mechanics*, Graduate Texts in Physics, https://doi.org/10.1007/978-3-031-04576-9

Fig. L.1 Three body impact
model

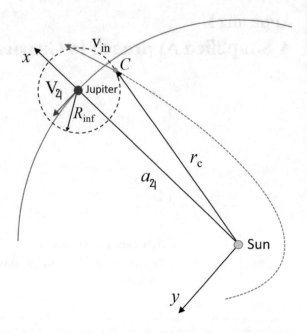

and $M_\jupiter/M_\odot \simeq 10^{-3}$ (Fig. L.1). This result is twice what would be obtained by imposing that the radius of SOI is where the gravitational forces of the Sun and Jupiter equalize (the reader will try to explain why there is such a large difference). Inside the sphere of influence (see also Sect. 2.11) we pretend that the gravitational attraction of the Sun on C becomes negligible compared to that of Jupiter and can be ignored. This dichotomous scheme introduces a clear discontinuity in the field of forces acting on C, but it is quite useful for understanding the phenomenon we are dealing with.

Suppose now that the orbit of C intersects that of Jupiter and that, at the time t_o, the body penetrates the planet SOI. Since the true anomaly of Jupiter varies marginally during the crossing time Δt (which is small compared to the orbital period $P_\jupiter$), the velocity of the planet can be assumed to be rectilinear and uniform, with modulus $V_\jupiter \approx \sqrt{G\,M_\odot/a_\jupiter}$. We are also legitimated to confuse the modulus of the heliocentric distance r of each point within the SOI with the value of the semimajor axis of Jupiter's orbit, $a_\jupiter$: therefore $r_c \equiv a_\jupiter$ (since $R_{\mathrm{SOI}}/a_\jupiter \simeq 0.07$).

Now consider a Cartesian orthogonal system centered at the Sun, with axes x and y coplanar to the orbits of C and Jupiter, and assume that, at t_o, the axis x contains the planet (Fig. L.1). According to the (2.68), the modulus of the heliocentric velocity $\mathbf{v}_{in}$ of C at the point where it penetrates the SOI of Jupiter (remember that $r_c = a_\jupiter$), is:

$$v_{in} = \sqrt{2\,G\,M_\odot \left[\frac{1}{a_\jupiter} - \frac{1}{2a_c}\right]}, \tag{L.1}$$

and has the following components:

Fig. L.2 Three body impact model: effects on the velocity vector

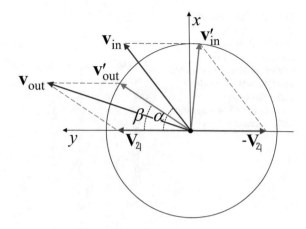

$$(v_{in})_x = v_{in} \sin \alpha,$$
$$(v_{in})_y = v_{in} \cos \alpha, \tag{L.2}$$

where the angle is that shown in Fig. L.2.

In our schematic impact model, when the body C enters the SOI, it starts to be subject to the action of Jupiter only. Therefore its orbit is again a conic, but with a new focus on Jupiter. It can be computed using, as initial conditions (at t_o), the radius vector $\mathbf{r}' = \mathbf{r} - \mathbf{a}_{2\!\!\!\perp}$ with modulus R_{SOI} and the velocity whose components are:

$$\left(v'_{in}\right)_x = v_{in} \sin \alpha,$$
$$\left(v'_{in}\right)_y = v_{in} \cos \alpha - V_{2\!\!\!\perp}. \tag{L.3}$$

Symmetry assures us that the velocity with which C leaves the SOI of Jupiter, $\mathbf{v}'_{out}$ has the same modulus of $\mathbf{v}'_{in}$, $|\mathbf{v}'_{out}| = |\mathbf{v}'_{in}|$; but it will form a different angle (Fig. L.2). Therefore, its components are:

$$\left(v'_{out}\right)_x = v'_{in} \sin \beta,$$
$$\left(v'_{out}\right)_y = v'_{in} \cos \beta, \tag{L.4}$$

At this point C leaves the SOI. Its velocity with respect to the Sun must be that relative to Jupiter plus the velocity of the planet with respect to the Sun:

$$(v_{out})_x = v'_{in} \sin \beta,$$
$$(v_{out})_y = v'_{in} \cos \beta + V_{2\!\!\!\perp}. \tag{L.5}$$

Figure L.2 helps us to understand better the meaning of this qualitative digression. We start from the velocity vector $\mathbf{v}_{in}$ relative to the Sun with which C meets the

Fig. L.3 The trend of
Voyager 2 heliocentric
velocity showing the
consequences of the
gravitational assists with the
outer planets. The yellow
curve shows the escape
velocity for a Solar System
body. Credit: NASA

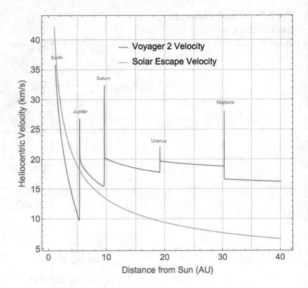

SOI of Jupiter. In order to obtain the input velocity of C relative to Jupiter, $\mathbf{v}'_{in}$, we subtract $\mathbf{V}_{2\!\!\!\!+}$ to $\mathbf{v}_{in}$, then rotate the resultant by an angle whose magnitude depends on the parameters of the impact, and finally add $\mathbf{V}_{2\!\!\!\!+}$ to it, obtaining the velocity vector relative to the Sun as it has been modified by the gravitational encounter. It is apparent that in this way both the modulus of $\mathbf{v}_{in}$ and it orientation (on the plane of motion, given our assumptions) can be changed.

The problem can be usefully treat as an elastic collision, remembering that in this case both kinetic energy and momentum are conserved (Fig. L.3).

Appendix M
The Brackets of Poisson and Lagrange

Siméon-Denis Poisson

Life is good for only two things, discovering mathematics and teaching mathematics.

The Lagrangian and Hamiltonian formalism provides an elegant and fast way to describe the equations of motion in the case of complex systems. The introduction of some operators, such as those described in detail in this appendix, allows us to further simplify this formalism.

M.1 Poisson Brackets

Let us consider the two functions: $f(q, p, t) = f(q_1, q_2, \ldots, q_n, p_1, p_2, \ldots, p_n, t)$ and $g(q, p, t) = g(q_1, q_2, \ldots, q_n, p_1, p_2, \ldots, p_n, t)$. We denote by (f, g) the operator:

$$(f, g) = \sum_{k=1}^{n} \frac{\partial f}{\partial p_k} \frac{\partial g}{\partial q_k} - \sum_{k=1}^{n} \frac{\partial f}{\partial q_k} \frac{\partial g}{\partial p_k}, \tag{M.1}$$

which is called Poisson bracket. Using the definition, it is immediate to verify the following identities:

$$(f, f) = 0,$$
$$(f, g) = -(g, f), \tag{M.2}$$
$$(f_1 + f_2, g) = (f_1, g) + (f_2, g).$$

Equally direct, though more laborious, is the proof of the Jacobi identity:

© The Editor(s) (if applicable) and The Author(s), under exclusive license to Springer Nature Switzerland AG 2022
E. Bannikova and M. Capaccioli, *Foundations of Celestial Mechanics*, Graduate Texts in Physics, https://doi.org/10.1007/978-3-031-04576-9

$$\big((f, g), h\big) + \big((h, f), g\big) + \big((g, h), f\big) = 0, \qquad (\text{M.3})$$

where h is a third function of q_k, p_k, and t. It is simple to remember this formula using of cyclic permutations rule. We start developing the expression $\big((f, g), h\big)$:

$$
\begin{aligned}
\big((f, g), h\big) = {} & \sum_{j=1}^{n}\sum_{k=1}^{n} \frac{\partial^2 f}{\partial p_j \partial p_k} \frac{\partial g}{\partial q_j} \frac{\partial h}{\partial q_k} + \sum_{j=1}^{n}\sum_{k=1}^{n} \frac{\partial^2 g}{\partial q_j \partial p_k} \frac{\partial f}{\partial p_j} \frac{\partial h}{\partial q_k} + \\
& - \sum_{j=1}^{n}\sum_{k=1}^{n} \frac{\partial^2 f}{\partial q_j \partial p_k} \frac{\partial g}{\partial p_j} \frac{\partial h}{\partial q_k} - \sum_{j=1}^{n}\sum_{k=1}^{n} \frac{\partial^2 g}{\partial p_j \partial p_k} \frac{\partial f}{\partial q_j} \frac{\partial h}{\partial q_k} + \\
& - \sum_{j=1}^{n}\sum_{k=1}^{n} \frac{\partial^2 f}{\partial p_j \partial q_k} \frac{\partial g}{\partial q_j} \frac{\partial h}{\partial p_k} - \sum_{j=1}^{n}\sum_{k=1}^{n} \frac{\partial^2 g}{\partial q_j \partial q_k} \frac{\partial f}{\partial p_j} \frac{\partial h}{\partial p_k} + \\
& + \sum_{j=1}^{n}\sum_{k=1}^{n} \frac{\partial^2 f}{\partial q_j \partial q_k} \frac{\partial g}{\partial p_j} \frac{\partial h}{\partial p_k} + \sum_{j=1}^{n}\sum_{k=1}^{n} \frac{\partial^2 g}{\partial p_j \partial q_k} \frac{\partial f}{\partial q_j} \frac{\partial h}{\partial p_k}. \quad (\text{M.4})
\end{aligned}
$$

We represent the set of the four double summations which contain the second partial derivatives of the function f by means of the auxiliary operator $\big(f, (g, h)\big)$:

$$
\begin{aligned}
\big(f, (g, h)\big) = {} & \sum_{j=1}^{n}\sum_{k=1}^{n} \frac{\partial^2 f}{\partial p_j \partial p_k} \frac{\partial g}{\partial q_j} \frac{\partial h}{\partial q_k} - \sum_{j=1}^{n}\sum_{k=1}^{n} \frac{\partial^2 f}{\partial q_j \partial p_k} \frac{\partial g}{\partial p_j} \frac{\partial h}{\partial q_k} + \\
& - \sum_{j=1}^{n}\sum_{k=1}^{n} \frac{\partial^2 f}{\partial p_j \partial q_k} \frac{\partial g}{\partial q_j} \frac{\partial h}{\partial p_k} + \sum_{j=1}^{n}\sum_{k=1}^{n} \frac{\partial^2 f}{\partial q_j \partial q_k} \frac{\partial g}{\partial p_j} \frac{\partial h}{\partial p_k}. \quad (\text{M.5})
\end{aligned}
$$

It is apparent that, by permuting the index of the summations, the operator just defined is symmetric with respect to the order of the functions within the square brackets, that is:

$$\big(f, (g, h)\big) = \big(f, (h, g)\big). \qquad (\text{M.6})$$

Expanding this notation to the other four summations of (M.4) which contain the second partial derivatives of the g function, we can put:

$$\big((f, g), h\big) = \big(f, (g, h)\big) - \big(g, (f, h)\big), \qquad (\text{M.7})$$

and, adding to these the analogous expressions related to the brackets $\big((g, h), f\big)$ and $\big((h, f), g\big)$, the identity (M.3) is immediately verified.

We now list some useful relations that follow directly from the definition (M.1):

$$\frac{d}{dt}(f, g) = (\frac{df}{dt}, g) + (f, \frac{dg}{dt}),$$

$$(f, q_k) = \frac{\partial f}{\partial p_k},$$ (M.8)

$$(f, p_k) = -\frac{\partial f}{\partial q_k},$$

where f is any function of q_k, p_k and t. Moreover:

$$(q_k, q_h) = 0,$$
$$(p_k, p_h) = 0,$$ (M.9)
$$(p_k, q_h) = \delta_{kh} = \begin{cases} 0 \text{ if } k \neq h \\ 1 \text{ if } k = h, \end{cases}$$

where δ_{kh} is the Kronecker[27] symbol. The relations (M.9) are called fundamental Poisson brackets.

We conclude this section by proving the following theorem: for given m functions $g_i(p, q, t)$ of $2n$ dynamic coordinates, if the function $f(g_1, g_2, \ldots, g_m)$ does not explicitly contain the coordinates and the moments, then:

$$(h, f) = \sum_{j=1}^{m}(h, g_j)\frac{\partial f}{\partial g_j},$$ (M.10)

where h is in turn a function of q_k, p_k and t. Indeed:

$$(h, f) = \sum_{k=1}^{n}\left(\frac{\partial h}{\partial p_k}\frac{\partial f}{\partial q_k} - \frac{\partial h}{\partial q_k}\frac{\partial f}{\partial p_k}\right) =$$

$$= \sum_{k=1}^{n}\left[\frac{\partial h}{\partial p_k}\sum_{j=1}^{m}\left(\frac{\partial f}{\partial g_j}\frac{\partial g_j}{\partial q_k}\right) - \frac{\partial h}{\partial q_k}\sum_{j=1}^{m}\left(\frac{\partial f}{\partial g_j}\frac{\partial g_j}{\partial p_k}\right)\right] =$$

$$= \sum_{j=1}^{m}\left[\frac{\partial f}{\partial g_j}\sum_{k=1}^{n}\left(\frac{\partial h}{\partial p_k}\frac{\partial g_j}{\partial q_k} - \frac{\partial h}{\partial q_k}\frac{\partial g_j}{\partial p_k}\right)\right],$$ (M.11)

that is:

$$(h, f) = \sum_{j=1}^{m}(h, g_j)\frac{\partial f}{\partial g_j}.$$ (M.12)

Similarly, it is immediately shown that:

---

[27] Leopold Kronecker (1823–1891): German mathematician and logician.

$$(f_1, f_2) = \sum_{i=1}^{m} \sum_{j=1}^{m} (g_i, g_j) \frac{\partial f_1}{\partial g_i} \frac{\partial f_2}{\partial g_j}, \tag{M.13}$$

where $f_1(g_1, g_2, \ldots, g_m)$ and $f_2(g_1, g_2, \ldots, g_m)$.

M.2 Lagrange Brackets

We now introduce a second operator, which is called Lagrange bracket. Given two functions f and g of $2n$ dynamic coordinates, and possibly of time, it is defined by the expression:

$$[f, g] = \sum_{k=1}^{n} \frac{\partial p_k}{\partial f} \frac{\partial q_k}{\partial g} - \sum_{k=1}^{n} \frac{\partial q_k}{\partial f} \frac{\partial p_k}{\partial g}. \tag{M.14}$$

From the definition it immediately follows that:

$$\begin{aligned}
[f, f] &= 0, \\
[f, g] &= -[g, f], \\
[q_i, q_j] &= 0, \\
[p_i, p_j] &= 0, \\
[p_i, q_j] &= \delta_{ij}.
\end{aligned} \tag{M.15}$$

Having said that, we will prove the following theorem: the Lagrange bracket between two constants of motion a_i and a_j:

$$[a_i, a_j] = \sum_{k=1}^{n} \frac{\partial p_k}{\partial a_i} \frac{\partial q_k}{\partial a_j} - \sum_{k=1}^{n} \frac{\partial q_k}{\partial a_i} \frac{\partial p_k}{\partial a_j}, \tag{M.16}$$

is independent of time and is a function of the constants themselves.[28] To carry on the demonstration, we derive the $2n$ Hamilton equations (4.45a) with respect to a_i. We obtain n pairs of expressions like:

$$\begin{aligned}
\frac{\partial}{\partial a_i}\left(\frac{dq_k}{dt}\right) &= \frac{\partial}{\partial a_i}\left(\frac{\partial \mathcal{H}}{\partial p_k}\right) = \sum_{h=1}^{n}\left(\frac{\partial^2 \mathcal{H}}{\partial q_h \partial p_k}\frac{\partial q_h}{\partial a_i} + \frac{\partial^2 \mathcal{H}}{\partial p_h \partial p_k}\frac{\partial p_h}{\partial a_i}\right), \\
\frac{\partial}{\partial a_i}\left(\frac{dp_k}{dt}\right) &= \frac{\partial}{\partial a_i}\left(-\frac{\partial \mathcal{H}}{\partial q_k}\right) = -\sum_{h=1}^{n}\left(\frac{\partial^2 \mathcal{H}}{\partial q_h \partial q_k}\frac{\partial q_h}{\partial a_i} + \frac{\partial^2 \mathcal{H}}{\partial p_h \partial q_k}\frac{\partial p_h}{\partial a_i}\right),
\end{aligned} \tag{M.17}$$

---

[28] The theorem has a general validity; nowhere in the proof it occurs the condition that the functions a are constants of motion.

for each value of the index k. Denoting a second constant of motion with a_j, we multiply both terms of the first of (M.17) by $-\partial p_k/\partial a_j$, and the second by $\partial q_k/\partial a_j$, and then we add the products. Summing up the result with respect to the index k, we obtain the expression:

$$\sum_{k=1}^{n}\left[\frac{\partial}{\partial a_i}\left(\frac{dp_k}{dt}\right)\frac{\partial q_k}{\partial a_j}-\frac{\partial}{\partial a_i}\left(\frac{dq_k}{dt}\right)\frac{\partial p_k}{\partial a_j}\right]=$$

$$=\sum_{k=1}^{n}\sum_{h=1}^{n}\left[-\frac{\partial^2\mathcal{H}}{\partial q_h\partial q_k}\frac{\partial q_h}{\partial a_i}\frac{\partial q_k}{\partial a_j}-\frac{\partial^2\mathcal{H}}{\partial p_h\partial q_k}\frac{\partial p_h}{\partial a_i}\frac{\partial q_k}{\partial a_j}+\right.$$
$$\left.-\frac{\partial^2\mathcal{H}}{\partial q_h\partial p_k}\frac{\partial q_h}{\partial a_i}\frac{\partial p_k}{\partial a_j}-\frac{\partial^2\mathcal{H}}{\partial p_h\partial p_k}\frac{\partial p_h}{\partial a_i}\frac{\partial p_k}{\partial a_j}\right], \quad \text{(M.18)}$$

whose second term is symmetric with respect to a_i and a_j. Consequently, it is zero the difference between the first terms of (M.18) and the equation obtained from this by inverting the index i with the index j, that is:

$$\sum_{k=1}^{n}\left[\frac{\partial}{\partial a_i}\left(\frac{dp_k}{dt}\right)\frac{\partial q_k}{\partial a_j}-\frac{\partial}{\partial a_i}\left(\frac{dq_k}{dt}\right)\frac{\partial p_k}{\partial a_j}+\right.$$
$$\left.-\frac{\partial}{\partial a_j}\left(\frac{dp_k}{dt}\right)\frac{\partial q_k}{\partial a_i}+\frac{\partial}{\partial a_j}\left(\frac{dq_k}{dt}\right)\frac{\partial p_k}{\partial a_i}\right]=0. \quad \text{(M.19)}$$

Since this identity coincides with the total derivative with respect to the time of (M.16), it proves the assumption.

M.3 Relations Between the Brackets of Poisson and Lagrange

The two operators are so closely related that they can be considered one the opposite of the other. To clarify this concept, we demonstrate that, for the set of $2n$ independent functions f_i of the variables q_k, p_k ($k = 1, \ldots, n$), and t, the following identity holds:

$$\sum_{i=1}^{2n}[f_i, f_j](f_i, f_h) = \delta_{jh}. \quad \text{(M.20)}$$

Applying the definitions (M.1) and (M.14) it results:

$$\sum_{i=1}^{2n} [f_i, f_j](f_i, f_h) =$$

$$= \sum_{i=1}^{2n} \sum_{k=1}^{n} \sum_{m=1}^{n} \left(\frac{\partial p_k}{\partial f_i} \frac{\partial q_k}{\partial f_j} - \frac{\partial q_k}{\partial f_i} \frac{\partial p_k}{\partial f_j} \right) \left(\frac{\partial f_i}{\partial p_m} \frac{\partial f_h}{\partial q_m} - \frac{\partial f_i}{\partial q_m} \frac{\partial f_h}{\partial p_m} \right). \quad (M.21)$$

In the development of the right-hand side there are four terms; the first can be written in the form:

$$\sum_{k=1}^{n} \sum_{m=1}^{n} \frac{\partial q_k}{\partial f_j} \frac{\partial f_h}{\partial q_m} \sum_{i=1}^{2n} \frac{\partial p_k}{\partial f_i} \frac{\partial f_i}{\partial p_m} = \sum_{k=1}^{n} \sum_{m=1}^{n} \frac{\partial q_k}{\partial f_j} \frac{\partial f_h}{\partial q_m} \frac{\partial p_k}{\partial p_m} =$$

$$= \sum_{k=1}^{n} \sum_{m=1}^{n} \frac{\partial q_k}{\partial f_j} \frac{\partial f_h}{\partial q_m} \delta_{km} = \sum_{k=1}^{n} \frac{\partial q_k}{\partial f_j} \frac{\partial f_h}{\partial q_k}. \quad (M.22)$$

The last of the four terms of (M.21) is identical to the first, except that q_k is replaced by p_k and vice versa, while the other two terms do not contribute to the sum. It turns out that:

$$\sum_{i=1}^{2n} [f_i, f_j](f_i, f_h) = \sum_{k=1}^{n} \left(\frac{\partial q_k}{\partial f_j} \frac{\partial f_h}{\partial q_k} + \frac{\partial p_k}{\partial f_j} \frac{\partial f_h}{\partial p_k} \right) = \frac{\partial f_h}{\partial f_j} = \delta_{jh}, \quad (M.23)$$

which is what we wanted to prove.

In the remaining part of this section, we shall interpret the relations between Poisson and Lagrange brackets in terms of matrices. Using $2n$ independent functions $f_i(q, p, t)$ of n coordinates q_k and n conjugate moments p_k, it is possible to build $(2n)^2$ Poisson brackets. These can be arranged in the form of an array of order $2n$:

$$\|A\| = \begin{Vmatrix} (f_1, f_1) & (f_1, f_2) & \cdots & (f_1, f_{2n}) \\ (f_2, f_1) & (f_2, f_2) & \cdots & (f_2, f_{2n}) \\ \cdots & \cdots & \cdots & \cdots \\ (f_{2n}, f_1) & (f_{2n}, f_2) & \cdots & (f_{2n}, f_{2n}) \end{Vmatrix}, \quad (M.24)$$

which, owing to (M.2), is antisymmetric. Similarly, from the $(2n)^2$ Lagrange brackets we obtain the antisymmetric matrix:

$$\|A'\| = \begin{Vmatrix} [f_1, f_1] & [f_1, f_2] & \cdots & [f_1, f_{2n}] \\ [f_2, f_1] & [f_2, f_2] & \cdots & [f_2, f_{2n}] \\ \cdots & \cdots & \cdots & \cdots \\ [f_{2n}, f_1] & [f_{2n}, f_2] & \cdots & [f_{2n}, f_{2n}] \end{Vmatrix}. \quad (M.25)$$

The relation (M.23) proves that the two matrices are reciprocal; hence their product, $\|U\| = \|A\| \cdot \|A'\|$, is the unit matrix. So, between the elements of $\|A\|$ and $\|A'\|$, there are the following relations:

$$(f_i, f_j) = \frac{1}{D'} \frac{\partial D'}{\partial [f_j, f_i]},$$

$$[f_i, f_j] = \frac{1}{D} \frac{\partial D}{\partial (f_j, f_i)}, \qquad \text{(M.26)}$$

where D and D' are the determinants of $\|A\|$ and $\|A'\|$ respectively, and:

$$\frac{\partial D'}{\partial [f_j, f_i]} = \frac{\partial \sum_{i=1}^{2n} (f_j, f_i) |F_{j,i}|}{\partial [f_j, f_i]}, \qquad \text{(M.27)}$$

with $|F_{j,i}|$ first minor of (f_j, f_i).

Appendix N
Some Special Functions

Pierre Simon de Laplace

Nature laughs at the difficulties of integration.

We call special functions a whole set of real or complex functions of real or complex variables, such as, for instance, the trigonometric functions, that have gained a proper name and a more or less standardized notation in virtue of a wide range applications in pure mathematics, statistics, and physics. Many, but not all, appear as solutions of differential or integral equations. Here we list some classes of special functions of interest for celestial mechanics, outlining their main characteristics (see [9] for an complete presentation).

N.1 Gamma Function

The function Γ is the extension of the factorial to real or complex numbers. Often used in statistics and to manipulate Legendre polynomials, fundamental in the description of the gravitational potential, it is defined by the integral[29]:

$$\Gamma(z) = \int_0^\infty t^{z-1} e^{-t} dt, \tag{N.1}$$

which, if the real part of z is positive, converges absolutely. Represented graphically in Fig. N.1, the $\Gamma(z)$ function turns into the usual definition of factorial for non-negative integers $z = n$, as it is easily verified. Indeed:

---

[29] The notation $\Gamma(z)$ is due to Legendre.

E. Bannikova and M. Capaccioli, *Foundations of Celestial Mechanics*, Graduate Texts in Physics, https://doi.org/10.1007/978-3-031-04576-9

Fig. N.1 Gamma function

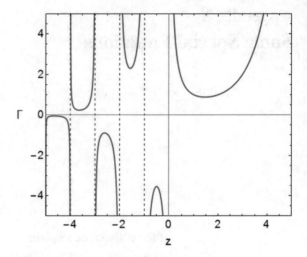

$$\Gamma(1) = \int_0^\infty e^{-t} dt = \lim_{k\to\infty} -e^{-t}\Big|_0^k = 1, \tag{N.2}$$

and, for all integers $n > 0$:

$$\Gamma(n+1) = n\,\Gamma(n) = \cdots = n!\,\Gamma(1) = n!, \tag{N.3}$$

which is the application to n of the more general property:

$$\Gamma(z+1) = z\Gamma(z), \tag{N.4}$$

obtained integrating (N.1) by parts.

Alternative definitions of $\Gamma(z)$ were provided by Euler and Weierstrass[30] respectively:

$$\Gamma(z) = \lim_{n\to\infty} \frac{n!\,n^z}{z\,(z+1)\cdots(z+n)}, \tag{N.5}$$

and:

$$\Gamma(z) = \frac{e^{-\gamma z}}{z} \prod_{n=1}^\infty \left(1 + \frac{z}{n}\right)^{-1} e^{z/n}, \tag{N.6}$$

where $\gamma \approx 0.5772$ is the Euler-Mascheroni constant.[31] By (N.5), we can now prove the (N.4):

---

[30] Karl Theodor Wilhelm Weierstrass (1815–1897): German mathematician, founder of modern mathematical analysis.

[31] Lorenzo Mascheroni (1750–1800): Italian mathematician and writer.

$$\Gamma(z+1) = \lim_{n \to \infty} \frac{n! \, n^{z+1}}{(z+1)\,(z+2)\cdots(z+1+n)} =$$

$$= \lim_{n \to \infty} \left(z \, \frac{n! \, n^z}{(z+1)\,(z+2)\cdots(z+n)} \, \frac{n}{(z+1+n)} \right) =$$

$$= z \, \Gamma(z) \lim_{n \to \infty} \frac{n}{(z+1+n)} = z \, \Gamma(z). \tag{N.7}$$

We report below some properties of the Γ function. Euler's reflection formula is:

$$\Gamma(1-z) \, \Gamma(z) = \frac{\pi}{\sin(\pi z)}, \tag{N.8}$$

and the multiplication theorem:

$$\Gamma(z) \, \Gamma\left(z+\frac{1}{m}\right) \Gamma\left(z+\frac{2}{m}\right) \cdots \Gamma\left(z+\frac{m-1}{m}\right) =$$

$$= (2\pi)^{(m-1)/2} \, m^{(1-2mz)/2} \, \Gamma(mz), \tag{N.9}$$

that, for $m = 2$, becomes the duplication formula:

$$\Gamma(z) \, \Gamma\left(z+\frac{1}{2}\right) = \frac{\sqrt{\pi}}{2^{2z-1}} \, \Gamma(2z). \tag{N.10}$$

Some typical values of Γ are:

$$\Gamma(1/2) = \sqrt{\pi} \approx 1.772,$$
$$\Gamma(1) = 0! = 1,$$
$$\Gamma(3/2) = \frac{\sqrt{\pi}}{2} \approx 0.886, \tag{N.11}$$
$$\Gamma(2) = 1! = 1,$$
$$\Gamma(3) = 2! = 2.$$

N.2   Beta Function

The Beta function (Fig. N.2), also called Euler integral of first kind, is defined by the integral:

$$B(x, y) = \int_0^1 t^{x-1}(1-t)^{y-1} \, dt, \tag{N.12}$$

for $\Re(x), \Re(y) > 0$. It is symmetric:

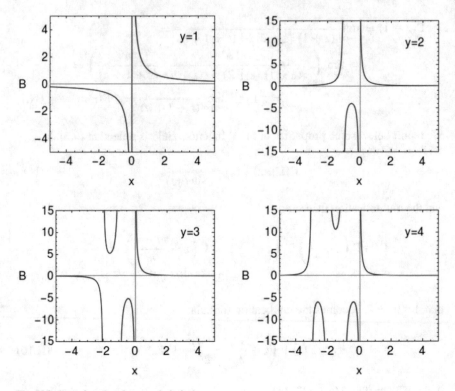

Fig. N.2 Beta function for $y = 1, 2, 3, 4$

$$B(x, y) = B(y, x), \tag{N.13}$$

and can be written in the following ways:

$$B(x, y) = \frac{\Gamma(x)\,\Gamma(y)}{\Gamma(x+y)}, \tag{N.14a}$$

$$B(x, y) = 2 \int_0^{\pi/2} \sin^{2x-1}\theta \cos^{2y-1}\theta\, d\theta, \tag{N.14b}$$

$$\Re(x) > 0,\ \Re(y) > 0$$

$$B(x, y) = \int_0^{\infty} \frac{t^{x-1}}{(1+t)^{x+y}}\, dt, \tag{N.14c}$$

The definition of the function B can be suitably expanded by introducing the generalized (incomplete) Beta function, where the definite integral of (N.12) is replaced by an indefinite one:

$$B(x; a, b) = \int_0^x t^{a-1}\,(1-t)^{b-1}\, dt. \tag{N.15}$$

Finally, we mention the regularized Beta function:

$$I_x(a, b) = \frac{B(x; \, a, b)}{B(a, b)} \, . \tag{N.16}$$

N.3 Bessel Functions

Defined by Daniel Bernoulli,[32] the Bessel[33] functions were generalized as solutions of Bessel's differential equation:

$$x^2 \frac{d^2 y(x)}{dx^2} + x \frac{dy(x)}{dx} + (x^2 - \alpha^2) y(x) = 0, \tag{N.17}$$

where α is a real or complex number. Special and interesting cases are obtained when α is an integer n (or half-integer), then called order, as they appear in the solution to Laplace's equation (5.21) in cylindrical coordinates. Being a second order differential equation, (N.17) gives two linearly independent solutions. We will obtain and study two such solutions, indicated with $J_n(x)$ and $Y_n(x)$ respectively, that are called Bessel functions of the first and of the second kind. They are widely used in various problems of wave propagation and static potentials, and also in the signal analysis [10]. Other useful functions can be expressed simply through them.

N.3.1 Bessel Functions of the First Kind

We obtain J_n (Fig. N.3) by an indirect procedure, proving only at the end that it is a solution of the differential equation (N.17). We develop the integral:

---

[32] Daniel Bernoulli (1700–1782): Swiss mathematician and physicist belonging to a family of scientists. Of Belgian origin, the dynasty of Bernoulli settled at Basel, Switzerland, at the end of the XVI century. The mathematical genius manifested with the progenitor, Jakob Bernoulli (1654–1705), who investigated the probability theory and the statistics. Son of Johann (1667–1748), Jakob's brother and a mathematician too, Daniel continued the studies of applied mathematics undertaken by the uncle, towering in the field of fluid dynamics. He was the one calling in St. Petersburgh his compatriot Euler, who was a student of his father. He is also known for the scarcely commendable scientific and academic conflict with his father Johann.

[33] Friedrich Wilhelm Bessel (1784–1846): German mathematician and astronomer, contemporary to Gauss. Protected by Olbers, although without a university education, he could start a fast-track career in astronomy. Thanks to Gauss, when just 26 years old he was appointed director of the Prussian Observatory of Königsberg. There, with an instrument produced by Joseph von Fraunhofer, in 1838 he made the first determination of the annual parallax of a star (61 Cygni), beating his competitors: the German Friedrich von Struve, director of the Dorpat Observatory in Tartu, Estonia, who was following the star Vega, and the Scottish Thomas Henderson who looked at the star αCen from the Cape of Good Hope Observatory.

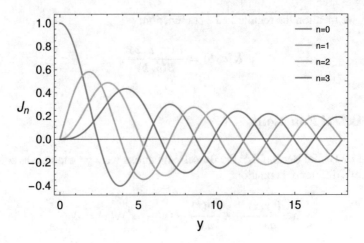

Fig. N.3 Bessel functions of the first kind

$$A = \frac{1}{2\pi} \int_0^{2\pi} e^{-i\,(n\,x - y\,\sin x)}\, dx, \tag{N.18}$$

using Euler's formula[34]

$$A = \frac{1}{2\pi} \int_0^{2\pi} \cos\left(n\,x - y\,\sin x\right) dx + \frac{i}{2\pi} \int_0^{2\pi} \sin\left(n\,x - y\,\sin x\right) dx. \tag{N.19}$$

By the substitution $x = 2\pi - x'$, the integrand of the second integral at the right-hand side of (N.19) becomes:

$$\sin\left(n\,x - y\,\sin x\right) = \sin\left(2\,n\pi - n\,x' + y\,\sin x'\right) =$$
$$= -\sin\left(n\,x' - y\,\sin x'\right), \tag{N.20}$$

showing that, for $x' = 2\pi - x$, it is equal and opposite to the same integrand as a function of x. The sum between 0 and π is therefore cancelled by that made between π and 2π. As a result, we get an expression equivalent to (F.13):

$$J_n(y) = \frac{1}{2\pi} \int_0^{2\pi} e^{-i\,(n\,x - y\,\sin x)}\, dx = \frac{1}{2\pi} \int_0^{2\pi} \cos\left(n\,x - y\,\sin x\right) dx. \tag{N.21}$$

Now we rewrite (N.19) in the form:

---

[34] Found by Euler in 1748, the the famous formula of the same name, "*the most remarkable formula in mathematics*" according to Richard Feynman, is: $e^{ix} = \cos x + i \sin x$. With $x = \pi$ you readily get the famous Euler's identity: $e^{i\pi} + 1 = 0$.

$$J_n(y) = \frac{1}{2\pi} \int_0^{2\pi} e^{-inx}\, e^{i\, y\, \sin x}\, dx, \tag{N.22}$$

which proves that $J_n(y)$ is the coefficient of e^{-inx} in the Fourier expansion (cf. Sect. F.2):

$$e^{i\, y\, \sin x} = \sum_{n=-\infty}^{+\infty} J_n(y)\, e^{inx}. \tag{N.23}$$

By introducing the new variable:

$$z = e^{ix}, \tag{N.24}$$

from where:

$$2i\,\sin x = z - \frac{1}{z}, \tag{N.25}$$

the expansion (N.23) takes the form:

$$\exp\left[\frac{y}{2}\left(z - \frac{1}{z}\right)\right] = e^{yz/2}\, e^{-y/2z} = \sum_{n=-\infty}^{+\infty} J_n(y)\, z^n. \tag{N.26}$$

Therefore, $J_n(y)$ is also the coefficient of z^n in the power series expansion of negative and positive powers of the function:

$$F(y, z) = e^{yz/2}\, e^{-y/2z} = \left(\sum_{k=0}^{\infty} \frac{(y/2)^k}{k!}\, z^k\right)\left(\sum_{j=0}^{\infty}(-1)^j \frac{(y/2)^j}{j!}\, z^{-j}\right) =$$

$$= \sum_{k=0}^{\infty}\sum_{j=0}^{\infty}(-1)^j \frac{(y/2)^{k+j}}{k!\, j!}\, z^{k-j}. \tag{N.27}$$

Finally, with the substitution $n = k - j$:

$$F(y, z) = \sum_{n=-\infty}^{\infty}\sum_{j=0}^{\infty}(-1)^j \frac{(y/2)^{n+2j}}{(n+j)!\, j!}\, z^n = \sum_{n=-\infty}^{+\infty} J_n(y)\, z^n. \tag{N.28}$$

Note that the lower limit of the sum has been extended from $k = 0$ to $n = -\infty$. This is because the added terms are null, as they contain in the denominator the factorial of a negative number that is infinite. The absence of terms with $n + k < 0$ can be considered equivalent to the fact that the power series of $y/2$:

$$J_n(y) = \sum_{j=0}^{\infty}(-1)^j \frac{(y/2)^{n+2j}}{(n+j)!\, j!}, \tag{N.29}$$

has only positive powers of $y/2$, as shown by the way we built the function $F(y, z)$. The characteristics of this latter function ensure that:

$$J_n(-y) = (-1)^n J_n(y),$$ (N.30a)

$$J_{-n}(-y) = (-1)^n J_n(y).$$ (N.30b)

In turn they prove that:

$$J_{-n}(-y) = J_n(y).$$ (N.30c)

It follows from all these relations that it is not necessary to consider negative values of the index n. For $n > 0$, the (N.29) gives:

$$J_n(y) = \frac{1}{n!} \left(\frac{y}{2}\right)^n \sum_{j=0}^{\infty} \frac{(y/2)^{2j}}{j!\,(n+1)(n+2)\cdots(n+j)}.$$ (N.31)

The ratio between two consecutive terms of the series in (N.31) with index j and $j+1$ is:

$$R = -\left(\frac{y}{2}\right)^2 \frac{1}{(j+1)(n+j+1)};$$ (N.32)

its numerical value is less than unity for each value of y, as long as j is chosen sufficiently large. So, based on d'Alembert's criterion of the ratio for series with positive terms, the series (N.31) is absolutely convergent for all values of y. It can be equally shown that the series (N.28) for $F(y, z)$ is absolutely convergent for each value of y and z but $z = 0$ and $z = +\infty$ (which are the poles of F in the complex plane). Differentiating the Eq. (N.26) with respect to z:

$$\frac{y}{2}\left(1 + \frac{1}{z^2}\right) \sum_{-\infty}^{+\infty} J_n(y)\, z^n = \sum_{-\infty}^{+\infty} n\, J_n(y)\, z^{n-1},$$ (N.33)

and equating the coefficients of z^{n-1} in the two sides of this equality, we have:

$$\frac{y}{2}\left[J_{n-1}(y) + J_{n+1}(y)\right] = n\, J_n(y).$$ (N.34)

Again, by differentiating the (N.26), but this time with respect to y:

$$\frac{1}{2}\left(z - \frac{1}{z}\right) \sum_{-\infty}^{+\infty} J_n(y)\, z^n = \sum_{-\infty}^{+\infty} z^n \frac{d\, J_n(y)}{dy},$$ (N.35)

and equaying the coefficients of z^{n-1} on the two sides, we have:

$$\frac{1}{2}\left(J_{n-1}(y) - J_{n+1}(y)\right) = \frac{dJ_n(y)}{dy}. \tag{N.36}$$

From (N.34) and (N.36) we obtain:

$$J_{n-1}(y) = \frac{n}{y} J_n(y) + \frac{dJ_n(y)}{dy}, \tag{N.37a}$$

$$J_{n+1}(y) = \frac{n}{y} J_n(y) - \frac{dJ_n(y)}{dy}. \tag{N.37b}$$

By changing n in (N.34) with $(n-1)$ and a second time with $(n+1)$, it is:

$$J_n(y) + J_{n-2}(y) = \frac{2(n-1)}{y} J_{n-1}(y), \tag{N.38a}$$

$$J_n(y) + J_{n+2}(y) = \frac{2(n+1)}{y} J_{n+1}(y), \tag{N.38b}$$

or, with (N.37a):

$$J_{n-2}(y) = \left(\frac{2n(n-1)}{y^2} - 1\right) J_n(y) + \frac{2(n-1)}{y} \frac{dJ_n(y)}{dy}, \tag{N.39a}$$

$$J_{n+2}(y) = \left(\frac{2n(n+1)}{y^2} - 1\right) J_n(y) - \frac{2(n+1)}{y} \frac{dJ_n(y)}{dy}. \tag{N.39b}$$

This procedure, which can be iterated endlessly, shows that J_{n+k}, where the integer $k \geq -n$, can always be expressed as a linear combination of $J_n(y)$ and $dJ_n(y)/dy$, through coefficients that are functions of the argument y.

Another important property of Bessel functions $J_n(y)$ is that they are solutions of a linear differential equation with non-constant coefficients. This equation is obtained by differentiating the (N.36):

$$\frac{d^2 J_n(y)}{dy^2} = \frac{1}{2}\left(\frac{dJ_{n-1}(y)}{dy} - \frac{dJ_{n+1}(y)}{dy}\right). \tag{N.40}$$

By replacing the derivatives of J_{n-1} and J_{n+1} through (N.36), we have:

$$\frac{d^2 J_n(y)}{dy^2} = \frac{1}{4}\left(J_{n-2}(y) - 2 J_n(y) + J_{n+2}(y)\right), \tag{N.41}$$

and, with the (N.39):

$$\frac{d^2 J_n(y)}{dy^2} = -\frac{1}{2} J_n(y) + \left(\frac{n^2}{y^2} - \frac{1}{2}\right) J_n(y) - \frac{1}{y}\frac{dJ_n(y)}{dy}, \tag{N.42}$$

that is:

Fig. N.4 Bessel functions of the second kind

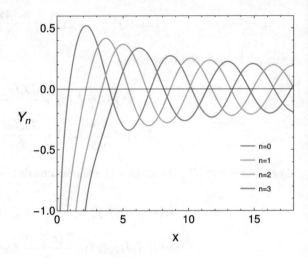

$$\frac{d^2 J_n(y)}{dy^2} + \frac{1}{y}\frac{d J_n(y)}{dy} + \left(1 - \frac{n^2}{y^2}\right) J_n(y) = 0. \tag{N.43}$$

Once multiplied by y^2, this differential equation is identical to (N.17) and can be used to define Bessel functions.

N.3.2 Bessel Functions of the Second Kind

These functions (see Fig. N.4) represent the second solution of the Eq. (N.17). They are singular at $x = 0$ and can be obtained from the functions of the first kind (Fig. N.5):

$$Y_n(x) = \frac{J_n(x)\cos(n\pi) - J_{-n}(x)}{\sin(n\pi)}. \tag{N.44}$$

The following relation holds:

$$Y_{-n}(x) = (-1)^n Y_n(x). \tag{N.45}$$

N.3.3 Modified Bessel Functions

Bessel functions can also be defined when the variable x is complex. A very special case is when x is purely imaginary. In this case the solutions of the Bessel equation

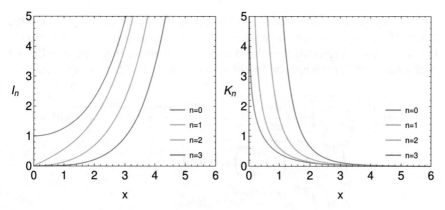

Fig. N.5 Modified Bessel functions of the first and the second kind

are called modified Bessel functions, or hyperbolic Bessel functions, of the first and the second kind:

$$I_n(x) = i^{-n} J_n(ix), \tag{N.46a}$$

$$K_n(x) = \frac{\pi}{2} \frac{I_{-n}(x) - I_n(x)}{\sin(n\pi)}. \tag{N.46b}$$

These two functions are linearly independent solutions of the equation:

$$x^2 \frac{d^2 y}{dx^2} + x \frac{dy}{dx} - (x^2 + n^2) y = 0. \tag{N.47}$$

Compared to Bessel's ordinary functions, which are oscillating functions of a real argument, I_n and K_n are exponentially increasing and decreasing respectively.

N.4 Elliptical Integrals

An elliptic integral is any function f that can be expressed in the form:

$$f(x) = \int_{x_0}^{x} r(t, p(t)) dt, \tag{N.48}$$

where r is a rational function and p the square root of a polynomial of degree 3 or 4 with no repeated roots, and x_0 is a constant. We consider here only the particular cases of the the elliptic integrals used in this book.

It is easy to meet the elliptical integral even in the two-body problem since the length of an ellipse cannot be expressed through elementary functions.[35] Taking into account the parametric expression (2.50) for an ellipse with major and minor semi-axes a and b, the length of the curve, $L = \int dl = \int \sqrt{(dx)^2 + (dy)^2}$, is:

$$L = \int_0^{2\pi} \sqrt{\left(\frac{dx}{d\theta}\right)^2 + \left(\frac{dy}{d\theta}\right)^2}\, d\theta = 4 \int_0^{\pi/2} \sqrt{a^2 \sin^2\theta + b^2 \cos^2\theta}\, d\theta =$$

$$= 4a \int_0^{\pi/2} \sqrt{1 - \left(1 - \frac{b^2}{a^2}\right) \cos^2\theta}\, d\theta =$$

$$= 4a \int_0^{\pi/2} \sqrt{1 - e^2 \cos^2\theta}\, d\theta =$$

$$= 4a \int_0^{\pi/2} \sqrt{1 - e^2 \sin^2\theta}\, d\theta = 4a\, E(e), \qquad (N.49)$$

since $\cos\theta = \sin\left(\frac{\pi}{2} - \theta\right)$. The function:

$$E(k) = \int_0^{\pi/2} \sqrt{1 - k^2 \sin^2\theta}\, d\theta, \qquad (N.50)$$

is the complete elliptical integral of the second kind with elliptical modulus $0 < k < 1$ (sometimes replaced with the parameter $m = k^2$). We can see from (N.49) that $E(e)$ is a quarter of the length of the ellipse. For the circle, the eccentricity is $e = 0$, so $E(0) = \pi/2$, and $L = 2\pi a$, as expected.

Since elliptical integrals cannot be expressed in terms of elementary functions, we must first expand the integrand in (N.50) by the same binomial series (5.35) used at Sect. 5.3, where this time the variable is $x = -k^2 \sin^2\theta$. The fact that by hypothesis $0 \le k < 1$ implies that $0 \le |x| = k^2 \sin^2\theta < 1$, which ensured that the series:

$$(1 + x)^{1/2} = \sum_{i=0}^{\infty} (-1)^i \binom{1/2}{i} x^i, \qquad (N.51)$$

with the coefficients:

$$(-1)^i \binom{1/2}{i} = (-1)^i \frac{(1/2)!}{i!(1/2 - i)!} =$$

$$= \frac{(1/2)}{1} \frac{(1/2 - 1)}{2} \frac{(1/2 - 2)}{3} \cdots \frac{(1/2 - i + 1)}{i} = -\frac{1\,1\,3\,5\,(2i - 3)}{2\,4\,6\,8\,\;\;\;\;2i}, \qquad (N.52)$$

converges absolutely for $0 \le \theta \le \pi/2$. Hence, the (N.50) becomes:

---

[35] Incidentally, we report without demonstration one of the many approximated formulas for the perimeter of an ellipse, due to Euler (1773): $L \approx 2\pi \sqrt{\dfrac{a^2 + b^2}{2}}$, which holds for small ellipticities.

$$E(k) = \int_0^{\pi/2} \left(1 - \sum_{i=0}^{\infty} \frac{1\,1\,3\,5}{2\,4\,6\,8} \frac{(2i-3)}{2i} k^{2i} \sin^{2i}\theta \right) d\theta, \qquad (N.53)$$

which can be integrated term by term since the series is absolutely convergent:

$$E(k) = \frac{\pi}{2} - \sum_{i=0}^{\infty} \frac{1\,1\,3\,5}{2\,4\,6\,8} \frac{(2i-3)}{2i} k^{2i} \int_0^{\pi/2} \sin^{2i}\theta \, d\theta. \qquad (N.54)$$

Since:

$$\int_0^{\pi/2} \sin^m \theta \, d\theta = \frac{1\cdot3\cdot5\cdots(m-1)}{2\cdot4\cdot6\cdots m} \frac{\pi}{2}, \qquad (N.55)$$

we finally obtain:

$$E(k) = \frac{\pi}{2}\left[1 - \left(\frac{1}{2}\right)^2 \frac{k^2}{1} - \left(\frac{1\cdot3}{2\cdot4}\right)^2 \frac{k^4}{3} - \left(\frac{1\cdot3\cdot5}{2\cdot4\cdot6}\right)^2 \frac{k^6}{5} - \cdots\right] =$$
$$= \frac{\pi}{2}\left[1 - \frac{1}{4}k^2 - \frac{3}{64}k^4 - \frac{5}{256}k^6 - \cdots\right], \qquad (N.56)$$

from which we deduce the obvious property that, with the same semi-major axis, the length of the perimeter decreases with increasing ellipticity.

The complete elliptical integral of the first kind:

$$K(k) = \int_0^{\pi/2} \frac{d\theta}{\sqrt{1 - k^2 \sin^2\theta}}, \qquad (N.57)$$

appears in the relativistic two-body problem (Sect. 4.14) and in the gravitational potential of an infinitely thin ring and a circular homogenous torus (see Sect. 5.13.1). It can be seen from (N.57) that $K(0) = E(0) = \pi/2$.

If the upper limit depends on the angle, we obtain the incomplete elliptical integral of the corresponding type. The incomplete elliptical integral of the first kind is:

$$K(\beta, k) = \int_0^{\beta} \frac{d\theta}{\sqrt{1 - k^2 \sin^2\theta}}, \qquad (N.58)$$

and that of the second kind:

$$E(\beta, k) = \int_0^{\beta} \sqrt{1 - k^2 \sin^2\theta}\, d\theta. \qquad (N.59)$$

There is a beautiful identity between the complete integrals of the first and the second kind called Legendre relation:

$$K(k)E(\sqrt{1-k^2}) + E(k)K(\sqrt{1-k^2}) - K(k)K(\sqrt{1-k^2}) = \frac{\pi}{2}. \qquad \text{(N.60)}$$

The symmetric notation $K' = K(k')$ and $E' = E(k')$ with the complimentary modulus $k' = \sqrt{1-k^2}$ is also used, with which the Legendre relation takes a compact form:

$$KE' + EK' - KK' = \frac{\pi}{2}. \qquad \text{(N.61)}$$

Functional expressions that are useful for analytical calculation (cf. [11]) are:

$$\frac{dK(k)}{dk} = \frac{E(k)}{k\,k'^2} - \frac{K(k)}{k}, \qquad \text{(N.62)}$$

$$\frac{dE(k)}{dk} = \frac{E(k) - K(k)}{k}, \qquad \text{(N.63)}$$

$$K\left(\frac{2\sqrt{k}}{1+k}\right) = (1+k)K(k), \qquad \text{(N.64)}$$

$$E\left(\frac{2\sqrt{k}}{1+k}\right) = \frac{1}{1+k}\left[2E(k) - k'^2 K(k)\right]. \qquad \text{(N.65)}$$

There is also the incomplete elliptical integral of the third kind:

$$\Pi(n; \beta, k) = \int_0^\beta \frac{d\theta}{(1 + n\sin^2\theta)\sqrt{1 - k^2\sin^2\theta}}, \qquad \text{(N.66)}$$

and the corresponding complete form:

$$\Pi(n, k) = \Pi(n; \pi/2, k) = \int_0^{\pi/2} \frac{d\theta}{(1 + n\sin^2\theta)\sqrt{1 - k^2\sin^2\theta}}. \qquad \text{(N.67)}$$

Appendix O
Orthogonal Functions

Carl Friedrich Gauss

It is not knowledge, but the act of learning, not posses-sion but the act of getting there, which grants the greatest enjoyment.

Orthogonal functions assume considerable importance in several mathematical and physical problems, from quantum mechanics[36] to astrophysics (for instance, in astro-seismology which studies oscillations in stars). We devote this appendix to a general description of these functions, paying particular attention to the Legendre polynomials for their key role in celestial mechanics. We start considering a classical case in which the need for orthogonal functions becomes apparent.

O.1 Least Squares

Given a basis of n real functions $f_i(x)$ of the real variable x in the domain $[x_o, x_1]$, we aim to find the linear combination that best approximates a function $g(x)$ within the same domain. In other words, we want to determine the 'best' set of coefficients a_i that, according to some given rule, satisfy the condition:

$$g(x) \approx \sum_{i=1}^{n} a_i \, f_i(x), \tag{O.1}$$

for $x_o \leq x \leq x_1$. The formulation of the problem becomes clearer and lends to a quantitative statement if we introduce the cumulative error function:

---

[36] For a description of the applications of orthogonal functions to quantum mechanics see [7, 8].

© The Editor(s) (if applicable) and The Author(s), under exclusive license to Springer Nature Switzerland AG 2022
E. Bannikova and M. Capaccioli, *Foundations of Celestial Mechanics*, Graduate Texts in Physics, https://doi.org/10.1007/978-3-031-04576-9

$$\epsilon(x, a_1, \cdots, a_n) = g(x) - \sum_{i=1}^{n} a_i \, f_i(x). \tag{O.2}$$

Usually, no set of coefficients a_i is capable of cancelling ϵ on the entire interval $[x_o, x_1]$; if this were the case, we would have the exact solution to our problem. In most cases we have to settle for minimizing the error function. The dilemma is the choice of a minimization criterion suitable for the application we are interested in. For example, following Chebyshev,[37] we can try to minimize the maximum of $|\epsilon|$ within the range of x; in this way we obtain a representation of $g(x)$ with a known tolerance at any point of the domain. Following Gauss,[38] we can instead minimize the integral value of some power m of ϵ:

$$E_m(a_1, \cdots, a_n) = \int_{x_o}^{x_1} [\epsilon(x, a_1, \cdots, a_n)]^m \, dx, \tag{O.3}$$

where $m > 0$ is an even integer, as we shall see (in this second case there is no guarantee that in the domain there are no high error peaks). The minimum of E_m is given by the zeros of all its partial derivatives with respect to the parameters a_i:

$$\frac{\partial}{\partial a_i} E_m(a_1, \cdots, a_n) = \int_{x_o}^{x_1} m \left\{ g(x) - \sum_{j=1}^{n} a_j \, f_j(x) \right\}^{m-1} f_i \, dx = 0; \tag{O.4}$$

note that E_m does not depend of x. It is evident that $m = 1$ does not impose any conditions on the parameters (it is the trivial solution, being the exponent $m - 1 = 0$). An intuitive explanation is provided in Fig. O.1, where we have chosen $g(x)$ to be a segment, $n = 2$, $f_1(x) = a$ and $f_2(x) = x$. The problem is in the sign of the residuals $\epsilon(x)$ at each abscissa; problem shared by any odd value of m. One might get around the problem by using the absolute value of ϵ in the integral (O.3), but modules are difficult tools to handle analytically. If $m = 2$, the (O.4) gives:

$$\frac{\partial}{\partial a_i} E_2(a_1, \cdots, a_n) = 2 \int_{x_o}^{x_1} \left\{ g(x) - \sum_{j=1}^{n} a_j \, f_j(x) \right\} f_i \, dx = 0, \tag{O.5}$$

i.e., n linear algebraic equations in the parameters a_i of the type:

---

[37] Pafnuty L. Chebyshev (1821–1894): Russian mathematician, whose main contributions are in the fields of probability, statistics, and number theory.

[38] The first publication of the least squares method was by Legendre in 1805, but it was Gauss in 1809 to give a complete account, linking the method to the principles of probability. The German genius also claimed to have known the method since 1795 and to have used it in predicting the position of Ceres (see also the note 29 at Sect. 2.9). The fact gave rise to a bitter dispute for the priority with Legendre.

Fig. O.1 Best representation of a segment *A-C* (in red) by means of a straight line. If the condition is that the algebraic sum of the differences is minimal, then the solution exists but it is not unique. Any straight lines though the median point *B* satisfies it

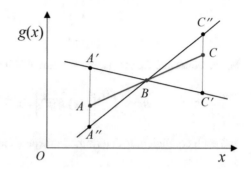

$$\sum_{j=1}^{n} a_j \int_{x_o}^{x_1} \left\{ f_j(x) f_i(x) \right\} dx = \int_{x_o}^{x_1} \left\{ g(x) f_i(x) \right\} dx \qquad (i = 1, \ldots, n),$$

(O.6)

If the system (O.6) can be inverted, i.e., if its determinant is not null, it provides one and only one solution and therefore uniquely determines the value of the parameters a_i (uniqueness does not exist for even integers $m > 2$).

In conclusion, the Gauss or least squares method consists in minimizing the standard deviation σ;

$$(x_1 - x_o) \sigma^2 = \int_{x_o}^{x_1} \left\{ g(x) - \sum_{j=1}^{n} a_j f_j(x) \right\}^2 dx.$$

(O.7)

The applications of this method to mathematics are numerous, but even larger are those to experimental sciences, where the least squares fit is commonly used for the analytic interpolation of measurements.

Before considering this aspect, however, let us return to the system of equations (O.6). It is apparent that the parameters a_i are intimately linked to all the n functions f_i. The consequence is that any modification of the set of functions f_i (for example, by adding or removing one function) affects the values of all the coefficients. This fact bears a practical difficulty: at any such change of the base of functions, we must recalculate the normal system (O.6) and solve it again (a drama when calculations were made by hand). This inconvenience is completely removed if the functions $f_i(x)$ satisfy the condition of orthogonality:

$$\int_{x_o}^{x_1} f_i(x) f_j(x) dx \propto \delta_{ij} \qquad (i, j = 1, \ldots, n).$$

(O.8)

When the proportionality turns into equality, the functions are called orthonormal. Any orthogonal family can be converted into an orthonormal family by dividing each function f_i by the normalization factor:

$$\left[\int_{x_o}^{x_1} \left\{ f_i(x) \right\}^2 dx \right]^{1/2}. \tag{O.9}$$

For orthonormal functions, the solution of (O.6) is reduced to n quadratures:

$$a_i = \int_{x_o}^{x_1} \left\{ g(x) f_i(x) \right\} dx \qquad (i = 1, \ldots, n), \tag{O.10}$$

and each coefficient is determined independently of the others.

O.2 A Family of Orthogonal Polynomials

We now examine a special class of orthogonal functions, the orthogonal polynomials. A system of polynomials $p_n(x)$ of degree n is said to be orthogonal in the interval $a \le x \le b$ with respect to the function $w(x)$, used for standardization purposes, if:

$$\int_a^b w(x) \, p_m(x) \, p_n(x) \, dx = 0, \qquad m \ne n \tag{O.11}$$

Then, placing:

$$\int_a^b w(x) \big[p_n(x) \big]^2 dx = h_n, \tag{O.12}$$

the system of polynomials $p_n(x)/\sqrt{h_n}$ is called orthonormal.

Orthogonal polynomial systems are obtained as solutions of the differential equation:

$$g_2(x) \frac{d^2 p_n(x)}{dx^2} + g_1(x) \frac{dp_n(x)}{dx} + a_n \, p_n(x) = 0. \tag{O.13}$$

The constant parameters a_n, dependent on the integer n, and the functions $g_1(x)$ and $g_2(x)$, independent of n, fix the family. Note that the family of polynomials formed by the derivatives of an orthogonal system retains the characteristics of orthogonality.

The following recurrence formulas apply:

$$p_{n+1}(x) = \big(a_n + x b_n \big) \, p_n(x) - c_n \, p_{n-1}(x), \tag{O.14}$$

as well as the formula of Rodriguez:

$$p_n(x) = \frac{1}{e_n w(x)} \frac{d^n}{dx^n} \left\{ w(x) \big[g(x) \big]^n \right\}. \tag{O.15}$$

The constants a_n, b_n, c_n, and e_n are specific to each family.

O.3 Legendre Polynomials

In Sect. 5.3 we showed how the function (5.25):

$$f(z, \mu) = (1 - 2\mu z + z^2)^{-1/2}, \tag{O.16}$$

can be expanded in powers whose coefficients are the Legendre polynomials:

$$f(z, \mu) = \sum_{k=0}^{\infty} P_k(\mu) z^k. \tag{O.17}$$

This development allowed us to obtain important information on the gravitational potential.

An alternative definition is possible if the Legendre functions are identified with the solutions of the differential equation:

$$\frac{d}{d\mu}\left[(1 - \mu^2)\frac{d}{d\mu}P(\mu)\right] + k(k+1)P(\mu) = 0, \tag{O.18}$$

named after Lagrange. This gives solutions in the form of a convergent series for $|\mu| < 1$. Converging solutions are also found for $\mu = \pm 1$, as long as the index k is a natural integer: in this case the above mentioned Legendre polynomials are found.

Legendre polynomials are orthogonal in the interval[39] $-1 \le \mu \le 1$:

$$\int_{-1}^{1} d\mu P_h(\mu) P_k(\mu) = \frac{2}{2k+1}\delta_{hk}, \tag{O.19}$$

and the Rodriguez formula (O.15) holds, which writes explicitly as:

$$P_k(\mu) = \frac{1}{2^k k!}\frac{d^k}{d\mu^k}\left[(\mu^2 - 1)^k\right]. \tag{O.20}$$

The first polynomials are:

$$\begin{aligned}
P_0(\mu) &= 1, \\
P_1(\mu) &= \mu, \\
P_2(\mu) &= \frac{1}{2}(3\mu^2 - 1), \\
P_3(\mu) &= \frac{1}{2}(5\mu^3 - 3\mu), \\
P_4(\mu) &= \frac{1}{8}(35\mu^4 - 30\mu^2 + 3).
\end{aligned} \tag{O.21}$$

---

[39] In the analysis done in Sect. 5.3, this region also coincided with the range of variability of the parameter μ, given the position: $\mu = \cos\gamma$.

Below we list and demonstrate some of the properties used in the calculus of gravitational potential, therein including the expressions (O.19) and (O.20).

(a) For each value of the index k it is: $P_k(1) = 1$.

In fact, if $\mu = 1$, then:

$$f(z, 1) = (1 - 2z + z^2)^{-1/2} = (1 - z)^{-1}, \tag{O.22}$$

which, expanded in binomial series, gives:

$$f(z, 1) = (1 - z)^{-1} = \sum_{k=0}^{\infty} z^k. \tag{O.23}$$

Equating (O.23) with (5.39), where $\mu = 1$, it is necessarily:

$$P_k(1) = 1. \tag{O.24}$$

(b) For each value of the index k it results: $P_k(-\mu) = (-1)^k P_k(\mu)$.

In fact, it is:

$$f(z, \mu) = \left(1 + 2(-\mu) z + z^2\right)^{-1/2} = \sum_{k=0}^{\infty} P_k(\mu) z^k, \tag{O.25}$$

but also:

$$f(z, \mu) = \left(1 + 2\mu(-z) + (-z)^2\right)^{-1/2} = \sum_{k=0}^{\infty} P_k(-\mu) (-z)^k. \tag{O.26}$$

By equating terms to terms these two series, we have:

$$P_k(\mu) z^k = P_k(-\mu) (-z)^k = (-1)^k P_k(-\mu) z^k, \tag{O.27}$$

and therefore, by eliminating z^k:

$$P_k(-\mu) = (-1)^k P_k(\mu). \tag{O.28}$$

Recalling the condition (O.24), from (O.28) we have as a corollary:

$$P_k(-1) = (-1)^k. \tag{O.29}$$

(c) For each value of the index k it results: $|P_k(\mu)| \leq 1$.

Using the Euler formula, we put:

$$\mu = \cos \gamma = \frac{e^{i\gamma} + e^{-i\gamma}}{2}, \tag{O.30}$$

with which the function (O.16) can be written as:

$$f(z, \mu) = (1 - 2z\mu + z^2)^{-1/2} = (1 - z\,e^{i\gamma})^{-1/2}(1 - z\,e^{-i\gamma})^{-1/2}, \tag{O.31}$$

and, due to (O.17):

$$f(z, \mu) = \sum_{k=0}^{\infty} P_k(\mu)\, z^k = (1 - z\,e^{i\gamma})^{-1/2}(1 - z\,e^{-i\gamma})^{-1/2}. \tag{O.32}$$

Having verified that the convergence condition $|z e^{\pm i\gamma}| < 1$ is satisfied, we expand in series[40] each binomial at the right-hand side of (O.31):

$$f(z, \mu) = \sum_{k=0}^{\infty} P_k(\mu)\, z^k = \sum_{k=0}^{\infty} \binom{-1/2}{k} z^k e^{ik\gamma} \sum_{k'=0}^{\infty} \binom{-1/2}{k'} z^{k'} e^{-ik'\gamma} =$$

$$= \left(1 + \frac{1}{2} z e^{i\gamma} + \frac{3}{8} z^2 e^{2i\gamma} + \dots \right) \times$$

$$\times \left(1 + \frac{1}{2} z e^{-i\gamma} + \frac{3}{8} z^2 e^{-2i\gamma} + \dots \right), \tag{O.33}$$

---

[40] The expansion of the binomial $(a + b)^\alpha$ is:

$$(a + b)^\alpha = \sum_{k=0}^{\infty} \binom{\alpha}{k} a^{\alpha-k} b^k,$$

where:

$$\binom{\alpha}{k} = \frac{\alpha(\alpha - 1)(\alpha - 2) \cdots (\alpha - k + 1)}{k!}.$$

and multiply the two series according to the Cauchy rule.[41] Grouping the coefficients of the same powers of z, we obtain[42]:

$$P_k(\mu) = \frac{(2k)!}{2^{2k} (k)!} \left[2\cos(k\gamma) + \frac{k}{(2k-1)} 2\cos\left((k-2)\gamma\right) + \cdots \right], \qquad (O.34)$$

with the series at the right-hand of (O.34) which ends with term containing $2\cos\gamma$ if k is odd, and with $\cos\left((k-k)\gamma\right) = 1$ if k is even. In other words, the expansion of $f(z, \mu)$ in a power series of z has coefficients given by the following functions:

$$P_k(\mu) = \sum_{h=0}^{n} a_{2h} \cos\left(\gamma(k-2h)\right) \begin{cases} n = \dfrac{k}{2} & \text{if } k \text{ is even,} \\[2mm] n = \dfrac{k-1}{2} & \text{if } k \text{ is odd,} \end{cases} \qquad (O.35)$$

where $a_{2h} \geq 0$ for each value of the index h. When γ is zero, i.e., when $\mu = \cos\gamma = 1$, the (O.35) reduces to:

$$P_k(1) = \sum_{h=0}^{n} a_{2h} = 1, \qquad (O.36)$$

as $P_k(1) = 1$ $\big($property a) on Sect. O.3$\big)$. Therefore:

---

[41] Given two series: $\displaystyle\sum_{k=0}^{\infty} a_k$ and $\displaystyle\sum_{k=0}^{\infty} b_k$, we call Cauchy product their discrete convolution:

$$\left(\sum_{k=0}^{\infty} a_k\right) \cdot \left(\sum_{k=0}^{\infty} b_k\right) = \sum_{k=0}^{\infty} c_n,$$

where:

$$c_n = \sum_{k=0}^{n} \binom{n}{k} a_k \, b_{n-k}.$$

A theorem due to the polish mathematician Franz Mertens (1840–1927) states that, if one series converges to S_a and the other to S_b, and if at least one of them converges absolutely, their Cauchy product converges to the product $S_a S_b$.

[42] This step is quite tricky. To reach the result, one must take into account the fact that, when performing the Cauchy product of the two series of the previous footnote 41, the following condition applies: $c_n = \displaystyle\sum_{k=0}^{\infty} a^k \, b^{n-k} = \sum_{k=0}^{\infty} a^{n-k} \, b^k$, i.e.:

$$c_n = \frac{1}{2} \left(\sum_{k=0}^{\infty} a^k \, b^{n-k} + \sum_{k=0}^{\infty} a^{n-k} \, b^k \right).$$

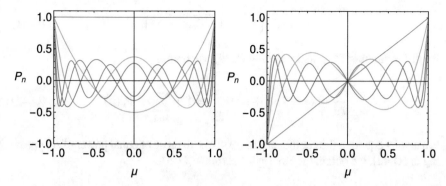

Fig. O.2 Legendre polynomials $P_n(\mu)$ up to the degree $n = 10$: the even polynomials are in left panel and the odd polynomials in the right one

$$|P_k(\mu)| = \left| \sum_{h=0}^{n} a_{2h} \cos\left(\gamma(k-2h)\right) \right| \le$$

$$\le \sum_{h=0}^{n} a_{2h} \left| \cos\left(\gamma(k-2h)\right) \right| \le \sum_{h=0}^{n} a_{2h} = 1, \qquad (O.37)$$

that is, $|P_k(\mu)| \le 1$. The trend of the Legendre polynomials up to the degree $n = 10$ is plotted in Fig. O.2.

(d) <u>For each value of the index k it results</u>: $P_k(\mu) = \dfrac{1}{2^k\, k!} \dfrac{d^k}{d\mu^k} (\mu^2 - 1)^k$.

To prove this property, we first consider the function:

$$\lambda = \mu + \frac{1}{2}\alpha\, (\lambda^2 - 1). \qquad (O.38)$$

Its dependent variable λ is equal to μ if the parameter α is zero. We transform the (O.38) through the Lagrange expansion of formula (F.2); it results:

$$\lambda = \mu + \sum_{k=1}^{\infty} \frac{\alpha^k}{k!} \frac{d^{k-1}}{d\mu^{k-1}} \left(\frac{\mu^2 - 1}{2}\right)^k, \qquad (O.39)$$

and, deriving with respect to μ:

$$\frac{d\lambda}{d\mu} = 1 + \sum_{k=1}^{\infty} \frac{\alpha^k}{k!} \frac{d^k}{d\mu^k} \left(\frac{\mu^2 - 1}{2}\right)^k. \qquad (O.40)$$

We solve now Eq. (O.38) with respect to λ. One solution is:

$$\lambda = \alpha^{-1}\left(1 - \sqrt{1 - 2\mu\alpha + \alpha^2}\right). \tag{O.41}$$

By differentiating for μ and recalling (5.39), we obtain:

$$\frac{d\lambda}{d\mu} = (1 - 2\mu\alpha + \alpha^2)^{-1/2} = 1 + \sum_{k=1}^{\infty} \alpha^k P_k(\mu). \tag{O.42}$$

Finally, by equating the coefficients of (O.40) with those of equal power of α appearing in (O.42), we obtain the Rodriguez formula (O.20).

(e) $P_k(\mu)$ has k and only k distinct zeros in the range $|\mu| < 1$.

Let us consider the Rodrigues formula (O.20). Being a polynomial of degree $2k$, the function $g(\mu) = (\mu^2 - 1)^k$ has $2k$ zeros, k of which at $\mu = -1$ and k at $\mu = 1$. According to the Rolle[43] theorem, its derivative, $dg(\mu)/d\mu$, has at least one zero within the interval $|\mu| < 1$. But, being a polynomial of degree $2k-1$, this derivative has a total of $2k-1$ zeros, of which $k-1$ at $\mu = -1$ and $k-1$ at $\mu = 1$. Therefore the zero internal to the interval $-1 < \mu < 1$ not only exists but is also unique. By the same procedure it is shown that the second derivative, $d^2g(\mu)/d\mu^2$, has two and only two zeros in the range $-1 < \mu < 1$. In general, therefore, $d^k g(\mu)/d\mu^k$ has k and only k zeros between -1 and $+1$. On the basis of Rodrigues' formula, it is then proved that, in the same interval, a Legendre polynomial $P_k(\mu)$ of degree k has k and only k zeros.

It remains to be proven that they are distinct. To this end we show that, if y is a non-trivial solution of the homogeneous linear differential equation:

$$a(x)\frac{d^2y}{dx^2} + b(x)\frac{dy}{dx} + c(x)\,y = 0, \tag{O.43}$$

with $a(x)$, $b(x)$, and $c(x)$, functions indefinitely derivable, it cannot have repeated zeros unless they satisfy the condition $a = 0$. Indeed, if $x_\circ$ is a repeated zero, it must be $y(x_\circ) = \left(\dfrac{dy}{dx}\right)_{x_\circ} = 0$, and therefore, based on (O.43), also $\left(\dfrac{d^2y}{dx^2}\right)_{x_\circ} = 0$, unless $a(x_\circ) = 0$. Furthermore, the derivative of (O.43) respect to x:

$$a(x)\frac{d^3y}{dx^3} + \left(\frac{d}{dx}a(x) + b(x)\right)\frac{d^2y}{dx^2} +$$

$$+ \left(\frac{d}{dx}b(x) + c(x)\right)\frac{dy}{dx} + y\,\frac{d}{dx}c(x) = 0, \tag{O.44}$$

---

[43] Michel Rolle (1652–1719), French mathematician, known for the following theorem: if a function is continuous in a closed interval $[a, b]$, differentiable at any point of the open interval (a, b), and with equal values at the extremes, i.e., $f(a) = f(b)$, there is at least one point inside (a, b) where the first derivative of $f(x)$ is zero.

shows that also the third derivative is null. Iterating the reasoning, it is proved that all the subsequent derivatives are null in x_o. But, based on the Taylor expansion theorem, this implies that the function $y(x)$ is identically null in a whole neighbourhood of x_o. This is therefore a trivial solution. Since, as we demonstrated in Sect. 5.5, the Legendre polynomials are non-trivial solutions of a differential equation of the type (O.43), with $a(x) = 1 - x^2 \neq 0$ in the open interval $(-1, +1)$, then in this same interval they cannot have repeated zeros.

(f) If the positive integers h and k are different from one another, it results:

$$\int_{-1}^{+1} P_h(\mu)\, P_k(\mu)\, d\mu = 0 \qquad (h \neq k). \tag{O.45}$$

If $h \neq k$, let $h > k$. By means of the Rodrigues formula:

$$I_{hk} = \int_{-1}^{+1} P_h(\mu) P_k(\mu)\, d\mu =$$

$$= \frac{1}{2^{h+k} h! k!} \int_{-1}^{+1} \frac{d^h}{d\mu^h} (\mu^2 - 1)^h \frac{d^k}{d\mu^k} (\mu^2 - 1)^k\, d\mu, \tag{O.46}$$

which, integrated by parts, provides:

$$I_{hk} = \frac{1}{2^{h+k} h! k!} \frac{d^{h-1}}{d\mu^{h-1}} (\mu^2 - 1)^h \frac{d^k}{d\mu^k} (\mu^2 - 1)^k \Big|_{-1}^{+1} +$$

$$- \frac{1}{2^{h+k} h! k!} \int_{-1}^{+1} \frac{d^{h-1}}{d\mu^{h-1}} (\mu^2 - 1)^h \frac{d^{k+1}}{d\mu^{k+1}} (\mu^2 - 1)^k\, d\mu. \tag{O.47}$$

The first element in the second term of (O.47) is cancelled by the presence of the factor $(\mu^2 - 1)$ in the $(h-1)$-th derivative of $(\mu^2 - 1)^h$. By repeating this operation $k - 1$ times, we obtain the expression:

$$I_{hk} = \frac{(-1)^k}{2^{h+k} h! k!} \int_{-1}^{+1} \frac{d^{h-k}}{d\mu^{h-k}} (\mu^2 - 1)^h \frac{d^{2k}}{d\mu^{2k}} (\mu^2 - 1)^k\, d\mu, \tag{O.48}$$

which integrates in:

$$I_{hk} = \frac{(-1)^k (2k)!}{2^{h+k} h! k!} \frac{d^{h-k-1}}{d\mu^{h-k-1}} (\mu^2 - 1)^h \Big|_{-1}^{+1}, \tag{O.49}$$

since $\dfrac{d^{2k}}{d\mu^{2k}} (\mu^2 - 1)^k = (2k)!$. Note that this relation holds because $(\mu^2 - 1)^k$ is a polynomial of degree $2k$.

It is easy to prove that I_{hk} is null. In fact, having assumed $h > k$, it is also $h - k - 1 \geq 0$. Therefore, the formal derivative of $(\mu^2 - 1)^h$ contains the factor $(\mu^2 - 1)$, which is zero at the extremes of integration. The property just proved can be summarized by saying that Legendre's polynomials form a family of orthogonal polynomials.

As a corollary, we observe that, if $h = k$, the (O.48) gives:

$$I_{kk} = \int_{-1}^{+1} \left[P_k(\mu) \right]^2 d\mu = \frac{(-1)^k (2k)!}{2^{2k} (k!)^2} \int_{-1}^{+1} (1 - \mu^2)^k \, d\mu, \tag{O.50}$$

from which, placing $\mu = 2x - 1$, it is:

$$\begin{aligned}
I_{kk} &= \int_{-1}^{+1} \left[P_k(\mu) \right]^2 d\mu = \frac{2(-1)^k (2k)!}{(k!)^2} \int_0^{+1} x^k (1-x)^k \, dx = \\
&= \frac{2(2k)!}{(k!)^2} B(k+1, k+1) = \frac{2(2k)!}{(k!)^2} \frac{\Gamma(k+1)\Gamma(k+1)}{\Gamma(2(k+1))} = \\
&= \frac{2(2k)!}{(k!)^2} \frac{\Gamma(k+1)\Gamma(k+1)}{\Gamma((2k+1)+1)} = \frac{2}{2k+1},
\end{aligned} \tag{O.51}$$

where B is the Beta function[44] (see Sects. N.1 and N.2). It follows immediately that:

$$\int_0^{+1} \left[P_k(\mu) \right]^2 d\mu = \frac{1}{2k+1}. \tag{O.52}$$

It can be proven that the moments of order $h \geq k$ of the Legendre polynomials are:

$$\int_0^{+1} \mu^h P_k(\mu) d\mu = \frac{h(h-1)(h-2) \cdots (h-k+2)}{(h+k+1)(h+k-1) \cdots (h-k+3)}. \tag{O.53}$$

g) Recurrence formula for Legendre function.

$$(k+1) P_{k+1}(\mu) - (2k+1)\mu P_k(\mu) + k P_{k-1}(\mu) = 0, \tag{O.54}$$

$$\mu \frac{d P_k(\mu)}{d\mu} - \frac{d P_{k-1}(\mu)}{d\mu} = k P_k(\mu), \tag{O.55}$$

$$\frac{d P_{k+1}(\mu)}{d\mu} - \frac{d P_{k-1}(\mu)}{d\mu} = (2k+1) P_k(\mu), \tag{O.56}$$

$$\frac{d P_{k+1}(\mu)}{d\mu} - \mu \frac{d P_k(\mu)}{d\mu} = (k+1) P_k(\mu), \tag{O.57}$$

$$(1 - \mu^2) \frac{d P_k(\mu)}{d\mu} = k P_{k-1}(\mu) - k\mu P_k(\mu). \tag{O.58}$$

---

[44] We may obtain the same result by integrating by parts k times.

In Sect. 5.5.3 we showed how the associated Legendre equation (5.82), which we rewrite here for clarity:

$$\frac{d}{d\mu}\left[(1-\mu^2)\frac{dy(\mu)}{d\mu}\right]+\left[n(n+1)-\frac{m^2}{1-\mu^2}\right]y(\mu)=0, \tag{O.59}$$

is solved by functions which depend on Legendre polynomials, called associated Legendre functions, defined through the Rodrigues formula:

$$P_n^m(\mu)=\frac{(\mu^2-1)^{m/2}}{2^n n!}\frac{d^{m+n}}{d\mu^{m+n}}(\mu^2-1)^n. \tag{O.60}$$

Note how these functions become Legendre polynomials for $m=0$. From the comparison with that for the Legendre polynomials, this expression proves that, apart from the multiplicative factor, $P_n^m(\mu)$, it is a Legendre polynomial of degree $(n-m)$. We will now show that these functions satisfy a relations of orthogonality. In fact, since both P_n^m and $P_{n'}^m$ are solutions of (O.59), it must be:

$$\frac{d}{d\mu}\left[(1-\mu^2)\frac{dP_n^m(\mu)}{d\mu}\right]+\left[n(n+1)-\frac{m^2}{1-\mu^2}\right]P_n^m(\mu)=0,$$
$$\frac{d}{d\mu}\left[(1-\mu^2)\frac{dP_{n'}^m(\mu)}{d\mu}\right]+\left[n'(n'+1)-\frac{m^2}{1-\mu^2}\right]P_{n'}^m(\mu)=0. \tag{O.61}$$

Subtracting the second equation, multiplied by $P_n^m(\mu)$, from the first one, multiplied by $P_{n'}^m(\mu)$, and integrating in the limits -1 and $+1$, we have:

$$(n-n')(n+n'+1)\int_{-1}^{+1}P_n^m(\mu)\,P_{n'}^m(\mu)\,d\mu+$$
$$+\int_{-1}^{+1}\left[P_{n'}^m(\mu)\frac{d}{d\mu}\left((1-\mu^2)\frac{dP_n^m(\mu)}{d\mu}\right)+\right.$$
$$\left.-P_n^m(\mu)\frac{d}{d\mu}\left((1-\mu^2)\frac{dP_{n'}^m(\mu)}{d\mu}\right)\right]d\mu=0. \tag{O.62}$$

Let us consider the second integral of this equation. Integrating by parts, we obtain:

$$\int_{-1}^{+1}\frac{d}{d\mu}\left[(1-\mu^2)\left(P_n^m(\mu)\frac{d}{d\mu}P_{n'}^m(\mu)-P_{n'}^m(\mu)\frac{d}{d\mu}P_n^m(\mu)\right)\right]d\mu=0, \tag{O.63}$$

which is zero because the multiplicative term $(1-\mu^2)$ vanishes at the limits of integration. It is therefore proven that:

$$(n-n')(n+n'+1)\int_{-1}^{+1}P_n^m(\mu)P_{n'}^m(\mu)\,d\mu=0. \tag{O.64}$$

In conclusion, if $n' \neq n$, the only way for (O.64) to be satisfied is that (5.99) holds. As a corollary, we give the following important result without demonstration:

$$\int_{-1}^{+1} \left[P_n^m(\mu) \right]^2 d\mu = (-1)^m \frac{(n+m)!}{(n-m)!} \frac{2}{2n+1}. \tag{O.65}$$

Summarizing, within the open interval $-1 < \mu < +1$, the functions (5.97) are solutions of the associated Legendre equation for integer values of the non-negative parameters n and m. Furthermore, since from (5.98) it turns out that $P_n^m \equiv 0$ for $m > n$, we can say that these solutions are non-trivial only if $m \leq n$. In conclusion, we have demonstrated that the associated Legendre polynomials form a family of orthogonal polynomials.

O.4 Spherical Harmonics

In Sect. 5.5 we introduced the concept of spherical harmonic. This appendix completes what was anticipated there.

Spherical harmonics (cf. [12]) are obtained as solutions of the angular part of the Laplace equation, that we rewrite here in spherical coordinates for convenience:

$$\nabla^2 f = \frac{1}{r^2} \frac{\partial}{\partial r} \left(r^2 \frac{\partial f}{\partial r} \right) + \frac{1}{r^2 \sin \theta} \frac{\partial}{\partial \theta} \left(\sin \theta \frac{\partial f}{\partial \theta} \right) + \frac{1}{r^2 \sin^2 \theta} \frac{\partial^2 f}{\partial \varphi^2} = 0. \tag{O.66}$$

The solutions of this equation, called U_n in Sect. 5.5 (but frequently met in the literature with the symbol Y_ℓ^m), are the product of a trigonometric function with the solution of the associated Legendre equation. In particular, there will be a solution such as (5.83):

$$U_n^m(\theta, \phi) = \left[A \cos(m\phi) + B \sin(m\phi) \right] T_n^m(\cos \theta), \tag{O.67}$$

known as spherical harmonic of degree n and order m, or another one having a more compressed form:

$$Y_\ell^m(\theta, \phi) = N e^{im\phi} P_\ell^m(\cos \theta). \tag{O.68}$$

When the Laplace equation is solved on the sphere, then it can be easily shown that n and m must satisfy the conditions $n \geq 0$ and $|m| \leq n$ (see also Appendix I).

O.4.1 Harmonics Representation

Going back to the formalism used in (5.83) and (O.67), we have shown in Sect. 5.5 that spherical harmonics can be classified into three subclasses: zonal ($m = 0$), tesseral ($0 < m < n$), and sectoral ($m = n$). We will now prove that a spherical harmonic of degree n is always a linear combination of the corresponding zonal, tesseral, and sectoral harmonics.

To this end, we start considering a homogeneous polynomial $V_n(x, y, z)$ of degree n in x, y, and z. Generally it contains $\frac{1}{2}(n+1)(n+2)$ arbitrary constants. But, if it is a solid spherical harmonic, i.e., a solution of the Laplace equation $\nabla^2 V_n = 0$, the number of arbitrary constants reduces to $(2n+1)$. In fact, the function $\nabla^2 V_n$ is itself a polynomial, but of degree $(n-2)$, with $\frac{1}{2}n(n-1)$ coefficients which are linear combinations of those of V_n and, for the condition posed by the Laplace equation, are null. In conclusion, if $V_n(x, y, z)$ is a solid spherical harmonic, only $\frac{1}{2}\left[(n+1)(n+2) - n(n-1)\right] = (2n+1)$ of its coefficients are linearly independent and can therefore be chosen arbitrarily. Consequently, $V_n(x, y, z)$ can be written in the general form:

$$V_n(x, y, z) = \sum_{k=1}^{2n+1} a_k \, (X_n)_k, \tag{O.69}$$

where a_k are constants. The $(2n + 1)$ functions $(X_n)_k$ are solid spherical harmonics of degree n. To prove it, we just cancel all the coefficients a_k but one. They are also linearly independent. If this were not the case, V_n would depend on a number of arbitrary constants less than $(2n+1)$.

We now denote by $\left(Y_n(\theta, \phi)\right)_k$ the surface spherical harmonic corresponding to $(X_n)_k$:

$$\left(X_n(x, y, z)\right)_k = r^n \left(Y_n(\theta, \phi)\right)_k. \tag{O.70}$$

Substituting in (O.69) we obtain:

$$r^{-n} V_n(x, y, z) = U_n(\theta, \phi) = \sum_{k=1}^{2n+1} a_k \left(Y_n(\theta, \phi)\right)_k, \tag{O.71}$$

where $U_n(\theta, \phi)$ is the surface spherical harmonic corresponding to V_n. Due to the polynomial nature of the latter, it is easy to realize that $U_n(\theta, \phi)$ is a homogeneous polynomial of integer degree n in the variables $\cos\phi \sin\theta$, $\sin\phi \sin\theta$, and $\cos\theta$. This structure is characteristic of each surface spherical harmonic of integer degree (5.101). The expression (O.69) shows that any spherical harmonic of integer degree n can always be a linear combination of $(2n+1)$ arbitrary spherical surface harmonics of equal degree n, provided that they are linearly independent of each other.

Based on this result, we now consider the following $(2n+1)$ harmonics:

$$\left(Z_n(\theta, \phi)\right)_1 = P_n(\cos\theta),$$

$$\cdots\cdots\cdots = \cdots\cdots\cdots$$

$$\left(Z_n(\theta, \phi)\right)_{1+m} = \cos(m\phi)\, T_n^m(\cos\theta), \qquad (O.72)$$

$$\cdots\cdots\cdots = \cdots\cdots\cdots$$

$$\left(Z_n(\theta, \phi)\right)_{1+n+m} = \sin(m\phi)\, T_n^m(\cos\theta),$$

$$\cdots\cdots\cdots = \cdots\cdots\cdots$$

which are obtained from (O.67) using the fact that the tesseral and sectoral harmonics have two arbitrary constants, A and B. For what it has been said, we may write $(2n+1)$ equations as:

$$\left(Z_n(\theta, \phi)\right)_j = \sum_{k=1}^{2n+1} (b_k)_j \left(Y_n(\theta, \phi)\right)_k \qquad (j = 1, \ldots, 2n+1). \qquad (O.73)$$

The system (O.72) can be solved because the functions $(Z_n)_j$ are linearly independent. We obtain:

$$\left(Y_n(\theta, \phi)\right)_k = \sum_{h=1}^{2n+1} (c_k)_h \left(Z_n(\theta, \phi)\right)_h \qquad (k = 1, \ldots, 2n+1), \qquad (O.74)$$

and therefore, substituting in (O.71):

$$U_n(\theta, \phi) = \sum_{k=1}^{2n+1} \left[a_k \left(\sum_{h=1}^{2n+1} (c_k)_h (Z_n)_h \right) \right] =$$

$$= \sum_{h=1}^{2n+1} (Z_n)_h \left(\sum_{k=1}^{2n+1} a_k\, (c_k)_h \right) = \sum_{h=1}^{2n+1} d_h \left(Z_n\right)_h =$$

$$= A_\circ P_n(\cos\theta) + \sum_{m=1}^{n} \left[A_m \cos(m\phi) + B_m \sin(m\phi) \right] T_n^m(\cos\theta), \qquad (O.75)$$

where $A_\circ$, A_m and B_m are constants. In other words, any surface spherical harmonic of integer degree n can be expressed as a linear combination of the corresponding zonal, tesseral, and sectoral harmonics. In compact notation:

$$U_n(\theta, \phi) = A_\circ P_n(\cos\theta) + \sum_{m=1}^{n} \frac{\cos(m\phi)}{\sin(m\phi)} T_n^m(\cos\theta), \qquad (O.76)$$

which is what we wanted to prove.

O.4.2 Orthogonality of Spherical Harmonics

The expression (O.76) allows us to show that two generic spherical harmonics of integer surface, $X_n(\theta, \phi)$ and $Y_{n'}(\theta, \phi)$, satisfy the condition of orthogonality:

$$\int_0^{2\pi} \int_0^{\pi} X_n(\theta, \phi)\, Y_{n'}(\theta, \phi)\, d(\cos\theta)\, d\phi = 0, \tag{O.77}$$

if $n' \neq n$. In fact, by developing the product through (O.76), we obtain $n(2n+1)$ integrals of the type:

$$\int_0^{2\pi} \int_0^{\pi} \frac{\cos(k\phi)}{\sin(k\phi)}\, T_{n'}^k(\cos\theta)\, \frac{\cos(h\phi)}{\sin(h\phi)}\, T_n^h(\cos\theta)\, d(\cos\theta)\, d\phi, \tag{O.78}$$

which, when $h = k$, are null because of (5.99). For $k \neq h$, they are equally zero because:

$$\int_0^{2\pi} \cos(k\phi)\, \cos(h\phi)\, d\phi = \int_0^{2\pi} \sin(k\phi)\, \sin(h\phi)\, d\phi = 0, \tag{O.79}$$

and in general:

$$\int_0^{2\pi} \sin(k\phi)\, \cos(h\phi)\, d\phi = 0. \tag{O.80}$$

If instead the two spherical harmonics have the same degree n, it can be shown that:

$$\int_0^{2\pi} \int_0^{\pi} X_n(\theta, \phi) X'_n(\theta, \phi)\, d(\cos\theta)\, d\phi =$$

$$= \frac{2\pi}{2n+1}\left[2A_\circ A'_\circ + \sum_{m=1}^{n} \frac{(n+m)!}{(n-m)!}\left(A_m A'_m + B_m B'_m \right) \right]. \tag{O.81}$$

In particular, if X'_n is a zonal harmonic P_n, i.e., if all the coefficients A'_m and B'_m but $A'_\circ = 1$ are null, the relation (O.81) becomes:

$$\int_0^{2\pi} \int_0^{\pi} X_n(\theta, \phi) P_n(\cos\theta)\, d(\cos\theta)\, d\phi = \frac{4\pi}{2n+1} A_\circ = \frac{4\pi}{2n+1} X_n(1), \tag{O.82}$$

since, at the pole, $T_n^m(\mu = 1)$ is cancelled due to the presence of the factor $(1 - \mu^2)^{1/2}$, and therefore $X_n(1) = A_\circ$.

O.4.3 Addition Theorem

Let γ be the angle between the radius vectors of two generic points $P(r, \theta, \phi)$ and $Q(r', \theta', \phi')$ relative to the origin of the reference system: $\gamma = \arccos\left(\mathbf{r}_P \cdot \mathbf{r}_Q\right)$ (see Fig. 5.2 at Sect. 5.3). Recalling (5.24):

$$\cos\gamma = \cos\theta\,\cos\theta' + \sin\theta\,\sin\theta'\,\cos(\phi - \phi'), \qquad (O.83)$$

we consider a Legendre polynomial of argument $\cos\gamma$. At Sect. 5.5 we have shown that $P_n(\cos\gamma)$, being a solution of equation (5.76), is a spherical surface harmonic of degree n; it can thus be developed in zonal, sectoral, and tesseral harmonics:

$$P_n(\cos\gamma) = A_\circ P_n(\cos\theta) + \sum_{k=1}^{n}\left[A_k\,\cos(k\phi) + B_k\,\sin(k\phi)\right] T_n^k(\cos\theta). \quad (O.84)$$

We want to prove the so-called addition theorem: the coefficients A_m and B_m are given by:

$$A_m = 2\,\frac{(n-m)}{(n+m)} \times \cos(m\phi')\,T_n^m(\cos\theta'), \qquad (O.85a)$$

$$B_m = 2\,\frac{(n-m)}{(n+m)} \times \sin(m\phi')\,T_n^m(\cos\theta'). \qquad (O.85b)$$

We multiply both terms of (O.84) by $\cos(m\phi)\,T_n^m(\cos\theta)$, and integrate on the unit sphere. Taking advantage of both the orthogonality of the Ferrers functions and the relations (O.65) and (5.76), and noticing that for every integer and non-null m it is:

$$\int_0^{2\pi} \cos^2(m\phi)\,d\phi = \pi, \qquad (O.86)$$

we have:

$$\int_0^{2\pi}\int_0^{\pi} \cos(m\phi)\,P_n(\cos\gamma)\,T_n^m(\cos\theta)\,d(\cos\theta)\,d\phi =$$

$$= A_m \int_0^{2\pi} \cos^2(m\phi)\,d\phi \int_0^{\pi} \left[T_n^m(\cos\theta)\right]^2 d(\cos\theta) =$$

$$= \frac{2\pi}{2n+1}\frac{(n+m)!}{(n-m)!} A_m. \qquad (O.87)$$

In order to explicit the first term inside the integral, we rotate the reference system, making the new z-axis coincide with the radius vector of the point Q, whose old angular coordinates are θ' and ϕ' (in this case the angle γ becomes the new coordinate

θ). We denote the new angular coordinates with the symbols ξ and η. Placed for convenience: $Y_n^m(\theta, \phi) = \cos(m\phi) T_n^m(\cos\theta)$, through (O.82) we have immediately:

$$\int_0^{2\pi}\int_0^{\pi} Y_n^m(\cos\theta, \phi) \, P_n(\cos\gamma) \, d(\cos\theta) \, d\phi =$$

$$= \int_0^{2\pi}\int_0^{\pi} Y_n^m(\xi, \eta) \, P_n(\cos\xi) \, d(\cos\xi) \, d\eta = \frac{4\pi}{2n+1} \, Y_n^m(1). \qquad (O.88)$$

Since in the new coordinate system the pole is the point Q, we have:

$$Y_n^m(1) = Y_n^m(\theta', \phi') = \cos(m\phi') \, T_n^m(\cos\theta'). \qquad (O.89)$$

By replacing (O.88) and (O.89) in (O.87), we obtain the first of the (O.85a); the second is readily obtained by a similar procedure.

In conclusion, the addition theorem proves that:

$$P_n(\cos\gamma) = P_n(\cos\theta) P_n(\cos\theta') +$$

$$+ \sum_{m=1}^{n} 2\frac{(n-m)!}{(n+m)!} \cos\left(m(\phi - \phi')\right) T_n^m(\cos\theta) \, T_n^m(\cos\theta'), \qquad (O.90)$$

if the angle γ is given by (O.83).

O.5 An Application of Spherical Harmonics

A modern application of spherical harmonics to astrophysics concerns the study of the anisotropies of cosmic background radiation (Cosmic Microwave Background = CMB): one of the tracers of the primordial properties of the universe capable of providing estimates of the cosmological parameters. The CMB, whose intensity peak falls in the microwave domain,[45] pervades the whole universe. Its spectrum is well described by a black body at a temperature of about 2.73 °K.

On large angular scales, the background radiation is isotropic, but it has some anisotropies[46] at some smaller scales. These are temperature fluctuations $C = \Delta T/T$ that must be measured by constructing the power spectrum. The operation is equivalent to expanding C in a Fourier series; or better, since the observational data are

---

[45] The microwaves cover a wavelength range from 1 mm to about 1 m, with frequencies between 0.3 and 300 GHz. The peak of the CMB is at 160.2 GHz, or 1.9 mm.

[46] Several observation campaigns have been launched to measure these anisotropies including Boomerang [13], COBE, WMAP [14], and Planck [15]. In 2006 the principal investigators of COBE, George Smoot and John Mather, received he Nobel Prize in physics for the discovery of CMB anisotropy.

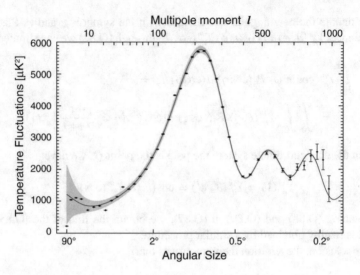

Fig. O.3 The CMB angular power spectrum. Credit: NASA/WMAP Science Team

collected on a sphere (the celestial vault), the expansion is done in terms of the angular variables:

$$\frac{\Delta T}{T} = \sum_l \sum_m a_{lm} Y_{lm}(\theta, \phi), \qquad (O.91)$$

where $l = 0$ indicates the monopole moment, $l = 1$ the dipole moment, $l = 2$ the quadrupole moment, and so on.[47] The different m for the same l correspond to different orientations with the same angular scale. It is customary to define the quantity[48] $C_l = \sum_m |a_{lm}|^2$ and represent the quantity $l(l + 1)C_l$ as a function of l, since the latter is approximately equal to the power per unit logarithmic interval of l:

$$\langle \frac{\delta T}{T} \rangle = \sum_l \frac{2l + 1}{4\pi} C_l. \qquad (O.92)$$

The final result is a function with primary and secondary peaks for different values of l, determined by the geometry of the universe. Therefore, the shape of this spectrum, and in particular the amplitude and the position of the peaks, allows to extract information on the cosmological parameters and on the spectrum of the initial perturbations of the universe, representing one of the most powerful tools of observational cosmology (see Fig. O.3).

---

[47] The multipole l is the counterpart of the angle θ, and the two are linked by a rough relation: $\theta \sim \pi/l$; therefore, as the angular scale increases, we are interested in smaller monopole moments.

[48] This is the average with respect to all observers in the universe; since the universe is homogeneous and isotropic, there are no preferred observing direction.

Appendix P
Some Theorems on Harmonic Functions

Heraclitus

The hidden harmony is better than the obvious.

The solid spherical harmonics, introduced at Sect. 5.5.1, are positively homogeneous integer functions which are also solutions of the Laplace equation. Here we present some theorems related to the complete family of solutions of the Laplace equation, still called harmonic functions. We require them to be regular within the domain D, limited by the surface L. The regularity of the function for the generic harmonic function $X(x, y, z)$ implies that it has:

1. first partial derivatives continuous within D and over L;
2. second partial derivatives continuous within D and limited on L.

In conclusion, it is essentially required that X is uniform in D.

P.1   Selected Theorems

Theorem No. 1

The flow of the gradient of a harmonic function X through the fixed surface S contained in D is zero:

$$\iint_S \nabla X \cdot d\mathbf{S} = 0. \tag{P.1}$$

By applying the Gauss theorem (Sect. 5.1) to the function ∇X, we have:

$$\iint_S \nabla X \cdot d\mathbf{S} = \iiint_\Sigma \nabla \cdot \nabla X \, d\Sigma, \tag{P.2}$$

E. Bannikova and M. Capaccioli, *Foundations of Celestial Mechanics*, Graduate Texts in Physics, https://doi.org/10.1007/978-3-031-04576-9

365

where Σ is the region enclosed by the surface S. Taking into account that, by assumption, X is a spherical harmonic, it is $\nabla \cdot \nabla X = \nabla^2 X = 0$, within the domain D, and consequently everywhere in Σ. Therefore, the second integral term of (P.2) is zero, with which the theorem is proven.

Theorem No. 2

If two harmonic functions X_1 and X_2 coincide in all points of the boundary S of a region inside the intersection of the domains D_1 and D_2, they are equal in all points of Σ.

It is sufficient to prove that the function $X = X_2 - X_1$ is null everywhere in Σ. From the Green[49] theorem applied to the function X^2, we have:

$$\iiint_\Sigma \nabla \cdot \nabla X^2 \, d\Sigma = \iint_S \nabla X^2 \cdot d\mathbf{S}. \tag{P.3}$$

On the surface S it results:

$$\nabla X^2 = 2 X \nabla X \equiv 0, \tag{P.4}$$

since $X \equiv 0$ by hypothesis. Furthermore, at any point of Σ it is:

$$\nabla \cdot \nabla X^2 = 2 \nabla X \cdot \nabla X + 2 X \cdot \nabla^2 X = 2 \left(\nabla X \right)^2, \tag{P.5}$$

as $\nabla^2 X = \nabla^2 (X_2 - X_1) = \nabla^2 X_2 - \nabla^2 X_1 \equiv 0$ within Σ. Therefore, the equality (P.3) becomes:

$$\iiint_\Sigma \left(\nabla X \right)^2 d\Sigma = 0, \tag{P.6}$$

which is satisfied only if the non-negative integrand function:

$$\nabla X = \frac{\partial X}{\partial x} + \frac{\partial X}{\partial y} + \frac{\partial X}{\partial z}, \tag{P.7}$$

vanishes identically in Σ. This is possible only if the partial derivatives of X are null everywhere in Σ, i.e., if the function X is constant there. Since by assumption X is null in S, it must be null at every point of Σ.

---

[49] George Green (1793–1841): English mathematician and physicist. Almost self-taught, he was a pioneer in exploring the theory of potential, anticipating the subsequent work of Maxwell and Thomson.

Fig. P.1 Scheme for theorem no. 6

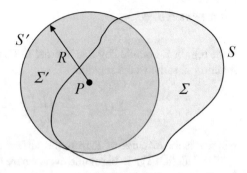

Theorem No. 3

If the harmonic function X_1 is null on the surface S which delimits the region Σ, it is null everywhere inside Σ.

Let X_2 be an identically null function in D_1. It is clearly a spherical harmonic; it also coincides with X_1 on the contour S and therefore, according to the previous theorem, at each point of Σ.

Theorem No. 4

If the gradients of the harmonic functions X_1 and X_2 coincide on the contour S of the region Σ inside the intersection of the domains D_1 and D_2, at any point of Σ the two functions differ at most by an additive constant.

As for theorem 4, it is proved that the first partial derivatives of the function $X = X_2 - X_1$ are null everywhere in Σ. In this case the relation (P.4) is satisfied because, by assumption, $\nabla X = 0$ on the whole surface S. The equality of the derivatives of X proves the constancy of the function; but the assumptions do not imply that the constant also vanishes.

Theorem No. 5

If the gradient of the harmonic function X_1 is null on the contour S of Σ, the function itself is constant at every point of Σ.

The gradient of a constant function X_2 is null everywhere, so it coincides with that of X_1 on S. According to the previous theorem, X_1 differs from the constant X_2 at most by an additive constant throughout Σ, and it is therefore itself a constant.

Theorem No. 6

If the region Σ bounded by S is internal to D, at each point P inside Σ the harmonic function X satisfies the relation:

$$X(P) = \frac{1}{4\pi} \iint_S \left(X\nabla\frac{1}{r} - \frac{1}{r}\nabla X \right) \cdot d\mathbf{S}, \tag{P.8}$$

where r is the distance of P from the surface element $d\mathbf{S}$.

The function $1/r$ is harmonic everywhere but at the point P where it is singular. Let S' be the surface of a sphere of volume Σ', radius R, and center P, inside Σ (Fig. P.1). We apply the Green theorem (no. 2) to the difference between the functions X and $1/r$ within the region limited by the surfaces S and S', i.e., within the region Σ deprived of that occupied by the sphere of radius R. It is:

$$\iint_{S-S'} \left(X\nabla\frac{1}{r} - \frac{1}{r}\nabla X \right) \cdot d\mathbf{S} =$$

$$= -\iiint_{\Sigma-\Sigma'} \left(X\nabla \cdot \nabla\frac{1}{r} - \frac{1}{r}\nabla \cdot \nabla X \right) d\Sigma = 0, \tag{P.9}$$

as $\nabla \cdot \nabla\left(1/r\right) = 0$ and $\nabla \cdot \nabla X \equiv 0$ everywhere in Σ. Considering that the vector $d\mathbf{S}$ is oriented outwards on S and inwards on S', the equality (P.9) allows to write:

$$\iint_S \left(X\nabla\frac{1}{r} - \frac{1}{r}\nabla X \right) \cdot d\mathbf{S} = \iint_{S'} \left(X\nabla\frac{1}{r} - \frac{1}{r}\nabla X \right) \cdot d\mathbf{S}. \tag{P.10}$$

On S' (sphere of radius R) the functions $\frac{1}{r}$ and $\nabla\frac{1}{r} \cdot \frac{d\mathbf{S}}{dS}$ are constants and equal to $\frac{1}{R^2}$ and $\frac{1}{R}$ respectively, and therefore the integral at the second term of (P.10) becomes:

$$\frac{1}{R^2} \iint_{S'} X \, dS - \frac{1}{R} \iint_{S'} \nabla X \cdot d\mathbf{S} =$$

$$= 4\pi \langle X(S') \rangle - 4\pi R \langle \left| \nabla X(S') \right|_n \rangle, \tag{P.11}$$

where the symbol $\langle \rangle$ indicates the average value of the function X and of the radial component of the gradient on the spherical surface S'. Substituting in (P.11) and (P.10) and passing to the limit for $R \to 0$, we finally obtain the formula (P.9).

Theorem No. 7

The value of the harmonic function X at any point P inside the domain D is equal to the average value that the function assumes on the surface and on the entire volume of any sphere of center P, provided that it is internal to D.

Apply the formula (P.9) to the surface S of a sphere of center P and generic radius $r \leq R$:

$$X(P) = \frac{1}{4\pi} \iint_S \left(X \frac{\mathbf{r}}{r} - \frac{1}{r} \nabla X \right) \cdot d\mathbf{S} =$$

$$= \frac{1}{4\pi r^2} \left\{ \iint_S X \, d\mathbf{S} - r \iint_S \nabla X \cdot d\mathbf{S} \right\}, \tag{P.12}$$

and, based on the theorem no. 1:

$$X(P) = \frac{1}{4\pi} \iint_S X \, d\mathbf{S} = \frac{\langle X(S) \rangle}{4\pi r^2} \iint_S d\mathbf{S} = \langle X(S) \rangle. \tag{P.13}$$

This shows that the value assumed by X in P is equal to the average value of the function on the surface of the sphere S. Multiplying both terms of (P.13) by dr, and integrating between 0 and R, we have:

$$\int_0^R X(P) \, dr = \frac{1}{4\pi} \iiint_\Sigma \frac{X}{r^2} \, d\Sigma, \tag{P.14}$$

that is:

$$\int_0^R X(P) \, dr = \frac{\langle X(\Sigma) \rangle}{4\pi} \iiint_\Sigma \frac{d\Sigma}{r^2} =$$

$$= \frac{\langle X(\Sigma) \rangle}{4\pi} \int_0^{2\pi} d\phi \int_0^\pi \sin\theta \, d\theta \int_0^R dr = R \langle X(\Sigma) \rangle. \tag{P.15}$$

Theorem No. 8

If X is harmonic and regular in the domain D, it has no relative maximum or minimum within D.

If P is a point internal to D where X has a maximum, there must be a subset D' about P such that at all the points Q, but P itself, the inequality $X(P) > X(Q)$ holds. Let Σ be a sphere of surface S and center P, entirely contained in D'. Then at all points of S, $X < X(P)$. But this is absurd because, owing to the previous theorem, $X(P) = \langle X(S) \rangle$. You can similarly prove the theorem for the minima.

P.2 Special Problems

In conclusion, we recall two famous problems dealing with the search for solutions of the Laplace equation under special boundary conditions. They are important for searching of the gravitational potential (see, as an example, Sect. 5.4).

Dirichlet Problem

We search for a harmonic and regular function in the domain D of frontier S that satisfies the condition of assuming values assigned continuously on S.

If such a function exists, theorem no. 2 ensures that it is unique. In fact, using the theory of integral equations, it can be proven that the problem of Dirichlet has a solution under convenient hypotheses on the domain D. We now show the nature of these solutions.

Let $X_\circ$ be a continuous function defined in S, which is the searched spherical harmonic. By the theorem no. 6, at any point P inside D:

$$X(P) = \frac{1}{4\pi} \iint_S \left(X_\circ \nabla \frac{1}{r} - \frac{1}{r} \nabla X_\circ \right) \cdot d\mathbf{S}. \tag{P.16}$$

Let g be a harmonic and regular function in D. By applying the Green (theorem no. 2 above) to the difference between X and g, we have:

$$\iint_S \left(X_\circ \nabla g - g \nabla X_\circ \right) \cdot d\mathbf{S} = 0. \tag{P.17}$$

The difference between (P.16) and (P.17) gives:

$$X(P) = \frac{1}{4\pi} \iint_S \left\{ X_\circ \left(\nabla \frac{1}{r} - \nabla g \right) - \left(\frac{1}{r} - g \right) \nabla X_\circ \right\} d\mathbf{S}. \tag{P.18}$$

Therefore, if the harmonic function g coincides with $1/r$ on S, the expression (P.18) becomes:

$$X(P) = \frac{1}{4\pi} \iint_S X_\circ \nabla \left(\frac{1}{r} - g \right) \cdot d\mathbf{S} = \frac{1}{4\pi} \iint_S X_\circ \nabla G \cdot d\mathbf{S}, \tag{P.19}$$

with the position $G = \frac{1}{r} - g$. This expression solves the Dirichlet problem as long as the function G is known. The latter is called Green function, and it is clearly a function of the points P inside D and the points Q belonging to S. Furthermore, it is harmonic and regular everywhere in D but in P, and shares the reciprocity relation: $G(P, Q) = G(Q, P)$. We can say that the latter property demonstrates the interchangeability between source and object point, i.e., the interchangeability between cause and effect.

Neumann[50] **problem.** We look for a harmonic and regular function in the domain D bordered by S, when the gradient on S is given.

According to theorem no. 1, this problem does not admit a solution when:

$$\iint \nabla X_\circ \cdot d\mathbf{S} \neq 0. \tag{P.20}$$

Furthermore, theorem no. 4 proves that, if the solution exists, it cannot be unique.

---

[50] Carl Gottfried Neumann (1832–1925): German mathematician, one of the forefathers in the study of integral equations.

Appendix Q
On the Principles of Analytical Mechanics

Dante Alighieri, *Paradise*
L'amor che move il sole e l'altre stelle
(Love that moves the Sun and other stars).

Q.1 Variational Formulation of Motion

The laws of motion can be derived starting from a principle formulated in the XVIII century by Maupertuis: in determining the motion of a system between two assigned extremes, nature chooses the most economic among all possible alternatives, where the adjective 'economic' refers to some criterion of judgment.

We want to describe the motion of a mechanical system B in a hyperspace (phase space) of n dimensions, as many as are the degrees of freedom (cf. Sect. 4.8.4). We consider a reference system where each axis represents one generalized coordinate q_j ($j = 1, \ldots, n$). At any time, the set of n coordinates of B, $q_j(t)$ identifies a point P in the hyperspace, and the curve that it draws while t varies is the trajectory of B.

Clearly, at any t a conjugated momentum $\dot{q}_j(t)$ is associated to the point P. Assuming that the configurations of the system are fixed at two epochs t_1, t_2 (which means knowing the coordinates of the two representative points in the hyperspace of the generalized coordinates), we try to determine the trajectory of the motion in the phase space in the time interval $\Delta t = t_2 - t_1$.

A priori, the trajectory can be any one of the infinite number of curves joining the two points of the hyperspace. But, if we accept a deterministic view of mechanics, we must pretend that the trajectory is one and only one. To identify it, all we have to do is to compare the various paths geometrically possible and take the most economic

E. Bannikova and M. Capaccioli, *Foundations of Celestial Mechanics*, Graduate Texts in Physics, https://doi.org/10.1007/978-3-031-04576-9

one. To this end we then define the action S:

$$S = \int_{t_1}^{t_2} \lambda(q_j, \dot{q}_j, t)dt, \tag{Q.1}$$

where λ is the function representing the '*cost*' of the operation in the interval dt. It depends on the particular path and therefore on the dynamic coordinates q_j and $\dot{q}_j$. Its minimum determines the true trajectory.

To better clarify what is behind this procedure, think of the optical principle devised by Fermat, which, in its strong form, states that light propagates in a non-homogeneous medium in such a way that the travel time:

$$t = \int_{s_o}^{s_1} \frac{ds}{v(t)}, \tag{Q.2}$$

is the least.

Minimizing S means placing $\delta S = 0$, where the symbol δ denotes the functional variation with respect to the overall generalized trajectories. It is used instead of d as trajectories are compared for different values of the coordinates q_j at the same time t. Let $q_j^\circ(t)$ be the true trajectory and $q_j^i(t)$ any other curve that joins the configurations identified by the two sets of n generalized coordinates $q_j^\circ(t_1)$ and $q_j^\circ(t_2)$. Then:

$$q_j^i(t) = q_j^\circ(t) + \delta q_j^\circ(t), \tag{Q.3}$$

where the difference $\delta q_j^\circ(t)$ is called variation. By hypothesis, the variation at the extremes of the trajectory must be zero, that is:

$$\delta q_j^\circ(t_1) = \delta q_j^\circ(t_2) = 0. \tag{Q.4}$$

Since the difference between q° and q^i is infinitesimal, the same happens for the relative actions, that is:

$$S^\circ = \int_{t_1}^{t_2} \lambda(q^\circ, \dot{q}^\circ, t)dt, \quad \text{and} \quad S^i = \int_{t_1}^{t_2} \lambda(q^i, \dot{q}^i, t)dt. \tag{Q.5}$$

So:

$$S^i - S^\circ = \delta S = \int_{t_1}^{t_2} \left[\lambda(q^i, \dot{q}^i, t) - \lambda(q^\circ, \dot{q}^\circ, t)\right]dt =$$

$$= \int_{t_1}^{t_2} \left[\lambda(q^\circ + \delta q^\circ, \dot{q}^\circ + \delta\dot{q}^\circ, t) - \lambda(q^\circ, \dot{q}^\circ, t)\right]dt. \tag{Q.6}$$

Expanding the argument of the integral in a power series of δq° and $\delta\dot{q}^\circ$, and truncating the expansion at the first order, we obtain:

$$\delta S = \int_{t_1}^{t_2} \left(\frac{\partial \lambda}{\partial q_j^{\circ}} \delta q_j^{\circ} + \frac{\partial \lambda}{\partial \dot{q}_j^{\circ}} \delta \dot{q}_j^{\circ} \right) dt. \tag{Q.7}$$

Since $\delta \dot{q} = \frac{d}{dt} \partial q$, integration by parts of the second term in brackets gives:

$$\delta S = \int_{t_1}^{t_2} \left(\frac{\partial \lambda}{\partial q_j^{\circ}} \delta q_j^{\circ} + \frac{\partial \lambda}{\partial \dot{q}_j^{\circ}} \frac{d}{dt} \delta q_j^{\circ} \right) dt =$$

$$= \int_{t_1}^{t_2} \frac{\partial \lambda}{\partial q_j^{\circ}} \delta q_j^{\circ} dt + \frac{\partial \lambda}{\partial \dot{q}_j^{\circ}} \delta q_j^{\circ} \Big|_{t_1}^{t_2} - \int_{t_1}^{t_2} \frac{d}{dt} \frac{\partial \lambda}{\partial \dot{q}_j^{\circ}} \delta q_j^{\circ} dt. \tag{Q.8}$$

Finally, observing that the second term in the last expression is null, we have:

$$\delta S = \int_{t_1}^{t_2} \left[\left(\frac{\partial \lambda}{\partial q_j^{\circ}} - \frac{d}{dt} \frac{\partial \lambda}{\partial \dot{q}_j^{\circ}} \right) \partial q_j^{\circ} \right] dt = 0. \tag{Q.9}$$

Since the variations δq_j° are completely arbitrary, the Eq. (Q.9) will always be true if and only if:

$$\frac{d}{dt} \frac{\partial \lambda}{\partial \dot{q}_j^{\circ}} - \frac{\partial \lambda}{\partial q_j^{\circ}} = 0. \tag{Q.10}$$

These equations are identical to Lagrange's equations when λ is identified with $\mathcal{L}$, which, as we already know, is a function of the dynamical coordinates and time.

Q.2 Conservation Laws

Assuming that the dynamical state function λ is the Lagrangian function, we will see how, under suitable hypotheses, conservation laws known by other way can be formulated and how their scope goes well beyond analytical demonstration. For a system with n degrees of freedom, the integration of the equations of motion leads to determine $2n$ constants that depend on the initial conditions and can be expressed as functions of q_j and $\dot{q}_j$. Such are the first integrals. But not all of them have the same importance; some, in fact, are additive.

This important property allows us to know the relative value of the whole system by means of a simple sum of the values pertaining to the single constituents, if they do not interact. It is therefore possible, given the state of the system before and after the interaction, to obtain information about the interaction itself. These conservation laws are based on the properties of isotropy and homogeneity of space and homogeneity of time, properties which may depend on the reference system (cf. Sect. 1.2). The systems in which they occur are called inertial reference systems.

Let us start by considering the homogeneity of time. In this case, the Lagrangian function, $\mathcal{L}$, of an isolated system is not an explicit function of time t, and therefore the differential:

$$\frac{d}{dt}\mathcal{L} = \sum_{j=1}^{n} \frac{\partial \mathcal{L}}{\partial q_j} \dot{q}_j + \sum_{j=1}^{n} \frac{\partial \mathcal{L}}{\partial \dot{q}_j} \ddot{q}_j, \tag{Q.11}$$

lacks the term:

$$\frac{\partial \mathcal{L}}{\partial t} = 0. \tag{Q.12}$$

Using the equations of Lagrange (Q.10), we have:

$$\frac{d}{dt}\mathcal{L} = \sum_{j=1}^{n} \dot{q}_j \frac{d}{dt}\frac{\partial \mathcal{L}}{\partial \dot{q}_j} + \sum_{j=1}^{n} \frac{\partial \mathcal{L}}{\partial \dot{q}_j} \ddot{q}_j = \frac{d}{dt} \sum_{j=1}^{n} \frac{\partial \mathcal{L}}{\partial \dot{q}_j} \dot{q}_j, \tag{Q.13}$$

or:

$$\frac{d}{dt} \left(\sum_{j=1}^{n} \frac{\partial \mathcal{L}}{\partial \dot{q}_j} \dot{q}_j - \mathcal{L} \right) = 0, \tag{Q.14}$$

from which:

$$\sum_{j=1}^{n} \frac{\partial \mathcal{L}}{\partial \dot{q}_j} \dot{q}_j - \mathcal{L} = \text{const} = \mathcal{E}, \tag{Q.15}$$

since the left term of this equation has the physical dimension of an energy.

The additivity of $\mathcal{E}$ follows directly from the additivity of the Lagrangian function, $\mathcal{L}$. The energy conservation law also applies to a non-isolated system as long as it is immersed in a constant external field, since the only hypothesis made is the independence of $\mathcal{L}$ from t. Systems where $\mathcal{E}$ is constant are called conservative.

Let us now consider the homogeneity of space. It allows us to arbitrarily choose the origin of the reference system, or better to be able to translate the latter without changing $\mathcal{L}$. If $\vec{\epsilon}$ is the radius vector joining O and O', origins of two inertial references T and T' translated from each other, then $\mathbf{r}'_j$, radius vector of the j-th point in T', will be:

$$\mathbf{r}'_j = \mathbf{r}_j + \vec{\epsilon}. \tag{Q.16}$$

Obviously $\vec{\epsilon}$ is constant and arbitrary, so:

$$\dot{\mathbf{r}}'_j = \dot{\mathbf{r}}_j. \tag{Q.17}$$

The variation of $\mathcal{L}$ for this translation will be:

$$\delta \mathcal{L} = \sum_{j=1}^{n} \frac{\partial \mathcal{L}}{\partial \mathbf{r}_j} \delta \mathbf{r}_j + \sum_{j=1}^{n} \frac{\partial \mathcal{L}}{\partial \dot{\mathbf{r}}_j} \delta \dot{\mathbf{r}}_j = \vec{\epsilon} \sum_{j=1}^{n} \frac{\partial \mathcal{L}}{\partial \mathbf{r}_j} = 0, \tag{Q.18}$$

from which:

$$\sum_{j=1}^{n} \frac{\partial \mathcal{L}}{\partial \mathbf{r}_j} = 0. \tag{Q.19}$$

The sense of this equality is that the sum of all the forces acting on all the particles of a closed system equals zero: $\sum_{j=1}^{n} \mathbf{F}_j = \sum_{j=1}^{n} \dot{\mathbf{p}}_j = 0$, where:

$$\dot{\mathbf{p}}_j = \frac{\partial \mathcal{L}}{\partial \mathbf{r}_j}, \tag{Q.20}$$

is the time derivative of the generalized momentum vector. Using the Lagrange equations, it is:

$$\sum_{j=1}^{n} \frac{d}{dt} \frac{\partial \mathcal{L}}{\partial \dot{\mathbf{r}}_j} = \frac{d}{dt} \sum_{j=1}^{n} \frac{\partial \mathcal{L}}{\partial \dot{\mathbf{r}}_j} = 0, \tag{Q.21}$$

and thus:

$$\sum_{j=1}^{n} \frac{\partial \mathcal{L}}{\partial \dot{\mathbf{r}}_j} = \sum_{j=1}^{n} \mathbf{p}_j = \overrightarrow{\text{const}} = \mathbf{P}. \tag{Q.22}$$

Due to its physical dimensions, $\mathbf{P}$ is the total momentum vector. Therefore, the momentum of the system is equal to the vector sum of the individual components even if they interact, differently from the case of the energy. In this respect it seems useful to stress that $\mathbf{P}$ is a vector while $\mathcal{E}$ is a scalar.

We come now to the isotropy of space. This property implies that a rotation of the reference system changes $\mathbf{r}_j$, but it leaves unchanged the $\mathcal{L}$ function. Let $\delta \overrightarrow{\varphi} = |\delta \varphi| \mathbf{k}$ be the rotation vector, with $|\delta \varphi|$ equal to the amplitude of the rotation. In this case:

$$\delta \mathbf{r}_j = \delta \overrightarrow{\varphi} \times \mathbf{r}_j,$$
$$\delta \dot{\mathbf{r}}_j = \delta \overrightarrow{\varphi} \times \dot{\mathbf{r}}_j, \tag{Q.23}$$

with which:

$$\delta \mathcal{L} = \sum_{j=1}^{n} \frac{\partial \mathcal{L}}{\partial \mathbf{r}_j} \delta \mathbf{r}_j + \sum_{j=1}^{n} \frac{\partial \mathcal{L}}{\partial \dot{\mathbf{r}}_j} \delta \dot{\mathbf{r}}_j = 0. \tag{Q.24}$$

With the usual substitutions and (Q.20), (Q.22):

$$\delta \mathcal{L} = \sum_{j=1}^{n} \left[\dot{\mathbf{p}}_j \left(\delta \overrightarrow{\varphi} \times \mathbf{r}_j \right) \right] + \sum_{j=1}^{n} \left[\mathbf{p}_j \left(\delta \overrightarrow{\varphi} \times \dot{\mathbf{r}}_j \right) \right] =$$
$$= \delta \overrightarrow{\varphi} \sum_{j=1}^{n} \left[\left(\mathbf{r}_j \times \dot{\mathbf{p}}_j \right) + \left(\dot{\mathbf{r}}_j \times \mathbf{p}_j \right) \right] = \delta \overrightarrow{\varphi} \frac{d}{dt} \sum_{j=1}^{n} \left(\mathbf{r}_j \times \mathbf{p}_j \right) = 0, \tag{Q.25}$$

from which:

$$\frac{d}{dt} \sum_{j=1}^{n} (\mathbf{r}_j \times \mathbf{p}_j) = 0, \tag{Q.26}$$

and therefore:

$$\sum_{j=1}^{n} (\mathbf{r}_j \times \mathbf{p}_j) = \overrightarrow{\text{const}} = \mathbf{J}, \tag{Q.27}$$

where $\mathbf{J}$ is the total angular momentum. This vector too is additive, regardless of any interactions.

In conclusion, for an isolated system there are 7 scalar first integrals: the energy $\mathcal{E}$, the three components of $\mathbf{P}$, and the three components of $\mathbf{J}$.

Q.3   Maupertuis's Principle

A mathematical formulation of the Maupertuis principle is now given which is very useful for subsequent applications.[51] To formulate the principle of least action we have considered the integral $S = \int_{t_1}^{t_2} \mathcal{L} dt$, along different trajectories that join the two fixed points of the space identified by the n coordinates $q_j(t_1)$ and $q_j(t_2)$, then looking for the minimum value of the action S.

The concept of action has a greater scope. It allows us to compare the trajectories that, although having the same origin $\left(\text{fixed values of } q_j(t_1)\right)$, yet have different extremes $\left(q_j(t_2)\text{free}\right)$, or pass through the same point $q_j(t)$ at different epochs.

In fact, by rewriting the variation of the action and integrating by parts, given that under these new hypotheses $\delta q_j(t_2) \neq 0$, we obtain a variation δS (or dS) which depends on the coordinates q_j or on time. Without going into details, we give the more general expression, valid even if the hypotheses made on the second integration term are also extended to the action. We find:

$$dS = \sum_{j=1}^{n} p_j^f dq_j^f - \mathcal{H}^f dt - \sum_{j=1}^{n} p_j^i dq_j^i + \mathcal{H}^i dt, \tag{Q.28}$$

where the superscripts i and f stay for initial and final. The expression is simplified if there is no explicit dependence on time; in this case $dt = 0$ and $d = \delta$. Then, the application of the principle of least action not only determines the shape of the trajectory but also the dependence on time of the position on the trajectory. It is

---

[51] "*L'action est proportionnelle au produit de la masse par la vitesse et par l'espace. Maintenant, voici ce principe, si sage, si digne de l'Être Suprême: lorsqu'il arrive quelque changement dans la Nature, la quantité d'action employée pour ce changement est toujours la plus petite qu'il soit possible*"; P.L.M. de Maupertuis, *Principe de la moindre quantité d'action pour la mécanique*, 1744.

therefore noted that the final state of a motion cannot be any function of the initial state but must be an exact differential. Suppose now that the Lagrangian system does not explicitly depend on time t, so that $\mathcal{H} = \mathcal{E} = $ const. Furthermore, the final instant is changed while the coordinates of position are kept fixed. In this case, from (Q.28) we have:

$$\delta S = -\mathcal{H}dt. \tag{Q.29}$$

Considering only trajectories for which energy conservation occurs, it follows that:

$$\delta S + \mathcal{E}dt = 0. \tag{Q.30}$$

We specify δS from (4.272):

$$S = \int_{q_j^i}^{q_j^f} \sum_{j=1}^{n} p_j dq_j - \mathcal{E}(t_f - t_i). \tag{Q.31}$$

Placing:

$$S_o = \int_{q_j^i}^{q_j^f} \sum_{j=1}^{n} p_j dq_j, \tag{Q.32}$$

and putting it in (Q.30), it follows:

$$\delta S_o = 0. \tag{Q.33}$$

This S_o is called abbreviated or Lagrangian action. Note that in this case the integration variables are the coordinates and not the time. Through the (Q.33), the abbreviated action has a minimum in relation to all the trajectories that satisfy the energy conservation law and that pass through the final configuration at an arbitrary time. However, in order to use the (Q.33), it is necessary to set everything according to the coordinates q_j (including the differential dt and $\mathcal{H}$ that appear in the (Q.33). After the appropriate substitutions and reductions, we have:

$$S_o = \int_{q^i}^{q_f} \left[2(\mathcal{E} - \mathcal{U}) \sum_{k,j=1}^{n} a_{k,j}(q) \, dq_k dq_j \right]^{1/2}, \tag{Q.34}$$

where:

$$\mathcal{L} = \frac{1}{2} \sum_{k,j=1}^{n} a_{k,j}(q) \, d\dot{q}_k d\dot{q}_j - \mathcal{U}(q), \tag{Q.35}$$

and the conjugated moments are:

$$p_k = \frac{\partial \mathcal{L}}{\partial \dot{q}_k} = \sum_{k,j=1}^{n} a_{k,j}(q) \, d\dot{q}_k. \tag{Q.36}$$

The expression (Q.33) is called Maupertuis's principle, although the analytical formulation was given by Lagrange and Euler. The importance of this principle lies in the fact that it determines the trajectories of the motion under suitable conditions.

Q.4 The Geodesic Lines

This section presents an application of what we have just seen in the previous one. We consider a system not subject to a potential ($\mathcal{U} \equiv 0$):

$$\mathcal{H} = \mathcal{E} = T = \frac{1}{2} \sum_{k,j=1}^{n} a_{k,j}(q_1, \ldots, q_n) \, \dot{q}_k \dot{q}_j. \tag{Q.37}$$

In the n-dimensional space $(q_1, \ldots, q_n)$, we introduce the metric:

$$ds^2 = \sum_{k,j=1}^{n} a_{k,j}(q_1, \ldots, q_n) \, dq_k dq_j. \tag{Q.38}$$

In this case the length l of the arc connecting two points $P_i(q_j^i)$ and $P_f(q_j^f)$ is given by:

$$l = \int_{q^i}^{q_f} ds = \int_{q^i}^{q_f} \left(\sum_{k,j=1}^{n} a_{k,j}(q_1, \ldots, q_n) \, dq_k dq_j \right)^{1/2}. \tag{Q.39}$$

Using (Q.34) we obtain:

$$S_\circ = \sqrt{2\mathcal{E}} \int_{q^i}^{q_f} \left(\sum_{k,j=1}^{n} a_{k,j}(q_1, \ldots, q_n) \, dq_k dq_j \right)^{1/2} = \sqrt{2\mathcal{E}} \, l, \tag{Q.40}$$

where $\mathcal{E}$ is the same constant for any line joining the two points P_i and P_f:

$$\delta S_\circ = \delta l = \delta \int_{q^i}^{q_f} \left(\sum_{k,j=1}^{n} a_{k,j}(q_1, \ldots, q_n) dq_k dq_j \right)^{1/2} = 0. \tag{Q.41}$$

In this way, the "true" trajectory is distinguished from the other possible trajectories in that the length of the corresponding curve arc is minimum with respect to the other arcs. The trajectories that satisfy (Q.41) are called geodesic lines.

Consider now a system for which $\mathcal{U} \neq 0$. From (Q.34) it follows:

$$S_{\circ} = \int_{q^i}^{q_f} \left(2(\mathcal{E} - \mathcal{U}) \sum_{k,j=1}^{n} a_{k,j} dq_k dq_j \right)^{1/2} , \tag{Q.42}$$

and the geodesic line will be such that:

$$\delta S_{\circ} = \delta \int_{q^i}^{q_f} \left(2(\mathcal{E} - \mathcal{U}) \sum_{k,j=1}^{n} a_{k,j}(q_1, \ldots, q_n) dq_k dq_j \right)^{1/2} . \tag{Q.43}$$

Comparing (Q.43) with (Q.41), we can state that the geodesics for a system subject to potential are rectilinear trajectories in a space with metric:

$$ds^2 = (\mathcal{E} - \mathcal{U}) \sum_{k,j=1}^{n} a_{k,j}(q_1, \ldots, q_n) \, dq_k dq_j . \tag{Q.44}$$

The concept of geodesic line is revolutionary as it transforms a dynamical into a geometric problem. At the basis of the general relativity theory, it allowed us to solve Mercury's perihelion precession problem even without mastering Einstein's theory in depth, just knowing the metric which, of course, is the solution of Einstein's equation (see Sect. 4.14).

Appendix R
Space-Time Invariant and Conservation Principles

Richard P. Feynman

It doesn't matter how beautiful your theory is, it doesn't matter how smart you are. If it doesn't agree with experiment, it's wrong.

R.1 Continuous Trajectories

The intimate connection between conservation laws and the existence of symmetries in the description of problems or of the invariance with respect to changes in the view point of both the observer and the reference system, has been stressed repeatedly. This connection becomes clear when it is examined in the context of the Hamilton variational principle. The strategy is to compare two contiguous trajectories (in space and time), both possible in the sense of the variational principle (they are therefore not virtual trajectories). We will represent them with the generalized coordinates $q(t)$ and $q'(t)$ respectively.

Consider the motion in the time interval (t_1, t_2) for the coordinate system q and $(t_1 + \Delta t_1, t_2 + \Delta t_2)$ for the q'. Unlike for virtual trajectories, in the present case it is not legitimate to admit that the coordinates q coincide with the q' at the ends of the trajectories. In fact, we will have:

$$\Delta q_i(t_{1,2}) = q_i'(t_{1,2} + \Delta t_{1,2}) - q_i(t_{1,2}) \qquad (i = 1, \ldots, N). \tag{R.1}$$

As we have done for virtual trajectories, we write $q_i'(t) = q_i(t) + \delta q_i(t)$ (but note that here δ does not represents a virtual variation, but an infinitesimal difference at

© The Editor(s) (if applicable) and The Author(s), under exclusive license to Springer   383
Nature Switzerland AG 2022
E. Bannikova and M. Capaccioli, *Foundations of Celestial Mechanics*, Graduate Texts
in Physics, https://doi.org/10.1007/978-3-031-04576-9

constant time between the same coordinates in two contiguous trajectories), with which:

$$\Delta q_i(t_1) = q_i(t_1 + \Delta t_1) + \delta q_i(t_1 + \Delta t_1) - q_i(t_1) \approx \delta q_i(t_1) + \dot{q}_i(t_1)\Delta t_1 , \quad (R.2)$$

valid for the lower extreme t_1 when $\Delta t_1 \to 0$. A similar expression applies to the upper extreme t_2. The difference between the variations in the two trajectories is:

$$\Delta S = \int_{t_1 + \Delta t_1}^{t_2 + \Delta t_2} \mathcal{L}(q + \delta q, \dot{q} + \delta \dot{q}, t) \, dt - \int_{t_1}^{t_2} \mathcal{L}(q, \dot{q}, t) \, dt =$$

$$= \int_{t_1}^{t_2} \Big(\mathcal{L}(q + \delta q, \dot{q} + \delta \dot{q}, t) - \mathcal{L}(q, \dot{q}, t) \Big) \, dt \ +$$

$$+ \int_{t_1 + \Delta t_1}^{t_1} \mathcal{L}(q + \delta q, \dot{q} + \delta \dot{q}, t) \, dt + \int_{t_2}^{t_2 + \Delta t_2} \mathcal{L}(q + \delta q, \dot{q} + \delta \dot{q}, t) \, dt =$$

$$= \int_{t_1}^{t_2} \Big(\mathcal{L}(q + \delta q, \dot{q} + \delta \dot{q}, t) - \mathcal{L}(q, \dot{q}, t) \Big) \, dt \ +$$

$$+ \mathcal{L}(q, \dot{q}, t)\Delta t \Big|_{t_1}^{t_2} , \qquad (R.3)$$

where, in deriving the last expression, the terms of the second order in δq_i and $\delta \dot{q}_i$ have been neglected. We now have to compute the integral in (R.3) from t_1 to t_2. It is:

$$\int_{t_1}^{t_2} \Big(\mathcal{L}(q + \delta q, \dot{q} + \delta \dot{q}, t) - \mathcal{L}(q, \dot{q}, t) \Big) \, dt =$$

$$= \int_{t_1}^{t_2} \sum_{i=1}^{N} \left(\frac{\partial \mathcal{L}}{\partial q_i} \delta q_i + \frac{\partial \mathcal{L}}{\partial \dot{q}_i} \delta \dot{q}_i \right) dt =$$

$$= \int_{t_1}^{t_2} dt \sum_{i=1}^{N} \delta q_i \left(\frac{\partial \mathcal{L}}{\partial q_i} - \frac{d}{dt} \frac{\partial \mathcal{L}}{\partial \dot{q}_i} \right) dt + \sum_{i=1}^{N} \frac{\partial \mathcal{L}}{\partial \dot{q}_i} \delta q_i \Big|_{t_1}^{t_2} . \qquad (R.4)$$

It must zero because the trajectories are both possible; therefore the Lagrange equations apply to them (4.24). On the other hand, the last term is not necessarily null (as it would be in the case of virtual trajectories). In conclusion:

$$\Delta S = \left(\mathcal{L} \Delta t + \sum_{i=1}^{N} \frac{\partial \mathcal{L}}{\partial \dot{q}_i} \delta q_i \right) \Big|_{t_1}^{t_2} . \qquad (R.5)$$

Replacing δq_i with the expression given by (R.2) and introducing the conjugated moments and the Hamilton function, we finally have:

$$\Delta S = \left(\sum_{i=1}^{N} \frac{\partial \mathcal{L}}{\partial \dot{q}_i} \Delta q_i - \left[\sum_{i=1}^{N} \frac{\partial \mathcal{L}}{\partial \dot{q}_i} \dot{q}_i - \mathcal{L} \right] \Delta t \right) \Bigg|_{t_1}^{t_2} =$$

$$= \sum_{i=1}^{N} p_i \Delta q_i - \mathcal{H} \Delta t \Bigg|_{t_1}^{t_2}. \qquad (R.6)$$

If $\Delta q_i(t_{1,2}) = 0$ and $\Delta t_{1,2} = 0$ (and therefore the two paths coincide), it is obviously $\Delta S = 0$. But the condition of coincidence is only necessary; in other words, there may be cases where, even if $\Delta q_i(t_{1,2}) \neq 0$ or $\Delta t_{1,2} \neq 0$, it is also $\Delta S = 0$. This is what we are going to verify.

R.2 Time-Invariance and Conservation of Energy

Assume that $\Delta q_i = 0$ and $\Delta t_1 = \Delta t_2 \neq 0$; this means admitting that the mechanical system follows an identical path, in the same time interval (t_2, t_1), both if the motion starts at t_1 (and ends at t_2) and if it starts at $t_1 + \Delta t_1$ (and ends at $t_2 + \Delta t_2 = t_2 + \Delta t_1$). Therefore, the two trajectories considered in the previous paragraph coincide entirely except in the origin of the time scale. The difference of the ΔS variation given by (R.6) must therefore be zero and equal to:

$$\Delta S = -\mathcal{H} \Big|_{t_1}^{t_2} \Delta t = 0. \qquad (R.7)$$

This can happen if and only if $\mathcal{H}(t_1) = \mathcal{H}(t_2)$, i.e., if the Hamiltonian does not explicitly depend on time. But, based on (4.101), we know that in this case the total derivative of $\mathcal{H}$ with respect to time is also null; therefore, the Hamiltonian is constant and coincides with the total energy of the system. In other words, the invariance with respect to a translation of the time coordinate implies the conservation of the (Hamiltonian) energy of the system.[52]

R.3 Invariance to Translations and Conservation of Moment

We now identify the two trajectories of (Sect. R.1) with the same one seen by two parallel and concordant spatial reference systems (with origins not necessarily coinciding). In other words, we want to see the effect of a translation in the Cartesian coordinate system. Then it is convenient to convert the coordinates q_i into Cartesian

---

[52] This statement also proves that an experiment is reproducible if and only if it is performed on a system with constant energy, provided that it is conveniently isolated.

vectors, $\Delta q_i \rightarrow \Delta \mathbf{r}_j = \Delta \mathbf{r}$ (the j index is suppressed because we are considering a translation). We will then indicate with $\mathbf{p}_j$ the conjugated moments (in the Lagrangian sense) to $\mathbf{r}_j$. It will also be $\Delta t_1 = \Delta t_2 = 0$, so the (R.6) is reduced to:

$$\Delta S = \left(\sum_{j=1}^{N'} \mathbf{p}_j \right) \Bigg|_{t_1}^{t_2} \cdot \Delta \mathbf{r}. \tag{R.8}$$

Being the same phenomenon, the invariance of the description ($\Delta S = 0$) with respect to a translation of the coordinates in the conventional space involves:

$$\sum_{j=1}^{N'} \mathbf{p}_j \Bigg|_{t_1}^{t_2} = 0, \tag{R.9}$$

that is the conservation of the total momentum.[53]

R.4 Rotational Invariance and Angular Momentum Conservation

We apply the same argument as in the previous section to the case where the two reference systems are rotated relative to each other. Denoting by $\mathbf{k}$ the vector of the direction chosen as the rotation axis, and with $\delta\phi$ the rotation angle (fixed), we have:

$$\Delta \mathbf{r}_j = \Delta\phi \, \mathbf{k} \times \mathbf{r}_j, \tag{R.10}$$

and:

$$\Delta S = \left(\sum_{j=1}^{N'} (\mathbf{k} \times \mathbf{r}_j) \cdot \mathbf{p}_j \right) \Bigg|_{t_1}^{t_2} \Delta\phi =$$

$$= \left(\sum_{j=1}^{N'} \mathbf{r}_j \times \mathbf{p}_j \right) \Bigg|_{t_1}^{t_2} \Delta\phi \, \mathbf{k} = \mathbf{M} \Big|_{t_1}^{t_2} \Delta\phi \, \mathbf{k}, \tag{R.11}$$

---

[53] In a general sense it can be stated that the constancy of the momentum derives from the homogeneity of space.

where $\mathbf{J} = \displaystyle\sum_{j=1}^{N'} \mathbf{r}_j \times \mathbf{p}_j$ is the total angular momentum with respect to the (fixed) origin of the two systems. Once again we will admit the invariance of the description ($\Delta S = 0$), with which we deduce the constancy of the total angular momentum. In fact, from (R.11), $\mathbf{J}(t_1)$ must be equal to $\mathbf{J}(t_2)$.

References

1. W.M. Smart, *Spherical Astronomy* (Cambridge University Press, Cambridge, 1949)
2. J. Binney, S. Tremaine, *Galactic Dynamics* (Princeton University Press, Princeton, 1987)
3. A.E. Roy, *Orbital Motion* (Adam Hilger Ltd, Bristol, 1982)
4. N. Grossman, A C^∞ lagrange inversion theorem. Am. Math. Mon. **112**(6), 512–514 (2005)
5. H.C. Plummer, *An Introductory Treatise on Dynamical Astronomy* (Dover Publications, New York, 1960)
6. G.P. Tolstov, *Fourier Series* (Dover, New York, 1976)
7. C. Cohen-Tannoudji, B. Diu, F. Laloë, *Quantum Mechanics*, vols. I and II (Wiley, New York, 1977)
8. A. Messiah, *Quantum Mechanics*, vol. I and II (Wiley, North Holland, 1966)
9. M. Abramowitz, I.A. Stegun (eds.), *Handbook of Mathematical Functions with Formulas, Graphs, and Mathematical Tables* (Dover, New York, 1972)
10. G.B. Arfken, H.J. Weber, F.E. Harris, *Mathematical Methods for Physicists* (Elsevier, Amsterdam, 2013)
11. I.S. Gradshteyn, I.W. Ryzhik, *Table of Integrals, Series and Products* (Academic, New York, 1966)
12. C. Müller, *Spherical Harmonics*. Lecture Notes in Mathematics, vol. 17 (Springer, Berlin, 1966)
13. P. de Bernardis et al., Nature **404**, 955 (2000)
14. D.N. Spergel et al., Astrophys. J. Suppl. Ser. **148**, 175 (2003)
15. Y. Akrami, et al. (Planck Collaboration), Planck 2018 results. I. Overview, and the cosmological legacy of Planck. Astron. Astrophys. **641**, A1 (2020)
16. V.V. Beletsky, *Essays on the Motion of Celestial Bodies* (Springer, Basel, 2001)
17. J. Binney, M. Merrifield, *Galactic Astronomy* (Princeton University Press, Princeton, 1989)
18. D. Brouwer, G.M. Clemence, *Methods of Celestial Mechanics* (Academic, London, 1961)
19. A. Celletti, A. Perozzi, *Celestial Mechanics. The Waltz of the Planets* (Springer, Berlin, 2007)
20. S. Chandrasekhar, *The Mathematical Theory of Black Holes* (Clarendon Press, Oxford, 1998)
21. A. Deprit, Celest. Mech. **30**, 181 (1983)
22. D.J.H. Garling, *A Course in Mathematical Analysis*, vol. 1 (Cambridge University Press, Cambridge, 2013)
23. C. Lanczos, *The Variational Principles of Mechanics* (Dover, Mineola, 1986)
24. A.I. Lurie, *Analytical Mechanics* (Springer, Berlin, 2002)
25. G.W. Marcy, et al., The Planet around 51 Pegasi. Astrophys. J. **481**, 926–935, 1379 (1997)
26. C.D. Murray, F.S. Dermott, *Solar System Dynamics* (Cambridge University Press, Cambridge, 2000)
27. H.C. Ohanian, R. Ruffini, *Gravitation and Spacetime* (Cambridge University Press, Cambridge, 2013)
28. H. Pollard, *Mathematical Introduction to Celestial Mechanics* (Prentice-Hall Inc., New Jersey, 1966)
29. W.H. Press et al., *Numerical Recipes: The Art of Scientific Computing* (Cambridge University Press, Cambridge, 1986)
30. E.T. Whittaker, G.N. Watson, *A Course in Modern Analysis* (Cambridge University Press, Cambridge, 1990)

Index

© The Editor(s) (if applicable) and The Author(s), under exclusive license to Springer
Nature Switzerland AG 2022
E. Bannikova and M. Capaccioli, *Foundations of Celestial Mechanics*, Graduate Texts
in Physics, https://doi.org/10.1007/978-3-031-04576-9

Printed in the United States
by Baker & Taylor Publisher Services